Particles and Fields

Particles and Fields

Edited by

DAVID H. BOAL

and

ABDUL N. KAMAL

University of Alberta
Edmonton, Alberta, Canada

Plenum Press · New York and London

Library of Congress Cataloging in Publication Data

Banff Summer Institute on Particles and Fields, 1977. Particles and fields.

Proceedings of the institute held at the Banff Center, Banff, Canada, Aug. 25–Sept. 3, 1977.

Includes index.

1. Particles (Nuclear physics)–Congresses. 2. Field theory (Physics)–Congresses. I. Boal, D. H. II. Kamal, A. N. III. Title.

QC770.B36 1977 539.7 78-2509

ISBN 0-306-31147-X

Proceedings of the Banff Summer Institute on Particles and Fields held at the Banff Center in Banff, Canada, August 25–September 3, 1977

Printed in the United States of America

PREFACE

This volume contains the invited lectures and seminars presented at the Banff Summer Institute on Particles and Fields held at the Banff Center in Banff, Canada, from 25 August to 3 September, 1977. The town is situated in the heart of the Canadian Rockies, and the observant reader may notice references in this volume to the bears which roam near the town.

The subject matter of the school was recent advances in particle physics and field theory. Lectures were given on such topics as extended objects, lattice gauge theories, quantum chromodynamics and Reggeon field theory. Experimental reviews were given of recent work in charmed particle and neutrino physics. Summaries of the theoretical implications of these experiments were also given. The format of the talks included eight lecture series (of three to four hours each) given by Profs. Abarbanel, Appelquist, Feldman, Gilman, 't Hooft, Jackiw, Mann and Weinstein, seven one-hour seminars given by Profs. Caianiello, Fujii, Johnson, Lam, Phillips, Sherry and Tze, and several short contributed seminars (which do not appear in this volume). There were also small informal seminar groups held at the Center and, we hope, many physics conversations on the hiking trails where most of the participants spent their afternoons. Not included in these proceedings are the banquet speeches by E. Caianiello and S. D. Drell, as well as (for copyright reasons) a seminar by K. Johnson.

We would like to thank the following members of the International Advisory Committee for their assistance and advice:

E. R. Caianiello, University of Salerno
S. D. Drell, Stanford Linear Accelerator Center
Y. Fujii, University of Tokyo
T. W. B. Kibble, Imperial College
C. S. Lam, McGill University
Y. Nambu, University of Chicago
R. J. N. Phillips, Rutherford Laboratory
A. Salam, I.C.T.P., Trieste

G. Sudarshan, University of Texas
M. K. Sundaresan, Carleton University.

Much of the detailed work was handled by the five man Local Organizing Committee, whose members, in addition to DHB and ANK were:

A. Z. Capri
Y. Takahashi
H. Umezawa

all of the University of Alberta. On the local level, our thanks also go to K. James, secretary at the conference, as well as G. Braun, L. Cech and M. Yiu, who typed the manuscript. Lastly, thanks are due to all those individuals at the University of Alberta who we coerced into folding letters, sealing envelopes and helping with the registration, and to the staff of the Banff Center who contributed to the smooth operation of the Institute.

Of course, the Institute was not self-supporting and we gratefully acknowledge the financial support we received from the following organizations:

National Research Council of Canada
University of Alberta
Institute of Particle Physics
Theoretical Physics Division, Canadian Association of Physicists
Atomic Energy of Canada Ltd.
Gulf Oil Canada Ltd.

The Institute was organized under the auspices of the Theoretical Physics Division of the Canadian Association of Physics, whose executive for 1976-77 was composed of three members of the Local Organizing Committee (DHB, ANK, HU). C.A.P. attempts to hold a summer school annually at Banff, and the royalties from the sale of the proceedings of this and previous schools will go toward the support of future Summer Institutes.

D. H. Boal,
A. N. Kamal

Edmonton, Canada
December, 1977.

CONTENTS

CONTEMPORARY REGGEON PHYSICS

Henry D.I. Abarbanel

Fermi National Accelerator Laboratory

Batavia, Illinois, U. S. A.

1. INTRODUCTION AND OUTLINE

This set of lectures aims to introduce the reader to the physics behind Reggeon field theories (RFT) and the developments in that subject. The goal of RFT is a coherent basis for the effects, in physical processes at large collision energies and small momentum transfers, of diffraction scattering subsumed in the form of the Pomeron Regge trajectory which has $\alpha(0) = 1$, the maximum allowed by unitarity.

Time, the patience of the listener, and the attention span of the lecturer did not permit a complete covering of all topics in RFT. Indeed, many have been left untouched. With extended apologies to the appropriate authors I refer the reader to the reviews in Reference 1. Here we treat the following topics:

1. Physical and Experimental Basis for Reggeon Physics

2. Theory of Regge Poles - The Multiperipheral Model

3. Reggeon Unitarity

4. A Suggestive Analogy

5. Critical Behavior in RFT and the "Ordered Phase" of RFT,

which should serve as an introduction to the literature and the subject as a whole.

2. EXPERIMENTAL AND PHYSICAL BASIS FOR REGGEON PHYSICS

We will be concerned at first with the elastic collision of equal mass, spinless hadrons as shown in Figure 1, where T(s,t) is the elastic scattering amplitude as a function of an energy variable $s = (p_A + p_B)^2 = 4E_{cm}^2$ and $t = (p_A - p_{A'})^2 = -2p_{cm}^2(1-\cos\theta)$; E_{cm}, p_{cm}, and θ are respectively the energy, momentum and scattering angle in the barycentric system. The differential cross section comes from T(s,t)

$$\frac{d\sigma(s,t)}{dt} = \frac{1}{16\pi^2 s^2} |T(s,t)|^2, \tag{1}$$

and via the optical theorem the total cross section is

$$\sigma_T(s) = \frac{\mathrm{Im}_s T(s,0)}{\sqrt{s(s-4m^2)}} \underset{s \text{ large}}{\approx} \frac{\mathrm{Im}_s T(s,t)}{s} . \tag{2}$$

Our interest will be in such processes at large s, fixed t. Since $d\sigma(s,t)/dt$ falls very rapidly as t moves away from 0 (typically as $\exp-[(8 \text{ or } 10)t/(\mathrm{GeV/c})^2]$), we are discussing the majority of hadron collision events.

Now the basic facts we need are as follows: If in the t-channel (the direction of $A\bar{A}'$ in Figure 1) the isospin is not zero; for example, $I_t = 1$ in $\pi^- p \to \pi^0 n$, then

$$T_I(s,t) \underset{\substack{s \text{ large} \\ t \text{ fixed}}}{\approx} s^{\alpha_I(t)} \frac{(\tau \pm e^{-i\pi\alpha_I(t)})}{\sin \pi\alpha_I(t)} , \tag{3}$$

where $\alpha_I(t)$, called the Regge trajectory, can be accurately approximated by

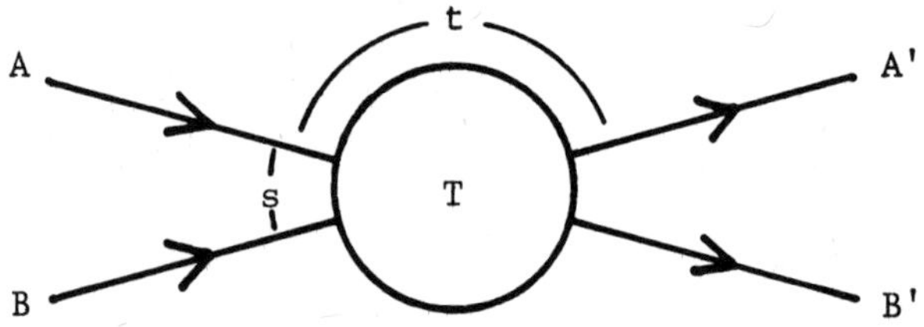

Fig. 1. The two-to-two hadron scattering amplitude T(s,t) as a function of energy $s = (p_A+p_B)^2$ and momentum transfer $t = (p_A-p_{A'})^2$. For equal mass particles the physical region is $s \geq 4m^2$, $t \leq 0$.

$$\alpha_I(t) = \alpha_o + \alpha_o' t , \tag{4}$$

and τ (the signature) depends on the particular quantum numbers in the t-channel. For the case when the allowed quantum numbers are those of the ρ-meson such processes have been studied over a large range of s and t and are accurately represented by $\tau_\rho = -1$ and

$$\alpha_\rho(t) = 0.53 + 0.91\, t/(\mathrm{GeV}/c)^2 , \tag{5}$$

which may approximately be read as $\alpha_\rho(t) = 0.5 + t$. When this straight line is projected from the scattering region ($t \leq 0$) where it is measured to the timelike region ($t > 0$) where the real ρ-meson lives one has to good accuracy the trajectory shown in Figure 2, which shows how a Regge trajectory connects scattering and resonances (here ρ and g mesons).

If the t-channel allows the quantum numbers of the vacuum (elastic scattering, for example), the situation is different

$$T_P(s,t) \underset{\substack{s \text{ large} \\ t \text{ fixed}}}{\approx} s^{\alpha_P(t)} \frac{(1 - e^{-i\pi\alpha_P(t)})}{\sin \pi\alpha_P(t)} (\log s)^\eta , \tag{6}$$

where the Pomeron or vacuum trajectory is

$$\alpha_P(t) = 1 + 0.287\, t/(\mathrm{GeV}/c)^2 , \tag{7}$$

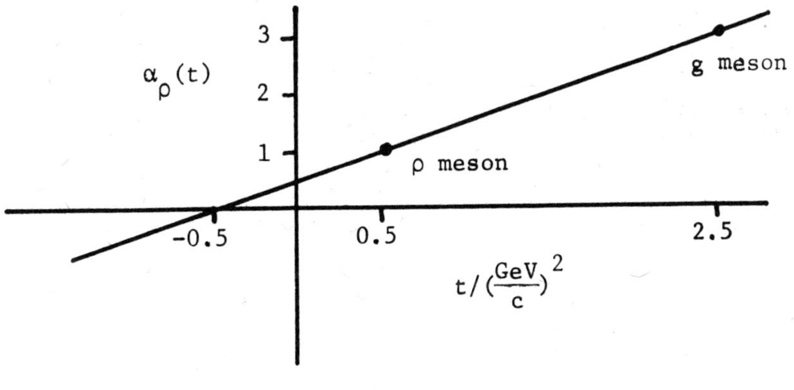

Fig. 2. The Chew-Frautschi $\alpha(t)$ vs. t plot for the ρ-meson trajectory.

and η may be as large as 2. A recent report on pp and $\bar{p}p$ total cross sections[2] gave the following fit

$$\sigma_{T\binom{pp}{\bar{p}p}}(s) = \sigma_0 + \sigma_1 \, [\log s/(\mathrm{GeV})^2]^{\eta} + c_1\left(\frac{s}{2m_p} - m_p\right)^{(\alpha_1-1)} \mp c_2\left(\frac{s}{2m_p} - m_p\right)^{(\alpha_2-1)} \quad (8)$$

with $\sigma_0 = 27.0 \pm 1.0$ mb, $\sigma_1 = 0.17 \pm 0.08$ mb, $\eta = 2.1 \pm 0.10$, $c_1 = 41.9 \pm 1.1$ mb, $c_2 = 24.2 \pm 1.1$ mb, $\alpha_1 = 0.63 \pm 0.03$, $\alpha_2 = 0.45 \pm 0.02$; $m_p = .938$ GeV/c^2. Clearly the logarithm, though not rapidly varying, adds a new dimension to the question of vacuum quantum numbers.

Now Amaldi, et al also report on the ratio $\rho(s) = \mathrm{Re}\, T(s,0)/\mathrm{Im}\, T(s,0)$ over the same range $5 \leq \sqrt{s}/\mathrm{GeV} \leq 62$ as for σ_{total}. As shown in Figure 3 (conveniently also Amaldi, et al's Figure 3) $\rho(s)$ is increasing over the measured range. By crossing symmetry a cross section $\sigma_T(s) \sim (\log s)^{\eta}$ must come from[3]

$$T(s,0) \approx s(\log s)^{\eta+1} + (-s)(\log(-s))^{\eta+1} , \quad (9)$$

which has

$$\rho(s) \approx \frac{\pi\eta}{2} \frac{1}{\log s} \quad (10)$$

for large s. Since $\rho(s)$ is still increasing at $\sqrt{s} = 62$ (lab equivalent momentum $\approx$ 2 TeV/c!), we can conclude that we are <u>not yet</u> observing the asymptotic diffraction (vacuum quantum number in t-channel) cross section. One's attention may focus on the peculiarly small coefficient of the $(\log s)^{\eta}$ term in (8) to guess that when $\sigma_0 \approx \sigma_1 (\log s)^2$ or $\log s \approx 13$ or $s \gtrsim 100\ (\mathrm{TeV})^2$ and the logarithmic term will be dominant and $\rho(s)$ will decrease. This corresponds to $\sqrt{s} \approx 300$ GeV which will be within the reach of the colliding beam device ISABELLE proposed at Brookhaven. Well, let's wait a few years and see.

The essence of this brief lesson is that

$$T(s,t) \underset{\substack{s\ \mathrm{large} \\ t\ \mathrm{fixed}}}{\sim} \beta(t)\, s^{\alpha(t)} (\log s)^{\eta} , \quad (11)$$

so almost power behavior in s with some log s. $\alpha(t) = \alpha_0 + \alpha_0' t$ so

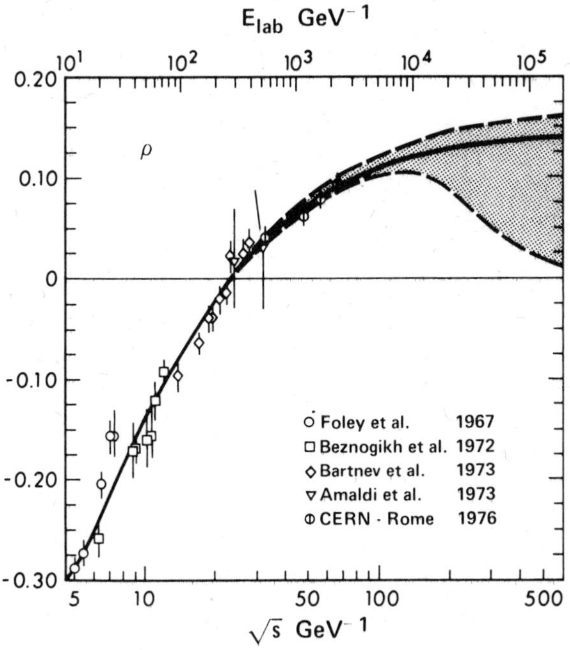

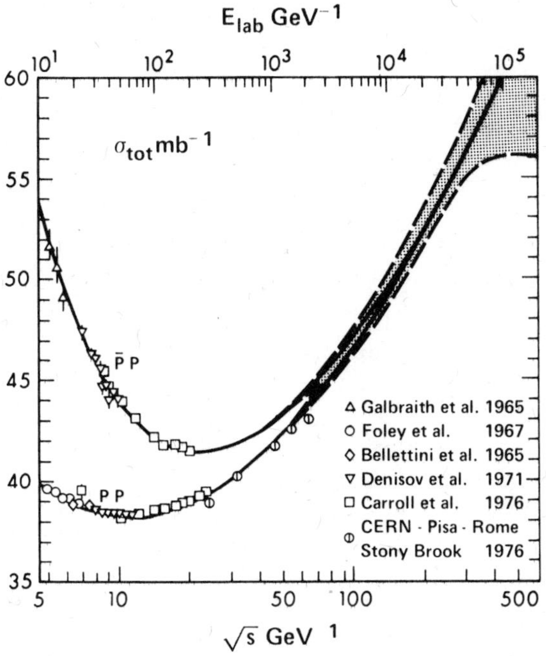

Fig. 3. $\rho(s) = \mathrm{Re}\, T(s,0)/\mathrm{Im}\, T(s,0)$ and $\sigma_T(s)$ for pp and $p\bar{p}$ collisions. Taken from Ref. 2 with permission from North-Holland Publishing Company, Amsterdam.

$$\frac{d\sigma(s,t)}{dt} \sim s^{2(\alpha_o - 1)} e^{-2\alpha_o'|t|\log s} \quad , \tag{12}$$

and the distribution in t becomes sharper as s increases.

These same Reggeons (quantities characterized by some $\alpha(t)$ and definite quantum numbers) appear wherever quantum numbers allow and whenever a process or sub-process has high s, small t. Several examples are discussed in Reference 1 and include multi-particle exclusive processes, inclusive processes ($A + B \to C +$ anything) - which are generalizations of σ_{total}, etc.

How can we find power behavior in s? If we turn to quantum field theory and calculate the amplitude for the exchange of a spin J quantum in our elastic scattering, Figure 4, there results

$$T(s,t) \underset{\substack{s \text{ large} \\ t \text{ fixed}}}{\approx} \frac{g_A g_B}{m_J^2 - t} (p_A \cdot p_B)^J \approx \frac{g_A g_B}{m_J^2 - t} s^J \quad . \tag{13}$$

This is indeed power behavior in s. It is, however, not acceptable because: (a) J is a fixed number and does not vary with t, (b) J may be any number, while the Froissart-Martin bound on T(s,0) which follows essentially from unitarity tells us that if

$$T(s,0) \approx s^{\alpha}(\log s) \quad , \tag{14}$$

then $\alpha \leq 1$ and if $\alpha = 1$, $\eta \leq 2$, and (c), the amplitude is real.

We learn from this example two important lessons: (i) $\alpha(t)$ is a generalized angular momentum of a t-channel quantum called a Reggeon. That is to say a Reggeon is a "state" off the mass shell and off the spin shell. (ii) We need more unitarity than is present in (13) to give our amplitude imaginarity.

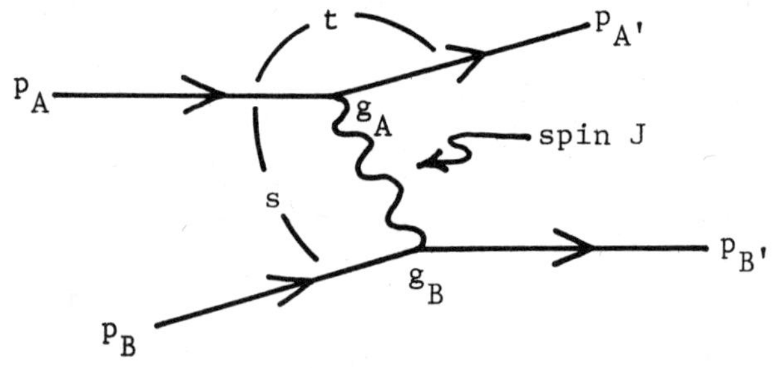

Fig. 4. The exchange of a spin-J quantum.

We will pursue these points in order. So first we wish to examine the question of amplitudes for arbitrary spin J. Let's begin in the t-channel center-of-momentum frame ($t \geq 4m^2$, $s \leq 0$) where $T(s,t)$ has the partial wave expansion

$$T(s,t) = \sum_{J=0}^{\infty} (2J+1) P_J(\cos\theta_t) F_J(t) \,, \tag{15}$$

$$\cos\theta_t = 1 + \frac{2s}{t - 4m^2} \,, \tag{16}$$

and

$$F_J(t) = \int_{-1}^{+1} d(\cos\theta_t) P_J(\cos\theta_t) T(s,t) \,. \tag{17}$$

This defines an amplitude for integer J for whatever t we like. The representation (17) is not adequate for general J[3] so we use a device introduced by Froissart and Gribov. Write a dispersion relation in s or z_t for fixed t

$$T(z_t,t) = \frac{1}{\pi} \int_{z_0(t)}^{\infty} dz' \frac{\mathrm{Im}\, T(z',t)}{z'-z_t} + \frac{1}{\pi} \int_{-\infty}^{-z_0(t)} dz' \frac{\mathrm{Im}\, T(z',t)}{z'-z_t} \,, \tag{18}$$

where

$$z_0(t) = 1 + 8m^2/(t - 4m^2). \tag{19}$$

Now $F_J(t)$ reads

$$F_J(t) = \frac{1}{\pi} \int_{z_0(t)}^{\infty} dz' Q_J(z') \{\mathrm{Im}\, T(z',t) - (-1)^J \,\mathrm{Im}\, T(-z',t)\}, \tag{20}$$

where the second kind Legendre function

$$Q_J(z) = \frac{1}{2} \int_{-1}^{+1} dx \frac{P_J(x)}{z - x} \,, \tag{21}$$

has appeared. But for the factor $(-1)^J$ in (20) we would be ready to continue $F_J(t)$ to arbitrary J; however, an ambiguity of the form $\sin \pi J$ which vanishes at $J = 0,1,...$ must be avoided so we define separate functions $F^{\tau}(J,t)$ for $\tau = \pm$ as the continuation into the J plane for J even and J odd

$$F^{\tau}(J,t) = \frac{1}{\pi}\int_{z_o(t)}^{\infty} \frac{dz'}{\pi} Q_J(z')\{\mathrm{Im}\ T(z',t) - \tau\,\mathrm{Im}\ T(-z',t)\}. \tag{22}$$

$\tau = \pm 1$ is called <u>signature</u>, and the $F(J,t)$ are called <u>signatured partial wave amplitudes</u>; they are well defined away from J = integer and coincide exactly with $F_J(t)$ for J even ($\tau = +1$) or J odd ($\tau = -1$). One reconstructs $T(s,t)$ by

$$T(s,t) = \sum_{\substack{J=0 \\ \tau=\pm 1}}^{\infty} (2J+1) F^{\tau}(J,t) \left\{\frac{P_J(z_t) + \tau P_J(-z_t)}{2}\right\} . \tag{23}$$

Since $F^{\tau}(J,t)$ is defined in the J-plane we may turn (23) into a contour integral first around the real axis and then parallel to the imaginary axis (Figure 5) so

$$T(s,t) = \frac{i}{2}\sum_{\tau}\int_{c-i\infty}^{c+i\infty} dJ\,(2J+1) F^{\tau}(J,t) \left[\frac{P_J(-z_t) + \tau P_J(z_t)}{\sin \pi J}\right] , \tag{24}$$

where from (14) we know Re $c > 1$.

At large s and fixed t we may approximate this transform pair for $T(s,t)$ and $F^{\tau}(J,t)$ by the more convenient forms

$$T(s,t) = \sum_{\tau}\int_{c-i\infty}^{c+i\infty} \frac{dJ}{2\pi i}\, s^J\, \frac{(\tau + e^{-i\pi J})}{\sin \pi J}\, F^{\tau}(J,t) , \tag{25}$$

and

$$F^{\tau}(J,t) = \int_{1}^{\infty} ds\ s^{-J-1}\, \frac{1}{2}\,[\mathrm{Im}\ T(s,t) - \tau\,\mathrm{Im}\ T(-s,t)] , \tag{26}$$

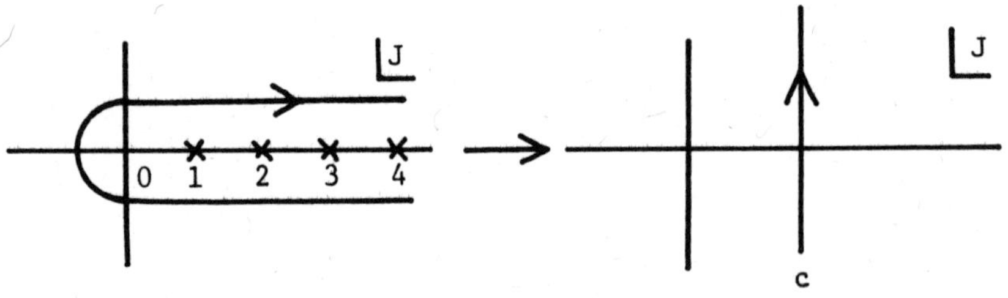

Fig. 5. The complex J-plane. The first contour reproduces the sum in Eq. (23) and the second uses the analyticity of $F^{\tau}(J,t)$ to open the contour parallel to the imaginary axis.

that is, a Mellin transform pair.

Now we want to translate powers of s and log s into the conjugate variable J. If $F^{\tau}(J,t)$ has a pole in J

$$F^{\tau}(J,t) = \beta_{\tau}(t)/(J - \alpha_{\tau}(t)), \tag{27}$$

then

$$T(s,t) = \sum_{\tau} \beta_{\tau}(t) \frac{(\tau + e^{-i\pi\alpha_{\tau}(t)})}{\sin \pi\alpha_{\tau}(t)} s^{\alpha_{\tau}(t)} . \tag{28}$$

So a pole in J is equivalent to power behavior in s. If we have a branch point in J

$$F^{\tau}(J,t) = \frac{1}{[J - \alpha(t)]^{\eta+1}} , \tag{29}$$

then

$$T(s,t) \sim s^{\alpha(t)} (\log s)^{\eta} . \tag{30}$$

This ends our lesson in J-plane technology. We now have the formal apparatus to pass from s dependence to the dependence on the conjugate variable J. Our path will be to consider how we can establish power dependence in s of T(s,t). We will utilize the physics we found lacking after Eq. (14), namely unitarity, to study poles in J. This study will not make us happy enough, but in its flaws we will discover that unitarity also demands branch points in J (read powers of log s in T(s,t)) and in following up on that we will become much happier.

3. THEORY OF REGGE POLES

Our task now is to inquire how we shall incorporate unitarity into theories with power behaved T(s,t). The framework for the discussion is that of the multiperipheral model and was essentially given 15 years ago by Amati, Fubini, and Stanghellini[4]. The first idea, which is a technical trick, is to use the unitarity relation to construct A(s,t) = Im T(s,t) and after that find Re T(s,t) via analyticity in s. Unitarity reads (Figure 6)

$$A(s,t) = \frac{(2\pi)^4}{2} \sum_N \prod_{j=1}^{N} \frac{d^4p_j \delta(m_j^2 - p_j^2)\theta(p_j^0)}{(2\pi)^3} T_{AB\to N}(p_A+p_B\to p_1+\ldots+p_N)$$
$$\times T^*_{A'B'\to N}(p_{A'}+p_{B'}\to p_1+\ldots+p_N)\delta^4(p_A+p_B-\sum_{j=1}^{N} p_j) \tag{31}$$

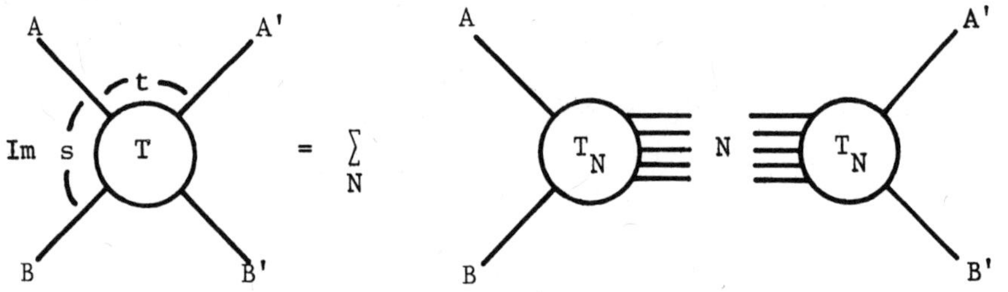

Fig. 6. Unitarity in the s-channel.

so the problem shifts to the study of the N particle production amplitude $T_{AB\to N}$. The approximation to this proposed by Amati, Fubini and Stanghellini was to treat it as closely as possible to the one quantum exchange process of Figure 4 with the exchange of light mass "pions". This peripheral form of interaction takes account of the longest range (~h/mc) forces. One essential ingredient in this multiperipheral model is to use the propagators $(m^2-t_j)^{-1}$ (see Figure 7) of the exchanged particles to yield small transverse momenta for the produced particles: $\langle p_T\rangle \approx \sqrt{-t_j} \approx m$, consistent with experimental observations on average transverse momentum. One may substitute for $(m^2-t_j)^{-1}$ any large inverse power $(m^2-t_j)^{-p}$ or even $\exp-|t_j|/m^2$ and the essential results to follow remain true. Now the multiperipheral construction for $T_{AB\to N}$ is

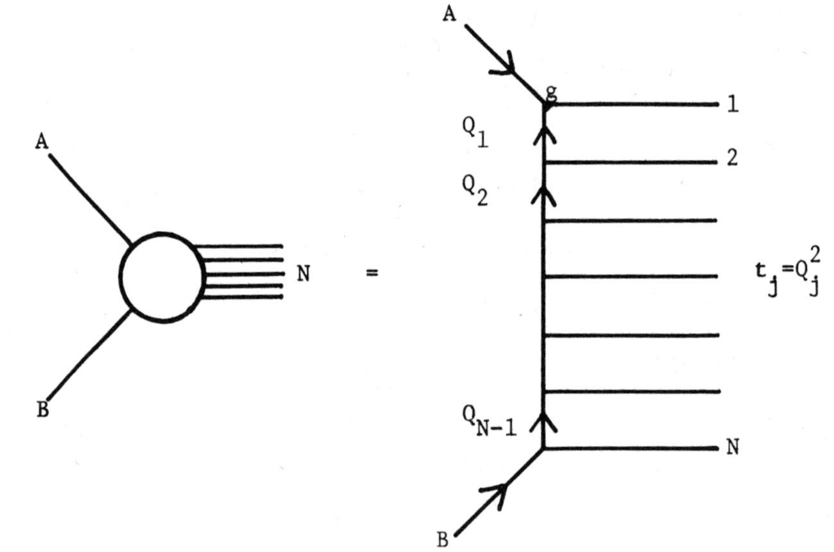

Fig. 7. The multiperipheral approximation to the N particle production amplitude.

$$T_{AB\to N} = g^N \prod_{j=1}^{N-1} \frac{1}{m^2 - t_j} \quad . \tag{32}$$

This yields for $A_N(s,t)$, the N particle contribution to $A(s,t)$ (Figure 8), from which we can deduce the recursion relation

$$A_N(s,t,t_1,t_2) = \int \frac{d^4p\, \delta^4(p^2-m_0^2)\theta(p^0)g^2A_{N-1}(s',t,t_1',t_2')}{(2\pi)^3\, (m_1^2-t_1')(m^2-t_2')} \quad , \quad N \geq 2 \tag{33}$$

which via the definition of $A(s,t)$

$$A(s,t) = A_1(s,t) + \sum_{N=2}^{\infty} A_N(s,t) \tag{34}$$

leads to the integral equation (Figure 9)

$$A(p,Q,k) = A_1(p,Q,k) + \int d^4p' A_1(p,Q,p')A(p',Q,k)G_0(t_+(p'),t_-(p')) \quad , \tag{35}$$

where

$$t_\pm(p') = (\frac{Q}{2} \pm p')^2 \quad , \tag{36}$$

and

$$G_0(t_1,t_2) = [(m^2-t_1)(m^2-t_2)]^{-1} \quad . \tag{37}$$

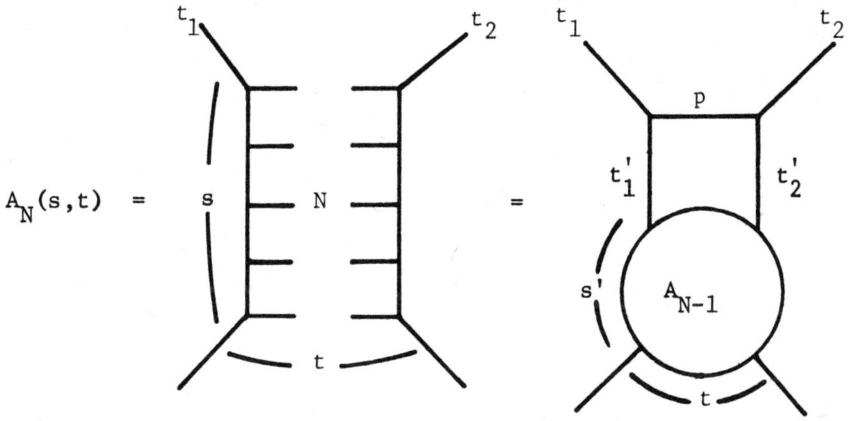

Fig. 8. The recursion relation.

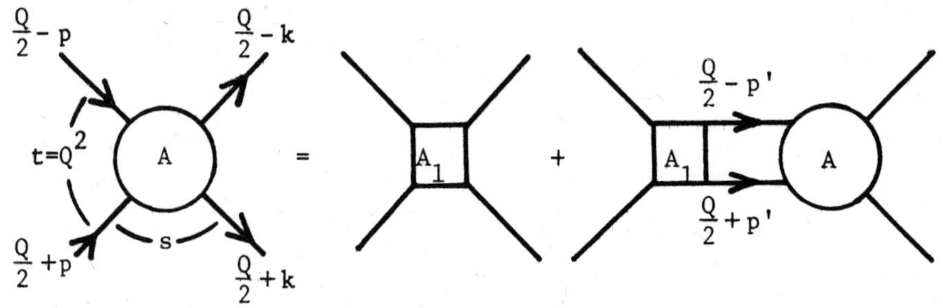

Fig. 9. The multiperipheral integral equation.

We can partially diagonalize this integral equation by going over to the signatured partial wave amplitude

$$F^{\tau}(J,t,t_+(p),t_-(p),t_+(k),t_-(k))$$

$$= \int_{z_0(t)}^{\infty} dz_t\, Q_J(z_t)[A(z_t,t) - \tau A(-z_t,t)] \tag{38}$$

$$= \int \frac{dt_+(p')dt_-(p')}{\sqrt{-\Delta(t,t_+(p'),t_-(p'))}} A_1(J,t,t_\pm(p),t_\pm(p'))$$

$$\times F^{\tau}(J,t,t_\pm(p'),t_\pm(k))G_0(t_\pm(p')) , \tag{39}$$

where

$$\Delta(x,y,z) = (x+y-z)^2 - 4xy . \tag{40}$$

This is an integral equation in two variables, namely the masses of the legs of $A(s,t)$, in which J and t, the two parameters we wish to correlate in $J = \alpha(t)$ are parametric. That two variables remain is due to our having integrated out a polar and an azimuthal angle from the full absorptive part $A(s,t)$ to form $F(J,t)$. So from four dimensions we have come to two remaining dynamical ones.

Now written as

$$F = A_1 + \int A_1 G_0 F \tag{41}$$

our integral equation is just like the Schrödinger equation

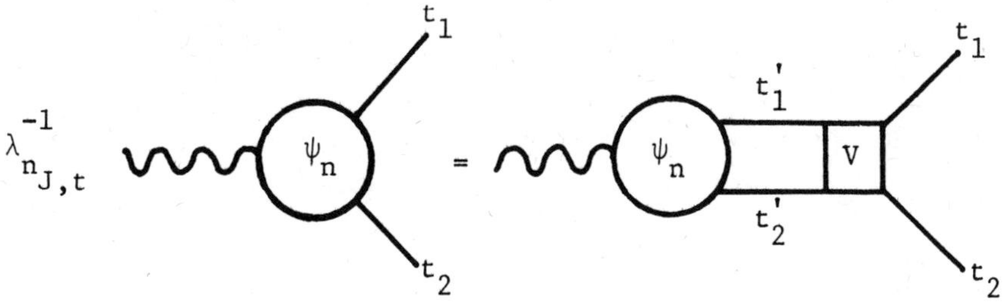

Fig. 10. The homogeneous equation for the eigenfunctions ψ_n and eigenvalues $\lambda_n(J,t)$ of the multiperipheral "potential" V. This defines the Reggeon as a hadron bound state for general J,t.

familiar from high school quantum mechanics. We solve it[5] by seeking the eigenfunctions and eigenvalues of the symmetrized kernel or "potential" $V = \sqrt{G_0}\, A_1 \sqrt{G_0}$

$$\lambda_n^{-1}(J,t)\psi_n(J,t,t_1,t_2) =$$

$$\int\frac{dt_1'dt_2'}{\sqrt{-\Delta(t,t_1,t_2)}}\, V(J,t,t_1,t_2,t_1',t_2')\psi_n(J,t,t_1',t_2') \,, \tag{42}$$

(see Figure 10)

and then we expand A(J,t,...) in these eigenfunctions

$$F^{\tau}(J,t,t_1,t_2,t_3,t_4) = \sum_n \frac{\psi_n(J,t,t_1,t_2)\psi_n^*(J,t,t_3,t_4)}{\lambda_n(J,t) - 1} \,. \tag{43}$$

When the eigenvalue $\lambda_n(J,t) = 1$ we have a singularity in F(J,t,...) along the trajectory $J = \alpha_n(t)$. Suppose the first λ_n to reach 1 is called λ_0 and near $J = \alpha_0(t)$

$$\lambda_0(J,t) = 1 + [J - \alpha_0(t)] \left.\frac{\partial\lambda}{\partial J}\right|_{J=\alpha_0(t)} + \ldots \,, \tag{44}$$

then

$$F(J,t,t_1,t_2,t_3,t_4) = \frac{\beta_0(t,t_1,t_2)\beta_0(t,t_3,t_4)}{J - \alpha_0(t)} \,, \tag{45}$$

(see Figure 11)

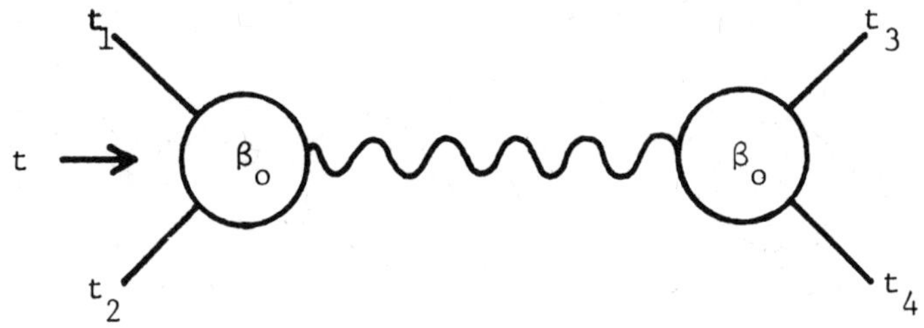

Fig. 11. The partial wave amplitude for one J-plane pole.

which is to be read as a vertex β_0 for hadrons of masses t_1 and t_2 to form a Reggeon of mass t, spin J, then the Reggeon propagates with propagator $(J - \alpha_0(t))^{-1}$, then emerges into hadrons of masses t_3 and t_4 with amplitude β_0.

We have achieved now the pole in J (or the power $s^{\alpha(t)}$) we set out to locate. Let us consider the physics of this before going on to let its faults guide us to the next level of Reggeon physics. First it is clear that we may generalize the "potential" embodied in A_1 to include any finite number of produced particles (see Figure 12). Second, the pole comes from a sum over an infinite number of produced particles. This is consistent with picturing the Reggeon as a bound state of hadrons with arbitrary J and t resulting from an infinite number of interactions via the potential A_1. Any finite number of interactions produces a $\sigma_{total}(s)$ which rises for small s and eventually falls away as $s^{-1}(\log s)^{power}$. Third, from our integral equation we can find the average number of "clusters" (Figure 12) produced at energy s:

$$\bar{N}(s) = \sum_N NA_N(s,0)/\sum_N A_N(s,0) \tag{46}$$

$$= \sum_N N\sigma_N(s)/\sigma_{total}(s) \tag{47}$$

$$= s^{-\alpha(0)} \sum_N N\sigma_N(s) . \tag{48}$$

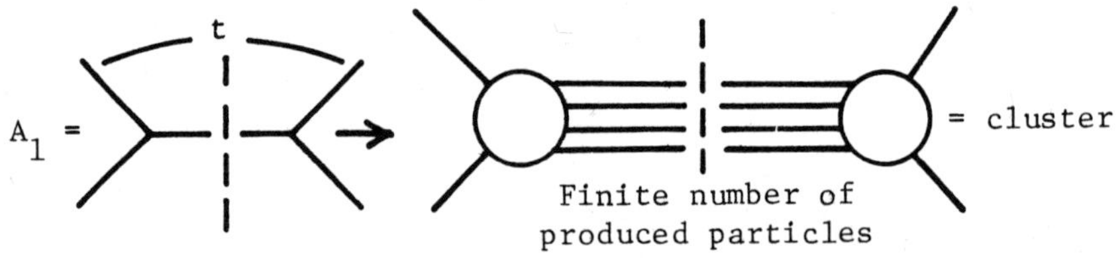

Fig. 12. The inhomogeneous term of the multiperipheral integral equation and its generalization to a cluster.

Using the recursion relation (33) we write

$$A_N(s) = \frac{1}{N} \sum_{j=1}^{N} A_{N-j}(s_1,t')A_j(s_2,t')dt' \tag{49}$$

as in Figure 13. So

$$\sum_N NA_N(s,0) = \int dt'A(s_1,t')A(s_2,t') , \tag{50}$$

which in the J plane means

$$\int_1^\infty ds\, s^{-J-1} \sum_N NA_N(s,0) = 1/(J - \alpha(0))^2 . \tag{51}$$

In s this means $\sum_N NA_N(s,0) = s^\alpha \log s$ according to our folklore around Eq. (29) and finally

$$\bar{N}(s) \sim \log s . \tag{52}$$

Just this average multiplicity of produced hadrons has been observed in reactions initiated by hadrons, in reactions initiated by photons (on mass shell and even far off mass shell), in jets associated with hadrons produced at large transverse momentum, and possibly elsewhere. This may be understood in a "back of the envelope" fashion by considering the phase space available for a produced object with longitudinal momentum p_L and transverse momentum p_T

$$\int \frac{d^2p_T\, dp_L}{\sqrt{m^2+p_T^2+p_L^2}} f(p_T^2) , \tag{53}$$

where $f(p_T^2)$ is sharply damped. Then (53) is essentially

$A_N(s,0) =$ 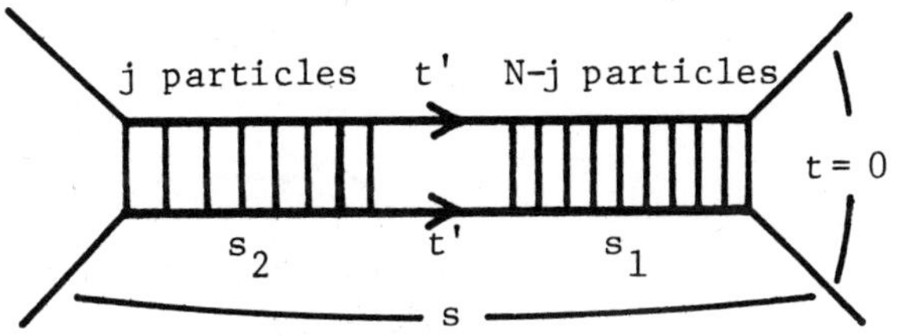

Fig. 13. The split of $A_N(s,0)$ for evaluating $\sum_N NA_N(s,0)$ used in $\bar{N}(s)$.

$$\int_{-\sqrt{s}/2}^{+\sqrt{s}/2} \frac{dp_L}{\sqrt{p_L^2+m^2+\langle p_T^2\rangle}} \sim \log s \,. \tag{54}$$

This leads one boldly to imagine that <u>any theory which exhibits bounded transverse momentum will have Regge behavior.</u>

4. REGGEON UNITARITY

After some effort we have arrived at poles in J - i.e. Reggeons. Now we are forced to face up to facts; namely, they won't do. The first argument against them is that there is no where in the theory of poles a restriction on $\alpha(t)$; indeed, there are examples of multiperipheral models which explicitly have $\alpha(t)$ as large as you like. Since unitarity requires $\alpha(0) \leq 1$, <u>we don't have enough unitarity</u>. Secondly, we have $T(s,t) \sim s^{\alpha(t)}$, but no $(\log s)^{\eta}$. We clearly haven't finished. Thirdly and most upsetting, <u>poles alone are inconsistent with unitarity</u>. In the demonstration of that we will also see how to eliminate all these diseases.

Let's return to the unitarity relation (31) and examine the N = 2 term (see Figure 14)

$$\mathrm{Im}\, T(s,t) = \int \frac{d^3p_1}{2E_1}\frac{d^3p_2}{2E_2}\,\delta^4(p_A+p_B-p_1-p_2)T(s,t_1)T^*(s,t_2) \tag{55}$$

$$= \frac{1}{\Delta^{\frac{1}{2}}(s,m^2,m^2)}\int \frac{dt_1 dt_2\,\theta(-\mathcal{J})}{\sqrt{-\mathcal{J}}}\, T(s,t_1)T^*(s,t_2) \tag{56}$$

where

$$\mathcal{J} = \Delta(t,t_1,t_2) + \frac{t\,t_1t_2}{\frac{s}{4}-m^2} \,. \tag{57}$$

Im T(s,t) =

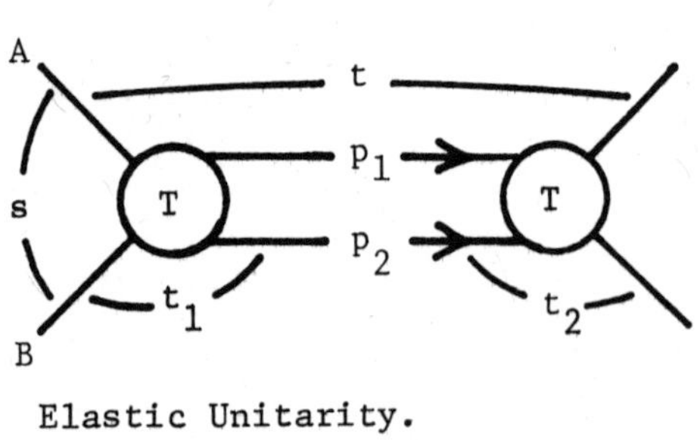

Fig. 14. Elastic Unitarity.

For s large, t fixed we may replace

$$T(s,t_i) = \beta(t_i)\, s^{\alpha(t_i)} \tag{58}$$

so

$$\mathrm{Im}\, T(s,t) = \int \frac{dt_1 dt_2}{\sqrt{-\Delta(t,t_1,t_2)}} \beta(t_1)\beta^*(t_2)\, s^{\alpha(t_1)+\alpha(t_2)-1} \quad , \tag{59}$$

yielding a partial wave amplitude[6]

$$F(J,t) = \int \frac{dt_1 dt_2}{\sqrt{-\Delta(t,t_1,t_2)}} \frac{\beta(t_1)\beta^*(t_2)}{(J-1)-(\alpha(t_1)-1)-(\alpha(t_2)-1)} \quad , \tag{60}$$

which leads to a branch point in J at

$$J_{BP}(t) - 1 = 2\left(\alpha(\tfrac{t}{4}) - 1\right) \quad . \tag{61}$$

So if we begin with poles, unitarity forces branch points on us. That, actually, is a most pleasant result for branch points give us log s terms in $\sigma_T(s)$, and we need them. Furthermore these branch points are important only when $\alpha(0) = 1$ since then $J_{BP}(0) = 1$. Only when we get up to the unitarity bound does the presence of branch points matter. If $\alpha(0) < 1$, then the pole gives us $s^{\alpha(0)-1}$ in the cross section and the branch point gives $s^{2(\alpha(0)-1)}$ log s which is unimportant for large s.

Next we write the expression (60) for the two Reggeon branch point in a suggestive fashion. Make the following change of variables: go from t_1 and t_2 to two momenta $\vec{q}_i$ and $\vec{q}$ such that

$$t_i = -|\vec{q}_i|^2 \quad , \qquad t = -|\vec{q}|^2 \quad , \tag{62}$$

define E = 1 - J (the Reggeon energy) and

$$\varepsilon(\vec{q}_i) = 1 - \alpha(\vec{q}_i) \quad , \tag{63}$$

and finally take the discontinuity across the branch line in E to find

$$\mathrm{disc}_E F(E,\vec{q}) = -2\pi i \int d^2q_1 d^2q_2 dE_1 dE_2 \delta^2(\vec{q}-\vec{q}_1-\vec{q}_2)\delta(E-E_1-E_2)$$
$$\times \delta(E_1-\varepsilon(\vec{q}_1))\delta(E_2-\varepsilon(\vec{q}_2))\beta(\vec{q}_1)\beta^*(\vec{q}_2) \quad . \tag{64}$$

Let us compare this with the two particle contribution to the usual unitarity relation

$$\text{disc}_s T(s,t) = \int d^4q_1 d^4q_2 \delta^4(q-q_1-q_2)\delta(q_1^2-m_1^2)\delta(q_2^2-m_2^2)$$
$$T(q\to q_1+q_2)T^*(q\to q_1+q_2) \ . \qquad (65)$$

This analogy suggests that each Reggeon lives in a phase space $d^2q\, dE$, just as Minkowski states live in a phase space d^4q. Each Reggeon in the Reggeon unitarity relation is on "mass shell", $E_i = \varepsilon(\vec{q}_i)$, just as an ordinary particle on "mass shell" has the energy-momentum relation $q^2 = m^2$ or $E_q = \sqrt{m^2+\vec{q}^{\,2}}$. Energy and momentum are conserved in Reggeon unitarity as in usual unitarity. Finally there is an amplitude to produce two Reggeons (called β here) and one to absorb them; these are the analogues of T in (65).

All this suggests we view a Reggeon as living in two space dimensions, $\vec{x}$ (conjugate to $\vec{q}$) and one time dimension, τ (conjugate to E). In this space it can propagate freely with amplitude $-(J-\alpha(t))^{-1} = (E-\varepsilon(\vec{q}))^{-1}$. If the energy momentum relation happens to be $\alpha(t) = \alpha_0+\alpha_0' t$, then $\varepsilon(\vec{q}) = \alpha_0'\vec{q}^{\,2} + (1-\alpha_0)$ and in $\vec{x},\tau$ space the amplitude is

$$\theta(\tau)\exp[-\tau(1-\alpha_0)]\ \frac{e^{-|\vec{x}|^2/4\alpha_0'\tau}}{4\pi\alpha_0'\tau} \ , \qquad (66)$$

that is the Reggeon diffuses in space to an average transverse distance $\sim\sqrt{\alpha_0'\tau}$ and decays in time τ with a decay length $\sim 1/(1-\alpha_0)$. If $\alpha_0 \approx 1$, then the Reggeon (now the Pomeron with vacuum quantum numbers) connects events over very long times.

To enforce general Reggeon unitarity we may now imagine introducing a Reggeon Field Theory (RFT) which associates a field amplitude $\phi(\vec{x},\tau)$ with each Reggeon. Then whatever interaction we may have, just kinematically the Reggeon unitarity relations will be satisfied. Such a field theory will give perturbative approximations to amplitudes for states with many intermediate Reggeons (we considered only two Reggeons up until now) which order by order satisfy Reggeon unitarity. Since Reggeon unitarity can be shown[1] to follow from t-channel unitarity, we are clearly now creating a tool for adding in more unitarity than we had with the multiperipheral model or equivalently, J-plane poles alone.

If we can devise non-perturbative schemes for studying our RFT, then we may have expressions for partial wave amplitudes with some approximation to full multiparticle t-channel unitarity built in. Now we know a case when we will be required to go beyond perturbation theory in our RFT. When $\alpha_0 = 1$, that is when we get up to the unitarity limit as far as powers of s go, N

Reggeon exchange gives a branch point in J at J = 1 when t = 0. In this case we have a conglomeration of an infinite number of branch points on top of a pole at E = 1-J = 0 and $\vec{q} = 0$. This regime of E,$\vec{q}$ space is precisely what we want to study for the small t, large s behavior of T(s,t). That very high energy behavior is thus an infrared problem in RFT.

5. A SUGGESTIVE ANALOGY

We have identified a Reggeon as a hadron bound state at general J and t (or E and $\vec{q}$) with a mass shell relation $J = \alpha(t)$ (or $E = \varepsilon(\vec{q})$), and then we have proposed for purposes of satisfying Reggeon unitarity writing a field theory to describe the propagation, interaction, emission, absorption, ... of these bound states. Such a field theory will in general be non-local and thus unfamiliar. Before launching into that we might well consider an analogous situation in ordinary field theory I trust the following example will illuminate the issues.

Suppose one were given the scalar field theory

$$\mathcal{L}(\phi) = \frac{1}{2}(\partial_\mu\phi)^2 - \frac{m_\phi^2}{2}\phi^2 - \frac{\lambda\phi^4}{4!} \tag{67}$$

and told that the ϕ quanta were the fundamental co-ordinates of the world. Furthermore you are informed there is a scalar bound state called χ which comes from solving the $\phi\phi$ Bethe-Salpeter equation (see Figure 15). How will you describe the properties of $\chi\chi$ scattering, production, etc.? Clearly χ is a composite of ϕ's and its field theory is likely non-local. It can scatter elastically, of course, via χ exchange (see Figure 16), and $T_{\chi\chi}(s,t)$ will satisfy elastic unitarity

$$\mathrm{Im}_s T_{\chi\chi}(s,t) = \int d^4p_1' d^4p_2' \delta^4(p_1+p_2\to p_1'+p_2')\delta(p_1'^2-m_\chi^2)\delta(p_2'^2-m_\chi^2)$$
$$\times\, T_{\chi\chi}(s+i\varepsilon)T_{\chi\chi}(s-i\varepsilon) \quad . \tag{68}$$

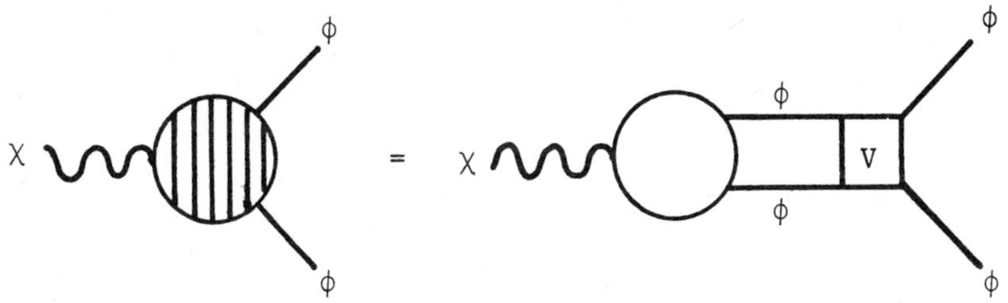

Fig. 15. χ as a $\phi\phi$ bound state.

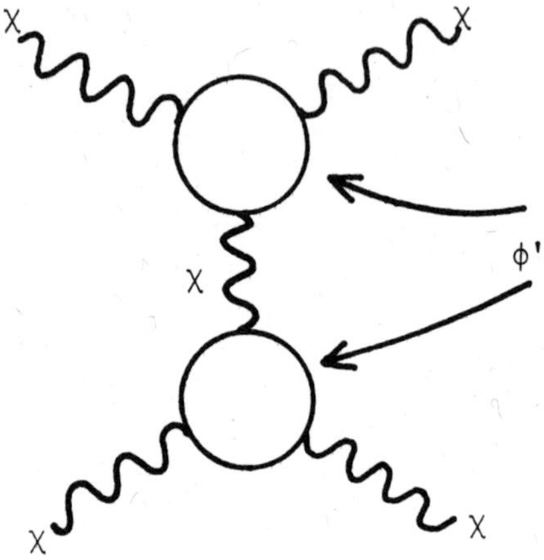

Fig. 16. χχ scattering via χ exchange.

So one might, despite the potential non-locality of the χ field theory, use it to investigate some physical questions.

One may arrive at the χ field theory in the following formal way. Begin with the generating functional W[J] for φ Green functions defined through the path integral

$$e^{iW[J]} = \int d\phi(x)\; e^{i\int d^4x[\mathcal{L}[\phi] + J(x)\phi(x)]} , \tag{69}$$

and note

$$\int d\chi(x)\; e^{i\int d^4x[-\frac{\chi(x)^2}{2} + \sqrt{\frac{\lambda}{12}}\,\chi(x)\phi(x)^2]}$$
$$= \exp[-i\int d^4x\; \frac{\lambda\phi(x)^4}{4!}] . \tag{70}$$

This introduces a field χ(x) as a "trade off" for $\phi(x)^2$. If we use (70) in the definition of W[J] and do the gaussian integration on φ(x) we arrive at

$$\exp\; iW[J] = \int d\chi(x)\exp i\left\{\int d^4x[-\frac{\chi(x)^2}{2} + \right.$$
$$\frac{1}{2}<x|\log(1 - \frac{1}{\partial^2 + m_\phi^2}\sqrt{\frac{\lambda}{12}}\,\chi(x))|x> +$$
$$\left. J(x)\int d^4y<x|(\partial^2 + m_\phi^2 - \sqrt{\frac{\lambda}{12}}\,\chi(x))^{-1}|y>J(y)]\right\} . \tag{71}$$

Now ϕ is "gone", though W[J] still generates connected ϕ Green functions.

We can now take the "effective" lagrangian

$$\mathcal{L}[\chi] = -\frac{\chi(x)^2}{2} + \frac{1}{2} \langle x|\log(1 - \frac{1}{\partial^2 + m_\phi^2}\sqrt{\frac{\lambda}{12}}\,\chi(x))|x\rangle \ , \tag{72}$$

to define χ interactions. Clearly this $\mathcal{L}[\chi]$ has an infinite number of χ interactions all of which are non-local. The cubic interaction is, for example,

$$\frac{1}{6}\left(\frac{\lambda}{12}\right)^{3/2} \int d^4y\ d^4z\ d^4x\ \chi(x)\chi(y)\chi(z)\ \times$$

$$\langle x|\frac{1}{\partial^2+m_\phi^2}|y\rangle\langle y|\frac{1}{\partial^2+m_\phi^2}|z\rangle\langle z|\frac{1}{\partial^2+m_\phi^2}|x\rangle \ . \tag{73}$$

When can we treat this as a local interaction? If $m_\phi \to \infty$, or more precisely when all momenta p_χ are small with respect to m_ϕ then

$$\langle x|\frac{1}{\partial^2+m_\phi^2}|y\rangle \approx \frac{1}{m_\phi^2}\,\delta^4(x-y) \tag{74}$$

and the theory becomes essentially local. That is to say, in an infrared limit where composite momenta are small with respect to constituent mass scales, the theory of composite interaction can be treated as local.

Lest we have gotten too obscure with our formal analogy let me note that hadrons are like the ϕ's and Reggeons, the χ's. If we are interested in the infrared structure of the RFT, we may treat it as local in $\vec{x},\tau$ space knowing by our analogy that non-infrared effects are where locality will bare its fangs.

6. CRITICAL POMERON AND ORDERED PHASE OF RFT[7]

We are finally prepared to consider a Reggeon Field Theory. Following the lesson of the previous section we will make it local in $\vec{x},\tau$ space and allow ourselves only to address infrared $(E,\vec{q} \to 0)$ questions. First, we must choose an energy momentum relationship for non-interacting Reggeons. There is no compelling choice but

$$\alpha_0(t) = \alpha_0 + \alpha_0' t \tag{75}$$

or

$$\varepsilon(\vec{q}) = \alpha_0' \vec{q}^{\,2} + \Delta_0 \tag{76}$$

with

$$\Delta_0 = 1 - \alpha_0 \quad , \tag{77}$$

the "mass" gap. This leads to the Lagrange density

$$\mathcal{L}_0 = \frac{i}{2} \tilde{\phi}(\vec{x},\tau) \frac{\overleftrightarrow{\partial}}{\partial\tau} \phi(\vec{x},\tau) - \alpha_0' \nabla\tilde{\phi}\cdot\nabla\phi - \Delta_0\tilde{\phi}\phi \quad , \tag{78}$$

where both $\phi(\vec{x},\tau)$ the Reggeon field amplitude and an "anti-amplitude" are necessary as always in lagrangian theories with linear energy, quadratic momentum relations (that is, Schrödinger or diffusion theories).

We will concentrate now on the vacuum Reggeon, the Pomeron, for which any power of fields $\tilde{\phi}$ and ϕ may enter the interaction lagrangian. For reasons outlined in Reference 1 only the lowest order theory with cubic interactions

$$\mathcal{L}_I = \frac{ir_0}{2} \tilde{\phi}(\tilde{\phi} + \phi)\phi \quad , \tag{79}$$

is important in the infrared limit. (Basically any higher powers of field in $\mathcal{L}_I$ are driven to zero by the renormalization process in the infrared limit.) Note the factor of i which appears before r_0, the real three Pomeron coupling, in $\mathcal{L}_I$. This arises from careful treatment[8] of the signature factors which enter the conjunction of three vacuum trajectories. Basically one finds that a Pomeron loop which comes from the second order of perturbation theory in r_0 must be negative. One could make Pomerons into fermions, I suppose, to get this (-1)/loop, but that seems extremely peculiar. Instead with each r_0 it is consistent to associate a factor i. In Reference 1 the phase of the n Pomeron → m Pomeron vertex is given.

With the total $\mathcal{L} = \mathcal{L}_0$ (Eq. (78)) + $\mathcal{L}_I$ (Eq. (79)) we want to evaluate the Reggeon Green function $G^{(n,m)}(E_i,\vec{q}_i)$ for n Reggeons to become m Reggeons. From that the signatured partial wave amplitude is given as in Figure 17, where N_k is the amplitude for two hadrons to produce k Reggeons and must be supplied to the theory from outside.

$$F(E,\vec{q}) = \sum_{n,m=1}^{\infty} E,\vec{q}$$

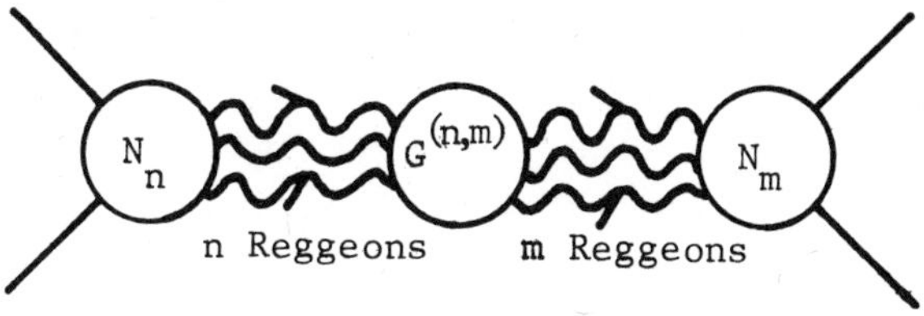

Fig. 17. The partial wave amplitude gotten from Reggeon Green functions, G, and Reggeon production amplitudes, N.

In the study of the Pomeron RFT we will discover the necessity to renormalize the quantities, ϕ, α_o', Δ_o, and r_o into new objects ϕ_R, α', $\Delta = 1 - \alpha(0)$ and r. Concentrate now on the renormalized intercept (or "mass") gap Δ. If $\Delta_o > \Delta_{oc}(\alpha_o', r_o, \ldots)$, it happens that $\Delta > 0$, which means the total cross section falls as a power of s

$$\sigma_T(s) \sim s^{-\Delta} = s^{-(1-\alpha(0))} \tag{80}$$

and we are in an uninteresting "phase" of the RFT. The interesting situation is $\Delta = 0$. This can come about through a miracle where all the constants $\alpha_o', r_o, \ldots$ are connected up to give $\Delta_o = \Delta_{oc}(\alpha_o', r_o, \ldots)$ (the critical gap) or $\alpha_o = \alpha_{oc}(\alpha_o', r_o, \ldots)$ such that renormalization just produces $\Delta = 0$. This unlikely event is called the critical Pomeron. The final possibility is that $\Delta_o < \Delta_{oc}$ (or $\alpha_o > \alpha_{oc}$) which is also of importance. It does not lead to $\Delta < 0$ $(\alpha(0) > 1)$ but to some unfamiliar other phenomena we will take up anon.

Just now let us focus on the critical Pomeron: $\Delta_o = \Delta_{oc}$ so $\Delta = 0$. This means the renormalized theory has no "mass" scale and there is a scale invariance which exhibits itself in scaling properties of the Green functions. (This is in direct analogy to scaling behavior[9] at the critical temperature in condensed matter physics.) These scaling properties are parametrized by an arbitrary energy scale, E_o, and two critical indices: η and z which reflect, respectively, how the field ϕ and the slope α' are renormalized. These critical indices can be calculated[1,10] in an ε-expansion about 4 dimensions for $\vec{x}$, in the analogue of high temperature expansions of many body theory, in loop expansions, and perhaps in other ways. After some effort the following approximate values emerge

$$\eta \approx 1/4 \quad , \quad z \approx 9/8 \quad . \tag{81}$$

An important example of such scaling laws occurs for the Pomeron propagator

$$G^{(1,1)}(E,q) = E_0\left(\frac{-E}{E_0}\right)^{\eta} H\left[\frac{\alpha'\vec{q}^{\,2}}{E_0}\left(\frac{-E}{E_0}\right)^{z}\right] , \tag{82}$$

where H is a function (not known in general) of the exhibited dimensionless argument. This and similar scaling laws for $G^{(n,m)}$ lead to the contributions to $\sigma_T(s)$ for A+B → anything (see Figure 18)

$$\sigma_T(s) \sim (\log s)^{\eta} g_A g_B - f_{AB}/\log s + h_{AB}/(\log s)^{2+\eta} + \ldots , \tag{83}$$

and the asymptotic differential cross section

$$\frac{d\sigma(s,t)}{dt} = (\log s)^{2\eta}\, B(\alpha' t(\log s)^{z}) . \tag{84}$$

These results indicate that $\sigma_T(s)$ should rise slowly for very large s; further calculations[1] of the scaling function B appearing in (84) show a structure of $d\sigma/dt$ quite consistent with experiment.

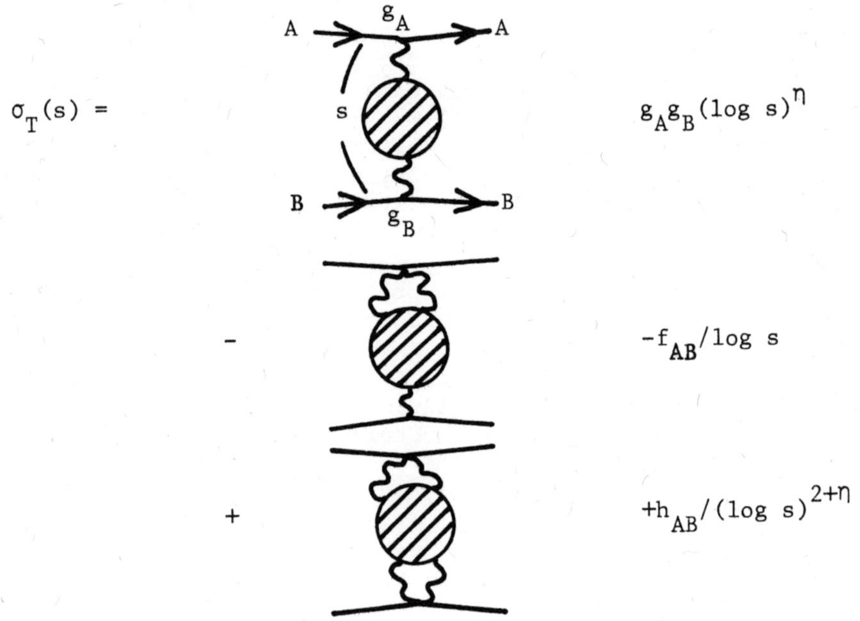

Fig. 18. The contributions to $\sigma_T(s)$ for the critical Pomeron. $\eta \approx 1/4$.

Whenever Pomerons appear the indices η and z will play a role. For example, the inclusive process $A+B \rightarrow C+$ anything has a cross section which depends on the total rapidity $Y = \log s$ and the rapidity of C, y_c,

$$\frac{d\sigma}{dy_c} = (Y - y_c)^\eta y_c^\eta \quad , \tag{85}$$

in the laboratory frame where $0 \le y_c \le Y$. This leads to an average multiplicity for particle c

$$\bar{n}_c(s) = (\log s)^{1+\eta} \quad , \tag{86}$$

which is a very mild departure from the $(\log s)^1$ of the multiperipheral model from which we began. That says that the presence of many, many critical Pomerons yields only a small change in $\bar{n}_c(s)$ and this must come from the alternating sign due to the ir_0 in $\mathcal{L}_I$ since each Pomeron by itself would try to produce log s secondaries.

With this solution of the Pomeron in hand one is able to consider Pomeron corrections to secondary quantum number carrying trajectories such as the ρ, corrections to fermion trajectories, properties of $\sigma_N(s)$, etc., etc.[1]. The picture that emerges is, remarkably enough completely consistent with s-channel unitarity. Only t-channel unitarity via the Reggeon unitarity relations was built in, you may recall, so it appears that we may finally have enough unitarity.

Unfortunately the critical Pomeron suffers from a logical ugliness; namely, where did we get the instruction to set $\Delta_o = \Delta_{oc}$ $(\alpha_o = \alpha_{oc})$? The bare Reggeon parameters are set by the underlying hadron theory (recall the ϕ,χ discussion of Section 5) and unless a miracle is working there it is unlikely in the extreme that $\alpha_o = \alpha_{oc}$. It is possible that $\alpha_o < \alpha_{oc}$ so $\alpha(0) < 1$, but that alternative appears to have been passed over by Nature (presumably she wished to use more of the limits allowed by unitarity). So we are invited to explore the regime $\Delta_o < \Delta_{oc}$ or $\alpha_o > \alpha_{oc}$.

The analysis of $\alpha_o > \alpha_{oc}$ has been carried out primarily by the CERN Reggeon team[11], and I will try to summarize their work and then sketch an instructive example due to Bronzan and Sugar[12].

One begins by treating $\vec{x},\tau$ space as a two dimensional square lattice of spacing a in $\vec{x}$ space and studies the field $\phi_j(\tau)$ at each space point j. This is a quantum mechanics problem since we deal with zero space dimensional fields on one time (i.e. point particles). At each lattice site one has the good old ground state $|0\rangle$ which is annihilated by $\phi_j(\tau)$

$$\phi_j(\tau)|0\rangle = 0 \tag{87}$$

and very close by a state $|1\rangle$ with energy

$$\varepsilon_1 = \frac{\Delta_o^2 a}{|r_o|} \exp -\left(\frac{2\Delta_o^2 a^2}{r_o^2}\right) , \tag{88}$$

where the now standard parameters Δ_o, r_o are connected with the hamiltonian

$$H = \int d^2x[\alpha_o' \nabla\tilde{\phi}\cdot\nabla\phi + \Delta_o\tilde{\phi}\phi + \frac{ir_o}{2}\tilde{\phi}(\tilde{\phi}+\phi)\phi] , \tag{89}$$

and on the lattice the α_o' term will couple sites. The two close states $|0\rangle$ and $|1\rangle$ are now split from the rest of the spectrum by order $|\Delta_o|$. This splitting with its non-analytic dependence on r_o is characteristic of a tunneling and occurs only for $\Delta_o < \Delta_{oc}$ (at this unrenormalized stage $\Delta_{oc} = 0$). In the limit $a \to 0$, as we go back to the continuum, $\varepsilon_1 \to 0$ and we have <u>two states $|0\rangle$ and $|1\rangle$ degenerate in energy</u>.

This is quite similar to the usual ϕ^4 theory

$$H = \frac{1}{2}(\nabla\phi)^2 + \lambda(\phi^2 - F^2)^2 \tag{90}$$

with spontaneous symmetry breaking where the field configurations

$$\phi = \pm F = \text{constant} \tag{91}$$

are degenerate in energy. However in the ϕ^4 theory one has (by considering the same lattice approach)

$$\langle 0|\phi|1\rangle = \langle 1|\phi|0\rangle = \Phi \neq 0 \tag{92}$$

and

$$\langle 0|\phi|0\rangle = \langle 1|\phi|1\rangle = 0 \tag{93}$$

so defining

$$|\Omega_\pm\rangle = \frac{1}{\sqrt{2}}(|0\rangle \pm |1\rangle) \tag{94}$$

we find

$$\langle\Omega_\pm|\phi|\Omega_\pm\rangle = \Phi , \tag{95}$$

$$\langle\Omega_\pm|\phi|\Omega_\mp\rangle = 0 , \tag{96}$$

and $<\Omega_{\pm}|\Omega_{\mp}> = 0$, (97)

that is, two ground states $|\Omega_{\pm}>$ in each of which the field is constant, between which there are no matrix elements of the field, and which are themselves orthogonal. That means one may choose <u>either</u> $|\Omega_{+}>$ or $|\Omega_{-}>$ as the ground state and build up a Hilbert space of excited levels on it. Those spaces will not communicate. Correlation functions

$$G(\vec{x}) = <\Omega_{\pm}|(\phi(\vec{x})-\Phi)(\phi(0)-\Phi)|\Omega_{\pm}> \quad (98)$$

for large $\vec{x}$ will behave as

$$G(\vec{x}) \sim e^{-|\vec{x}|(E_1 - E_0)} \quad (99)$$

where E_0 is the energy of $|\Omega_{\pm}>$ and E_1 the energy of the first level above it. Short range order is reestablished below the critical point in ϕ^4 theory.

In RFT the situation is quite different. Now one finds

$$<0|\phi|1> = <1|\tilde{\phi}|0> \quad , \quad (100)$$

$$<1|\phi|1> = <1|\tilde{\phi}|1> \quad , \quad (101)$$

and $<1|\phi|0> = <0|\phi|0> = 0$. (102)

We can no longer define disjoint, non-communicating ground states $|\Omega_{\pm}>$. We will always have state $|1>$ appearing as an allowed intermediate state in all Green functions

$$<0|\tilde{\phi} \;....\; \phi|0> \quad , \quad (103)$$

and the equivalent energy gap (E_1-E_0) of Eq. (99) is zero, and we will have long range order (in τ, the Reggeon time-really rapidity).

This means that the Reggeon energy gap is zero for $\Delta_o \leq \Delta_{oc}$ or equivalently $\Delta = 0$, $\alpha(0) = 1$ for $\alpha_o \geq \alpha_{oc}$ (see Figure 19) and the power behavior in s of $\sigma_T(s)$ is s^0 for $\alpha_o \geq \alpha_{oc}$. The log s behavior of $\sigma_T(s)$ is still $(\log s)^{\eta}$, and estimates of η are as yet unreliable for $\alpha_o > \alpha_{oc}$ but may be $3/2 \lesssim \eta \lesssim 2$.

Let us turn to an instructive example now due to Bronzan and Sugar. They work directly in continuum space at two dimensions for x and add to the hamiltonian, Eq. (89), the term $\int \frac{\lambda_o}{4} \tilde{\phi}^2\phi^2 d^2x$, $\lambda_o > 0$. This will not affect the $E,\vec{q} \to 0$ behavior

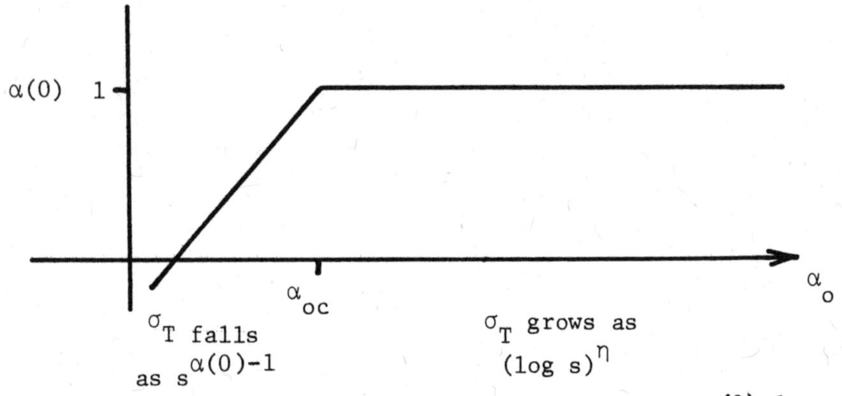

Fig. 19. The behavior of the power term in $\sigma_T(s) \sim s^{\alpha(0)-1}$ in RFT as the input parameter α_o passes its critical value. For $\alpha_o < \alpha_{oc}$, $\sigma_T(s)$ falls as a power of s. For $\alpha_o \geq \alpha_{oc}$, $\sigma_T(s) \sim (\log s)^{\eta}$, $\eta > 0$.

of the theory. Now they choose $\Delta_o = -r_o^2/\lambda_o \equiv \Delta_{os} < 0$, which means we are in the $\Delta_o < \Delta_{oc} = 0$ (unrenormalized still) or "ordered phase" of RFT. Just at this special combination of couplings one can find two states $|1\rangle$ and $\langle I|$ in addition to $|0\rangle$, the state for which $\phi|0\rangle = 0$, which satisfy

$$H|1\rangle = 0 \tag{104}$$

and
$$\langle I|H = 0 \quad . \tag{105}$$

Focus on $|1\rangle$ (clearly the analogue of the CERN $|1\rangle$ state) and form the zero energy state

$$|\Omega\rangle = iN[|1\rangle - |0\rangle] \ , \tag{106}$$

where the factor N is

$$N = \left[1 - \exp - \frac{4\Delta_{os}^2}{r_o^2} \times \text{Volume of } \vec{x} \text{ space}\right]^{-1/2} \quad . \tag{107}$$

Note now that if r_o is real (that is, ir_o coupling in the hamiltonian) as the volume of $\vec{x}$ space is taken to infinity, $N \to 1$. If r_o were imaginary (that is, a real coupling in the hamiltonian as in the ϕ^4 theory of before), $N \to 0$.

The important observation is now that

$$<0|\phi(\vec{x})|\Omega> = -\frac{2\Delta_{os}}{r_o} N \tag{108}$$

$$\sim \frac{4\Delta_o^2/r_o}{1 - \exp - \frac{4\Delta_o^2}{r_o^2} \times \text{Volume of } \vec{x} \text{ space}}, \tag{111}$$

$$\xrightarrow[\text{Volume}\to\infty]{} \begin{cases} -\frac{2\Delta_{os}}{r_o} & r_o \text{ real} \\ 0 & r_o \text{ imaginary} \end{cases}, \tag{109}$$

which demonstrates the key role played by the imaginary triple Reggeon coupling in establishing the properties (100), (101), and (102) of RFT.

If one examines the Reggeon propagator $G(\vec{x},t)$ for $\tau \to \infty$ $(E \to 0)$ then from the states $|0>$ and $|\Omega>$ alone

$$G(\vec{x},\tau) = <0|\phi(\vec{x},0)e^{-H\tau}\tilde{\phi}(0,0)|0> \tag{110}$$

$$\xrightarrow[\text{Volume}\to\infty]{} \begin{cases} 4\Delta_o^2/r_o & r_o \text{ real} \\ 0 & r_o \text{ imaginary} \end{cases} \tag{112}$$

So if r_o is real, the power behavior in s of $\sigma_T(s)$ is s^0, if r_o were imaginary, $\sigma_T(s) \to 0$ as a power of s as $s \to \infty$.

This construction of the properties of RFT for $\alpha_o > \alpha_{oc}$ is very attractive. It is not yet complete. The index η (and other indices) need to be precisely established, and questions of s-channel unitarity must be thoroughly investigated. It does, however, present a very physical tantalizing picture: as one varies the inputs to RFT which come from the underlying theory of hadrons we find for an infinite range of parameters: $\alpha_o \geq \alpha_{oc}$ $(r_o, \alpha_o', \ldots)$ a $(\log s)^\eta$ behaved $\sigma_T(s)$ with $\eta > 0$.

REFERENCES AND FOOTNOTES

1) H.D.I. Abarbanel, J.B. Bronzan, R.L. Sugar, and A.R. White, Physics Reports 21C, 119 (1975); H.D.I. Abarbanel, Reviews of Modern Physics 48, 435 (1976); M. Baker and K.A. Ter-Martirosyan, Physics Reports (1976); M. Moshe, Physics Reports (to be published).

2) U. Amaldi, et al, Phys. Lett. 66B, 390 (1977).

3) See R.J. Eden, 1967, High Energy Collisions of Elementary Particles (Cambridge University Press).

4) D. Amati, A. Stanghellini, and S. Fubini, Nuovo Cimento 26, 896 (1962).

5) One can properly only do this when the kernel is Fredholm but that can always be arranged with sufficient damping in transverse momentum.

6) Signature is treated rather sloppily here. See Ref. 1.

7) The lectures were interrupted at this juncture by the speaker's off-the-cuff attempt at humor with regard to the management of bears in Banff National Park. This was in response to S.D. Drell's urging at the previous night's banquet that physicists involve themselves in problems of society. He suggested we contemplate the issues of proliferation of nuclear weapons and disarmament. The virtues of that are unassailable, however, in the time allowed only the more mundane question of bears admitted of a solution. Hardly in response to popular request, the management is able to reproduce the theoretical physicist's solution to bears on Banff trails:

(i) Always travel in pairs on the trail.

(ii) One person/pair must carry chalk and blackboard.

(iii) Upon encountering a bear, do not run, but

(iv) Set up the blackboard and begin lecturing about field theory (bare field theory, of course).

(v) When the bear falls asleep, walk calmly away.

Upon close questioning it was revealed that the speaker had not been able to address the multi-bear problem nor the exceptional case when the bear was already versed in field theory. These cases were left as home assignments for the dedicated student.

8) V.N. Gribov, Sov. Phys. JETP 26, 414 (1968).

9) K.G. Wilson and J. Kogut, Phys. Rep. 12C, 75 (1974); M.E. Fisher, Rev. Mod. Phys. 46, 597 (1974).

10) J.L. Cardy, CERN preprint TH-2265 (1976) and references therein.

11) D. Amati, M. Ciafaloni, M. Le Bellac, and G. Marchesini, CERN preprint, TH-2152 (1976) and D. Amati, M. Ciafaloni, G. Marchesini, and G. Parisi, CERN preprint, TH-2185 (1976).

12) J.B. Bronzan and R.L. Sugar, U.C. Santa Barbara preprint, TH 77-1.

CHROMODYNAMIC STRUCTURE AND PHENOMENOLOGY

Thomas Appelquist*

Yale University

New Haven, Connecticut 06520, U.S.A.

1. INTRODUCTION

In these three lectures, I will discuss some of the formal and phenomenological features of quantum chromodynamics (QCD). This local gauge theory based on the color group SU(3) has stood as probably the most attractive candidate for a theory of strong interactions since the discovery of asymptotic freedom in 1973. There has been an enormous amount of theoretical work since then learning how to confront the theory with experiment in the short distance, weak coupling regime and trying to deal with the theory at larger distances(say $1/m_{\pi}$) where it presumably becomes strongly coupled.

These lectures are not intended to be a general review of all this work and the overall status of QCD[1]. What I have done is to pick a few topics which have interested me during the past year and which I feel are important and instructive. The material divides naturally into two parts, the first part more formal and the second more phenomenological. In the first part, I will be concerned with the computation of the static potential between two heavy color sources in quantum chromodynamics. This problem is an interesting theoretical laboratory for studying the structure of QCD and it has also taken on phenomenological importance since the discovery of charmed quarks.

I will start by reviewing the structure of the Yang-Mills theory in Coulomb gauge, the natural gauge for studying a static problem. Some time will be devoted to explaining the physical mechanism behind asymptotic freedom. One feature of the static potential which is properly studied with perturbation theory is the limit $M_Q \to \infty$. It will be shown that this limit is finite and the consequences of this fact for heavy quark phenomenology will be discussed.

There seems to be no immediate phenomenological reason for actually computing the static potential in higher order perturbation theory. The $c\bar{c}$ system probes quark separation distances of order $1/m_\pi$ and the linear potential that describes this system rather well, almost surely has a completely non-perturbative origin. Nevertheless, there are some interesting and potentially important things to be learned from such a study. For example, it has been shown that the color singlet potential is infrared finite through low orders. It is expected that this result is true to all orders and it is important to prove this. The potential in fact does not have a simple power expansion in $\alpha_s = g^2/4\pi$. The color correlations between the quark sources and the gluons lead to non-analyticity in α_s at $\alpha_s = 0$, but through low orders at least, these effects can be dealt with by selective resummation. I will summarize what is known about these and other features of the static potential in perturbation theory.

Some of the phenomenology of quantum chromodynamics is presented in sections 5 and 6. I will restrict myself to experimental situations involving currents probing time-like momenta. The more familiar application to deep inelastic scattering has been reviewed several times during the past year.[1,2] I will start by describing a useful technique suggested by Shankar[3], which draws on the old finite energy sum rules of Regge pole theory. It finds a very natural application in e^+e^- annihilation into hadrons and in the decay of heavy leptons, both of which I will discuss.

Lastly, I will review the status of the asymptotic freedom explanation of the OZI rule for heavy quarks[4]. The status of this somewhat speculative use of asymptotic freedom is still uncertain but there have been a few experimental and theoretical developments during the past year. I will describe both the theoretical considerations and the current experimental status.

The Lagrangian of quantum chromodynamics is

$$\mathcal{L}(x) = \bar{q}(x)[i\not{D} - M_0]q(x) - \frac{1}{4} F^a_{\mu\nu}(x)F^{a\mu\nu}(x) \quad . \tag{1.1}$$

The quark spinor $q(x)$ contains $4 \times 3 \times N$ components corresponding to the three colors and unknown number N of quark flavors. M_0 is the bare mass matrix, a product of Dirac and color unit matrices and a diagonal flavor matrix M_0^{AB}.[5] The gauge field covariants are

$$D_\mu = \partial_\mu + igT^a A^a_\mu$$
$$F^a_{\mu\nu} = \partial_\mu A^a_\nu - \partial_\nu A^a_\mu + gf^{abc} A^b_\mu A^c_\nu \quad . \tag{1.2}$$

The T matrices are normalized in the conventional way

$$T_a = \frac{1}{2}\lambda_a \;, \qquad \mathrm{Tr}\;\lambda_a\lambda_b = 2\delta_{ab} \quad . \tag{1.3}$$

For future reference we record, for a general SU(N) color group, the commutation relations among the T matrices and the quadratic Casimir operators of the fundamental and adjoint representations:

$$[T_a, T_b] = if_{abc}T_c \quad , \quad \{T_a, T_b\} = d_{abc}T_c + \frac{1}{N}\delta_{ab} \tag{1.4}$$

$$(T^a)_{ik}(T^a)_{kj} = C_F\delta_{ij} \;, \qquad C_F = \frac{N^2 - 1}{2N} \tag{1.5}$$

$$f_{acd}f_{bcd} = C_A\delta_{ab} \quad , \qquad C_A = N \quad . \tag{1.6}$$

The local gauge invariance of the QCD Lagrangian is

$$A_\mu(x) \equiv A^a_\mu(x)T^a \to U(x)A_\mu(x)U^{-1}(x) + \frac{i}{g}\,U(x)\partial_\mu U^{-1}(x) \tag{1.7}$$
$$q(x) \to U(x)q(x)$$

where

$$U(x) = e^{i\theta^a(x)T^a} \quad .$$

2. THE YANG-MILLS THEORY IN COULOMB GAUGE

To simplify the discussion, I will consider only one quark flavor Q with mass M. In the discussion of asymptotic freedom, M will be taken to be large. The Lagrangian density is

$$\mathcal{L}(x) = -\frac{1}{4}\,\vec{F}_{\mu\nu}(x)\cdot\vec{F}^{\mu\nu}(x) + \bar{\psi}(x)[i\partial\!\!\!/ - g\vec{T}\cdot\vec{A\!\!\!/}(x)]\psi(x)$$
$$- M\bar{\psi}(x)\psi(x) \tag{2.1}$$

where

$$\vec{F}_{\mu\nu} = \partial_\mu\vec{A}_\nu - \partial_\nu\vec{A}_\mu + g\vec{A}_\mu \times \vec{A}_\nu \quad . \tag{2.2}$$

For color SU(3), the vectors have eight components

$$\vec{T} = T_1 \ldots T_8 \quad , \quad (\vec{a} \times \vec{b})_i = f_{ijk} a_j b_k \quad . \tag{2.3}$$

The equations of motion are

$$\partial_\mu \vec{F}^{\mu\nu} + g\vec{A}_\mu \times \vec{F}^{\mu\nu} - g\bar{\psi}\gamma^\nu \vec{T}\psi = 0 \tag{2.4}$$

and the canonical momenta are

$$\vec{\pi}^0 = \frac{\partial \mathcal{L}}{\partial \dot{\vec{A}}_0} = 0$$

$$\vec{\pi}^i = \frac{\partial \mathcal{L}}{\partial \dot{\vec{A}}_i} = - \vec{F}^{0i} \equiv \vec{E}^i \tag{2.5}$$

$$\pi_\psi = i\psi^\dagger \quad .$$

It is clear from Eq. (2.5) that A_0 cannot be a dynamical variable. It must either be eliminated or expressed in terms of the independent dynamical variables. Eq. (2.4) contains both true equations of motion and a constraint equation which reads

$$\nabla^i \vec{E}^i - g\vec{A}^i \times \vec{E}^i - g\psi^\dagger \vec{T}\psi = 0 \quad . \tag{2.6}$$

The constraint means that all the components of $\vec{E}^i$ cannot be taken to be independent and therefore neither can all the components of $\vec{A}^i$.

To eliminate one of these components, we impose the Coulomb gauge condition

$$\nabla^i \vec{A}^i = 0 \tag{2.7}$$

which is always possible because of the gauge invariance of the theory. $\vec{E}^i$ can be separated into transverse and longitudinal parts

$$\vec{E}^i = \vec{E}^i_t + \nabla^i \vec{\phi} \tag{2.8}$$

with $\nabla^i \vec{E}^i_t = 0$. The constraint equation, Eq. (2.6), then becomes an equation for $\vec{\phi}$:

$$[\nabla^2 - g\vec{A}^i \times \nabla^i]\vec{\phi} = g\vec{A}^i \times \vec{E}^i_t - g\psi^\dagger\vec{T}\psi \equiv \vec{\rho} \quad . \tag{2.9}$$

From the definition of $\vec{F}^{oi}$, an equation for $\vec{A}^o$ is also obtained

$$[\nabla^2 - g\vec{A}^i \times \nabla^i]\vec{A}^o = -\nabla^2\vec{\phi} \quad . \tag{2.10}$$

The Hamiltonian is now easily written in terms of the dynamical variables and $\vec{\phi}$

$$H = \int d^3x \left\{ \frac{1}{2}\vec{E}^i_t \cdot \vec{E}^i_t + \frac{1}{4}\vec{F}^{ij} \cdot \vec{F}^{ij} + \bar{\psi}[i\nabla^i\gamma^i + m]\psi - g\bar{\psi}\vec{T}\gamma^i\psi \cdot \vec{A}^i - \frac{1}{2}\vec{\phi} \cdot \nabla^2\vec{\phi} \right\}. \tag{2.11}$$

Surface terms have been dropped when integrating by parts. The last term is the energy of the longitudinal electric field. It is given in terms of the source $\vec{\rho}$ if Eq. (2.9) can be solved uniquely for $\vec{\phi}$. A formal expression for the Green's function solution to Eq. (2.9) is

$$\vec{\phi} = \frac{1}{1 - g\frac{1}{\nabla^2}\vec{A}^i \times \nabla^i} \frac{1}{\nabla^2}\vec{\rho} \quad . \tag{2.12}$$

For purposes of doing perturbation theory in the coupling constant, this expression can be expanded and the Hamiltonian becomes an infinite series in the independent dynamical variables $\vec{A}^i$ and $\vec{E}^i_t$.

The fact that Eq. (2.9) does not determine $\vec{\phi}$ uniquely for large enough field strength has recently been emphasized by V.N. Gribov[6] and S. Mandelstam[7]. I will have very little to say about this during my lectures except to touch on it briefly while discussing the physical origin of asymptotic freedom.

Feynman rules may be derived in the Coulomb gauge using either path integral or operator methods through the interaction representation[8]. The gauge propagator breaks into the usual instantaneous Coulomb part and transverse part. There is a Fadeev-Popov ghost required but it couples only to the transverse gauge field and is itself instantaneous. Its propagator is $1/\vec{k}^2$ and it has nothing directly to do with unitarity.

3. ASYMPTOTIC FREEDOM AND OTHER FEATURES OF PERTURBATION THEORY

I will discuss the origin and physical content of asymptotic freedom in the context of the static potential between two heavy color sources. The sources will be taken to be a $Q\bar{Q}$ pair with a large mass M and I will first show that the limit $M \to \infty$ is free of divergences. The subsequent discussion of asymptotic freedom can then be carried out using infinitely massive, fixed sources. The Coulomb gauge is particularly convenient for studying this limit since the coupling of transverse gluons to the heavy quarks may be neglected when $M_Q \to \infty$. Thus attention may be focussed on the fermion-Coulomb gluon vertex which satisfies a Ward identity similar to QED. This identity can be derived by functional methods[9] and is most easily expressed in momentum space. In a standard notation

$$\begin{aligned} &-q^i\Gamma_i^a(q,p)(1+B(q))+q^o\Gamma_o^a(q,p)(1+A(q)) \\ &= (gT^a-B^a(q,p))S^{-1}(p)-S^{-1}(p+q)(gT^a-\hat{B}^a(q,p)) \ . \end{aligned} \tag{3.1}$$

$S(p)$ is the fermion propagator and the functions $A(q)$, $B(q)$, $B^a(q,p)$ and $\hat{B}^a(q,p)$ are two and three point functions involving the ghost. $B(q)\,\vec{q}^{\,2}$ is the ghost self energy while $A(q)$, $B^a(q,p)$ and $\hat{B}^a(q,p)$ are artificial constructs in which an external ghost line terminates.

As the ghost couples only to the transverse gluon, it is possible to factorize the A and B^a amplitudes as follows:

$$B^a(q,p) = q_i B_i^a(q,p) \quad , \qquad A(q) = \frac{1}{\vec{q}^{\,2}}\mathcal{A}(q,q_o) \ . \tag{3.2}$$

It then follows from simple power counting considerations that B_i^a and $\mathcal{A}$ are primitively convergent. The function $\mathcal{A}$ in fact vanishes to three loops and we conjecture this to be true to all orders. The primitive convergence of B_i^a and $\mathcal{A}$ can be used along with the Ward identity (3.1) to show that the only ultraviolet divergent renormalization of the Coulomb gluon-quark coupling constant comes from corrections to the Coulomb gluon propagator.

We now consider the infinite mass limit. By mass here, I mean the renormalized mass. All self energy insertions are to be expanded in powers of $(\not{p} - M_Q)$ with the zeroth order term being absorbed into mass renormalization. Letting the mass tend to infinity then, the Ward identity assures us that the combination of fermion vertex Γ_o^a and self energy $S(p)$ parts is finite in this limit. This result follows from the observation that in this limit, $\Gamma_i^a(q,p)$ vanishes while $A(q)$ and $B^a(q,p)$ remain primitively convergent. We omit the details of the power counting analysis required to demonstrate this. It is worth commenting that the finiteness of this limit is not at all apparent in a graph by graph analysis of Γ^a_o and S(p). The individual graphs contain log M_Q terms which sum to zero in each order as the Ward identity demands.

It is easy to see that the only possible sources of divergence as $M_Q \to \infty$ are the vertex Γ^a_o and the propagator S(p). Apart from these, the computation of the potential involves diagrams which are clearly finite as $M_Q \to \infty$. Diagrams with closed loops of heavy quarks in fact vanish in this limit[10]. Having shown that the combined effect of Γ^a_o and S(p) is finite as $M_Q \to \infty$, it follows that the static potential itself is finite. This result is perhaps expected intuitively, especially in an asymptotically free theory. Our proof involved only power counting and did not explicitly involve asymptotic freedom. On the other hand, it is only in an asymptotically free theory that one can be sure that anomalous short distance behavior does not develop nonperturbatively and destroy the naive power counting arguments we have employed.

Because the limit is finite, the $Q\bar{Q}$ potential at separation R is insensitive to the quantum fluctuations at distance scales $1/M_Q \ll R$. The same phenomenological potential can be used to describe charmonium and its heavier imitations. I emphasize that this result is quite general. Even though the confining potential describing these systems can almost surely not be derived from perturbation theory, the $M_Q \to \infty$ limit <u>is</u> properly studied this way, at least in an asymptotically free theory.

Because of the Ward identity, the two diagrams shown in Fig. 1 are the only important one loop contributions. They have already been computed in the literature[11] and I will describe them here from a physical point of view[12]. Fig. 1a is vacuum polarization, a shielding effect which should reduce the renormalized coupling strength. Computation bears this out, giving in the instantaneous approximation ($q_o = 0$) the following correction to the Coulomb propagator $\delta_{ab} \frac{i}{\vec{q}^2}$:

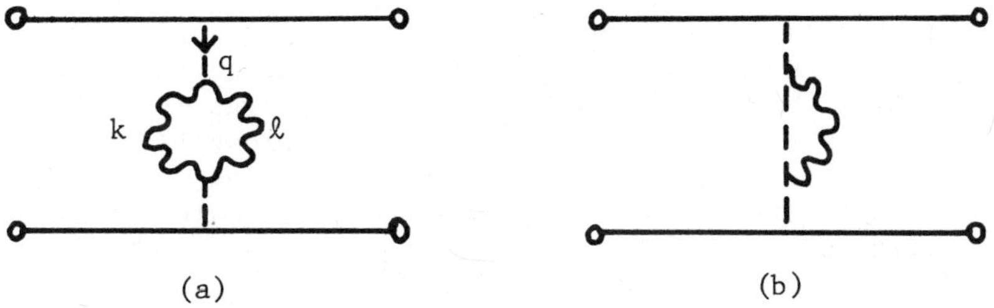

Fig. 1. One loop contributions to the potential.

$$D^{VP}_{ab}(q) = \delta_{ab} \frac{i}{\vec{q}^2} [1 + \frac{5}{24} \frac{g^2}{\pi^2} C_A \log q/\mu] , \quad (3.3)$$

where μ is the renormalization scale and $q \equiv |\vec{q}|$. Although vacuum polarization is not instantaneous, it appears as effectively instantaneous to the quarks. This can be seen by noting that the energy denominator $|\vec{k}| + |\vec{\ell}|$ in old fashioned perturbation theory is of order q if the loop is subtracted at $\mu \approx q$. Thus this process takes place over a typical time scale of order 1/q which is not instantaneous but which is small compared to the time scale for quark recoil (of order M/q^2).

The Coulomb field self energy contribution (Fig. 1b) is, by contrast, completely instantaneous. The full propagator can be written in the form

$$\delta_{ab} \frac{i}{\vec{q}^2} \left(2g^2 C_A \frac{q_i q_j}{\vec{q}^2} \int \frac{d^3k}{(2\pi)^3} \frac{\delta_{ij} - k_i k_j/\vec{k}^2}{|\vec{k}|} \frac{1}{|\vec{q} + \vec{k}|^2} \right) . \quad (3.4)$$

Momentum integration and subtraction at $q = \mu$ gives the resulting contribution to the renormalized Coulomb propagator:

$$D^{SE}_{ab}(q) = \delta_{ab} \frac{i}{\vec{q}^2} [- \frac{16}{24} \frac{g^2}{\pi^2} C_A \log q/\mu] . \quad (3.5)$$

The asymptotic freedom of the Coulomb field self energy can be seen from the positivity and ultraviolet divergence of the square bracketed correction in expression (3.4). It is a q dependent and cutoff dependent increase in the magnitude of the field strength over and above the bare Coulomb field. The renormalized Coulomb propagator shows the attendant weakening of the effective coupling strength at short distances. The asymptotic freedom of the theory follows from the dominance of this contribution over the polarization of the vacuum, a computational result without an apparent simple explanation. Adding (3.3) and (3.5) gives the usual running coupling constant of the pure gauge sector of the theory.

The physics of asymptotic freedom is most apparent in configuration space. Suppose that a quark source is located at the origin and an antiquark probe is at $\vec{R}$. The Fourier transform of expression (3.4) is a positive addition to the 1/R Coulomb potential. The dominant term in the transverse propagator is the δ_{ij}, so we exhibit only that part for simplicity. One of the external Coulomb propagators (e.g. the one connecting to the quark source) is then cancelled and the addition to the 1/R potential is

$$2g^2 C_A \int_{\Lambda^{-1}} d^3x \, \frac{1}{x^2} \frac{1}{x} \frac{1}{|\vec{R} - \vec{x}|} \quad . \tag{3.6}$$

Integration gives an expression of the form $\frac{1}{R} \log R\Lambda$. This is a scale dependent increase (for $R > 1/\Lambda$) of the attractive (repulsive) force between the quark and antiquark if they are in a color singlet (octet) state. The source of the increase is the attractive potential $1/R^2$ which operates between the Coulomb field lines at the position of the source and at the point $\vec{R}$. It arises from the instantaneous exchange of the octet transverse gluon between these points and is attractive because the two points of the Coulomb field are in a relative singlet state. The field lines connecting the source to the point $\vec{R}$ are thus drawn together, increasing the magnitude of the energy stored in the field. This mechanism of logarithmic flux collimation is clearly unique to Yang-Mills theories.

It is instructive to plot out the electric field lines between the quark and antiquark. It can then be seen directly that the additive piece in the electric field due to the Coulomb self energy has the effect of bending the electric field lines inward, making them more collimated. I leave this as an exercise.

To proceed beyond one loop, a general scheme must be developed to separate contributions to the static potential from potential iterations. It is most convenient to do this by working directly in configuration space. I will first consider a simpler problem than QCD in which the sources are classical, that is, uncorrelated with the quantum fields. The simplest such example is electrodynamics where the sources are static point-like electric charges, but this model is trivial in this context since the only interactions are between the sources and the field. More generally interactions among $A_\mu(x)$ and other quantum fields can be introduced as a prototype for the Yang-Mills theory.

A gauge invariant starting point for the computation of the potential is the Wilson loop integral

$$\langle P \, e^{ig\oint dx_\mu A^\mu} \rangle_o \quad .$$

The integral is taken about the rectangle of width R and length $T \gg R$ and P is the path ordering symbol. The expectation value is taken in the vacuum, which for a perturbation theory computation is built up perturbatively from the bare vacuum. As pointed out by Fischler[13], if the ends of the rectangle can be shown not to contribute in the limit $T \to \infty$, the loop integral becomes

$$\langle T \exp[ig \int_{-T/2}^{+T/2} A_o(0,t)dt - ig \int_{-T/2}^{+T/2} A_o(\vec{R},t)dt]\rangle_o \quad .$$

This then should give the exponential of the static potential $e^{iV(R)T}$ in the limit $T \to \infty$. Thus

$$V(R) = \lim_{T\to\infty} \frac{1}{iT} \log\langle P\ e^{ig\oint dx_\mu A^\mu}\rangle_o \quad . \tag{3.7}$$

The vanishing of the contributions from the ends of the rectangle is expected since $F_{\mu\nu} \to 0$ in the limit $T \to \infty$ and therefore the potential A_i becomes gauge equivalent to $A_i = 0$. This fact is easily seen in the perturbation expansion of Eq. (3.7). Diagrams with lines attaching to either end of the rectangle vanish as $T \to \infty$.

The expansion of Eq. (3.7) in perturbation theory produces, after the usual cancellation of vacuum disconnected graphs, two classes of diagrams which attach to the sides of the rectangle (the sources). One class is "connected" where connectivity is defined without the inclusion of the source lines, i.e., using only the quantum fields. The other class is "disconnected" in the same sense.

By iterating the connected diagrams in all possible ways along the sides of the rectangle, all the disconnected graphs will be generated. A given diagram will correspond to a definite time ordering of the various interactions with the source. These are specified by the source propagators which are θ functions in time. The effect of ordering one connected diagram C_1 in all possible ways with respect to another C_2 is to eliminate the θ functions involving the relative time intervals in C_1 with respect to those in C_2. Thus the two sets of time integrations range independently between $-T/2$ and $+T/2$. It is of course crucial here that the sources be uncorrelated. In adding together all time permutations, an overcounting by a factor of $n!$ will occur where n is the number of identical connected diagrams involved. This must be compensated by multiplying by $1/n!$.

The upshot of this analysis, which we have only sketched, is that the potential V(R) will be given by the connected diagrams. The disconnected diagrams are iterations which build up the exponential of Eq. (3.7). We note in passing that the class of connected diagrams includes those which connect only to one side of the rectangle (only one of the sources). These are source self mass contributions which, as we shall discuss further in the next section, can be regarded as cancelled by an appropriate mass counter term. They do not depend upon quark separation and make no contribution to the potential.

If it can be shown that the large spatial integrations and large relative time integrations in each connected diagram are convergent (infrared finiteness), then this procedure will give a finite result for the potential. The overall time integration, corresponding to sliding each connected structure as a unit along the rectangle, will give one overall factor of T. It will then follow that the limit $T \to \infty$ in Eq. (3.7) exists and the potential will be calculable to any order in the coupling constant. The infrared finiteness for connected diagrams can, in fact, be proven for a wide variety of field theories. In particular, for renormalizable theories even with all the fields massless, a simple power counting analysis can establish the result to all orders.

It is worth commenting that this type of connectivity result is familiar in many body theory. It is very similar to the Brueckner-Goldstone linked cluster expansion for the energy of a system of particles at zero temperature. The literature on this problem and the closely related finite temperature expansion is extensive[14].

Turning now to the Yang-Mills theory, we again start with the Wilson integral

$$\langle \mathrm{Tr}\, P\, e^{ig\oint dx_\mu \vec{T}\cdot\vec{A}^\mu} \rangle_o$$

where the contour is the same rectangle of width R and length $T \gg R$. In perturbation theory, it can again be shown[15] that diagrams with gluon lines attaching to the ends of the rectangle vanish in the limit $T \to \infty$. This makes it plausible that in this limit, the Wilson integral again takes the form $e^{iV(R)T}$ where V(R) is the static potential energy of a <u>color singlet</u> pair of color sources separated by a distance R. Then, as before,

$$V(R) = \lim_{T\to\infty} \frac{1}{iT} \log \langle \mathrm{Tr}\, P\, e^{ig\oint dx_\mu \vec{T}\cdot\vec{A}^\mu} \rangle_o \quad . \tag{3.8}$$

To actually demonstrate this and to give a direct prescription for computing V(R) analogous to the "connectivity" prescription with uncorrelated sources, is now a rather more difficult problem. All the complexities stem from the fact that the sources are no longer classical. Color is a quantum label and the color state of the sources is correlated to the gluonic sector. This leads to a number of interesting consequences.

I can now summarize the two loop structure of the potential where two new features, not seen at one loop, emerge. First, contributions appear which are more complex than the exchange of a single dressed gluon. To be specific, the contribution of Fig. 2 is proportional to $C_F C_A^2 \alpha_s^3$ 1/R. Second, infrared divergences appear in the individual graphs shown in Fig. 3. In the case of Fig. 3a, there is a leading (potential iteration) piece proportional to T^2 and residual pieces proportional to T log T as well as T. The other three diagrams all contain T log T and T pieces. The T log T parts are the infrared divergences, coming from integrations over large times and large spatial distances. In an Abelian model, both the T log T and T pieces would completely cancel pairwise, (a) with (c) and (b) with (d), leaving only the potential iteration part of Fig. 3a. In QCD, the T log T pieces cancel for a color singlet $Q\bar{Q}$ pair. The cancellation is again pairwise but now (a)

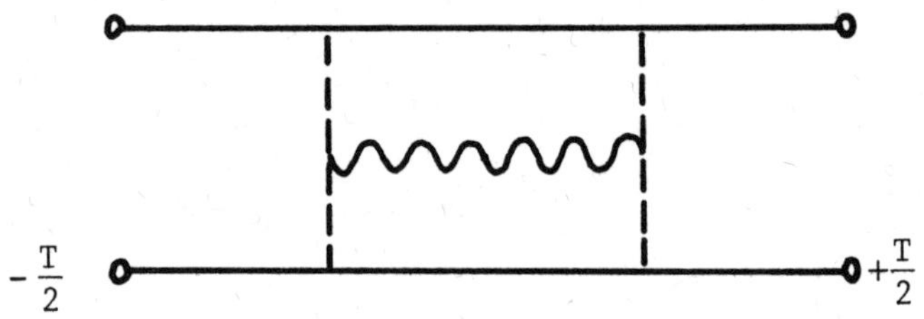

Fig. 2. A two loop contribution to the potential.

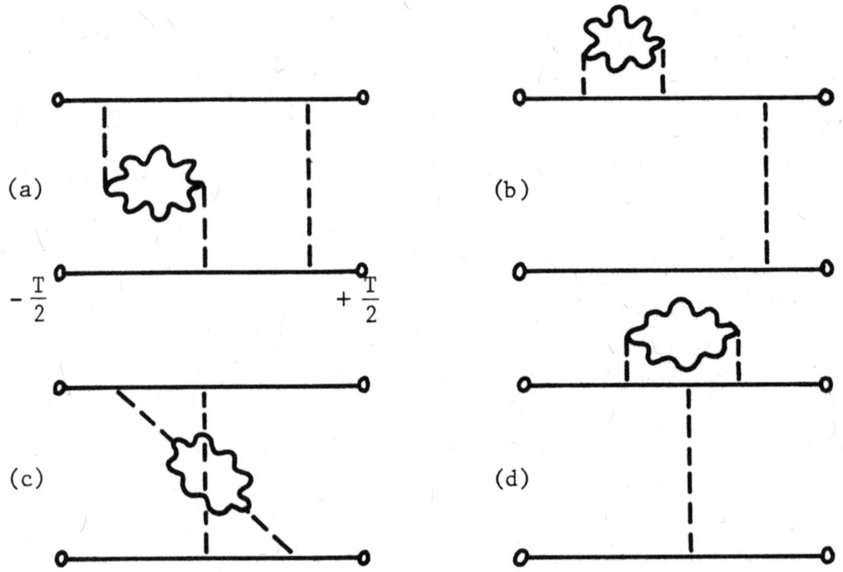

Fig. 3. Two loop diagrams with infrared singularities.

with (b) and (c) with (d). The remaining piece proportional to T is a contribution to the potential which must be added to other two loop contributions. From the full two loop result, one can for example extract the two loop β function. From the foregoing discussion, it is clear that although the potential is well defined through two loops (order α_s^3), it is no longer given only by the connected diagrams.

4. THE BREAKDOWN OF PERTURBATION THEORY

In higher orders, the disconnected diagrams introduce an interesting new feature. Before describing it, I should remark that infrared divergences of the type already encountered will continue to exist. It is important to prove that they cancel to all orders for a color singlet $Q\bar{Q}$ since only then can one be sure, for example, that the short distance potential ($R \lesssim 1\ \mathrm{GeV}^{-1}$) is independent of the confinement mechanism and truly described by asymptotic freedom.

The new feature is a divergence coming from large time ($\sim T$) but finite distance ($\lesssim R$) integrations. It does not cancel and in fact causes a breakdown of simple perturbation theory. We will illustrate the general problem by examining the simplest (lowest order) situation where it occurs. There the problem can be overcome by selective resummation. We conjecture that this technique can be systematically applied to any order but the combinatoric problems become formidable in higher orders.

Consider the disconnected diagram shown in Fig. 4, Without the middle rung, the remaining connected piece is easily evaluated. After doing the x_1 and x_2 spatial integrations, it is proportional to

$$a_s^3 \int_{-T/2}^{+T/2} dt_1 dt_2 \frac{C_F C_A^2}{R^2 + (t_2 - t_1)^2} = T\, C_F C_A^2 \alpha_s^3 \int_0^T \frac{dt}{R^2 + t^2} \quad . \tag{4.1}$$

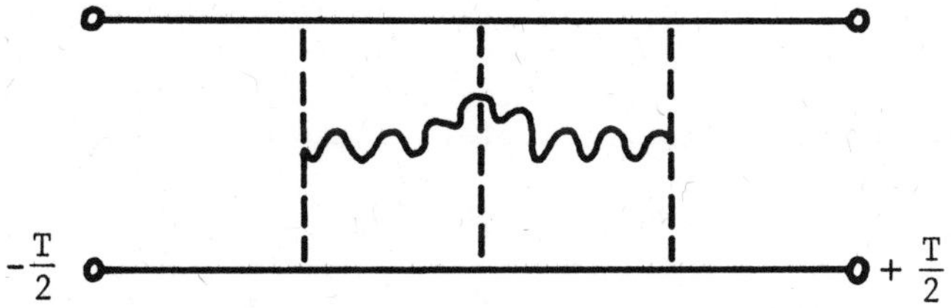

Fig. 4. A contribution to the Wilson loop integral behaving like T log T.

The upper limit on the relative time integration can be taken to infinity and the result is proportional to $T(\alpha_s^3/R)$. For simplicity, we have Wick rotated to Euclidean space-time in writing this expression. The presence of the additional rung leads to an extra numerator factor of $\alpha_s t/R$ in the relative time integration in Eq. (4.1). The result is proportional to $T \log T\, \alpha_s^4/R$ and it arises from the integration region $t \sim T$ but $x_1, x_2 \lesssim R$.

In an Abelian model, this would of course be cancelled by contributions from graphs with the extra rung placed outside the connected piece. In the Yang-Mills theory however, there is a left over piece. If the extra rung is outside the connected piece, then the group theory factor resulting from the trace is $C_F \times C_F C_A^2$. If it is inside as in Fig. 4, the result is proportional to $(C_F - C_A/2) \times C_F C_A^2$ and therefore a piece of the form $(-C_A/2) \times C_F C_A^2 \times T \log T$ remains uncancelled. According to Eq. (3.8), it would give an infinite contribution to the potential.

This divergence is a signal that the perturbation expansion for the potential has broken down. The breakdown is associated with the presence of infinite range fields and can be dealt with by selective resummation. Suppose we sum over any number of single Coulomb rungs both inside and outside the connected structure in Fig. 4. The result will be proportional to

$$\int_{-T/2}^{T/2} dt_1\, dt_2 \exp\left[C_F \frac{\alpha_s}{R}\left(\frac{T}{2} - t_2\right)\right] \exp\left[(C_F - C_A/2)\frac{\alpha_s}{R}(t_2 - t_1)\right]$$

$$\times \exp\left[C_F \frac{\alpha_s}{R}\left(t_1 + \frac{T}{2}\right)\right] \frac{C_F C_A^2 \alpha_s^3}{(t_1 - t_2)^2 + R^2}$$

$$= \exp\left[C_F \frac{\alpha_s}{R} T\right] \times C_F C_A^2 \alpha_s^3 T \int_0^\infty \frac{dt}{t^2 + R^2} \exp\left[-\frac{C_A}{2} \alpha_s \frac{t}{R}\right] \quad . \tag{4.2}$$

The multiplicative exponential factor is simply the iteration of the zeroth order potential. The resummation has also produced an exponential factor in the relative time integration which has already been extended to infinity. The result is proportional to T and is therefore a finite contribution to the potential. Note however that the exponential $\exp[-C_A/2\, \alpha_s\, t/R]$ cannot be re-expanded before doing the t integration. That would lead back to the $T \log T$ divergences of perturbation theory.

A contribution to the potential has been identified which is proportional to

$$C_F C_A^2 \frac{\alpha_s^3}{R} \int_0^\infty \frac{dx}{1 + x^2} \exp[- \frac{C_A}{2} \alpha_s x] \ .$$

The exponential integral is non-analytic in α_s at $\alpha_s = 0$. It contains a logarithmic discontinuity and can be written in the form

$$\int_0^\infty \frac{dx}{1 + x^2} e^{-ax} = \log a \sin a + g(a) \tag{4.3}$$

where g(a) is analytic at a = 0. To leading order in α_s, this contribution to the potential contains the α_s^3 term mentioned in section 3 (Fig. 2). In next order there is a term proportional to $C_F C_A^3 (\alpha_s^4/R) \log \alpha_s$ plus a term of order α_s^4. While there are many other contributions of order α_s^4, I believe this is the only $\alpha_s^4 \log \alpha_s$ term.

The existence of log α_s terms is familiar in quantum electrodynamics where they enter at the level of relativistic corrections in bound state problems. The physical processes involved are somewhat similar in the two cases. In either case, if the singularities are only logarithmic, it seems likely that perturbation theory can still be used. While there is no experimentally important reason for computing such terms in QCD, it is still interesting to know whether or not the computation could in principle be done[16] to arbitrary accuracy for small α_s. We conjecture that this is the case, that is, that the computation can be organized into a double expansion in $\alpha_s^n (\log \alpha_s)^m$ with $m < n$.

The proof of this conjecture appears to be quite difficult. The log α_s terms are associated with disconnected diagrams and it must be shown how to deal with them in general. The selective summation of diagrams must be systematized, perhaps leading to some simple prescription for the computation of V(R). Such a prescription will certainly be more complicated than the connected diagram prescription for the case of uncorrelated sources.

The physical content of the resummation is clear. In the simple example we considered, it corresponds to reorganizing the expansion about a Coulomb state rather than a state with non-interacting sources. In the ground state, the sources are in a color singlet state. The intermediate state shown in Fig. 4 consists of one gluon and the sources in an octet state. Since the Coulomb force is attractive in the singlet channel and repulsive in the octet channel, there is an energy gap of order $C_A \frac{\alpha_s}{R}$ between the ground state and the intermediate state. By the uncertainty principle, the intermediate state can therefore live at most a time of order $R(C_A\alpha_s)^{-1}$. That is the origin of the dying exponential on the RHS of Eq. (4.2) and, in fact, the $\alpha_s^4 \log \alpha_s$ term comes from $t \approx R(C_A\alpha_s)^{-1}$. It is worth pointing out that in a bound state, this will be a time on the order of the period of the motion and it is perhaps misleading at this level to say that one is still computing an instantaneous potential.

5. SOME APPLICATIONS OF QCD

I will begin by describing a technique suggested by Shankar[3] for the confronting of QCD with e^+e^- annihilation data. Asymptotic freedom allows the perturbation theory computation of the hadronic vacuum polarization $\Pi(s)$ in the deep Euclidean region, $s = E_{CM}^2 < -1$ GeV^2. Its absorptive part for $s > 0$ is $R(s) = \sigma(e^+e^-\to\text{hadrons})/\sigma(e^+e^-\to\mu^+\mu^-)$. Thus for $s < -1$ GeV^2,

$$\Pi^{\text{theor}}(s) = \Pi^{0}(s) + \frac{4}{3}\alpha_s(-\mu^2)\Pi^1(s) + \ldots \tag{5.1}$$

where $\alpha(-\mu^2)$ is the QCD running coupling constant at $-\mu^2 < -1$ GeV^2. The expressions for $\Pi^0(s)$ and $\Pi^1(s)$ are, for our purposes, most conveniently given in terms of their absorptive parts R^0 and R^1 as computed in perturbation theory. We have

$$\begin{aligned} R^{\text{theor}} &\equiv R^0(s) + \frac{4}{3}\alpha(-\mu^2)R^1(s) \\ &= 3 \sum_{\substack{i \\ \text{flavors}}} Q_i^2\{v_i \frac{3-v_i^2}{2}\}[1+\frac{4}{3}\alpha_s(-\mu^2)f(v_i)+\ldots] \end{aligned} \tag{5.2}$$

where

$$f(v) = \frac{\pi}{2v} - \frac{3+v}{4}\left(\frac{\pi}{2} - \frac{3}{4\pi}\right) \quad . \tag{5.3}$$

Here Q_i is the quark charge and v_i the quark velocity $= (1-4M_i^2/s)^{\frac{1}{2}}$.

It has been conjectured the R^{theor} represents a reliable direct computation of R^{exp} away from strong singularities such as charm threshold and for $s > 1$ GeV^2. There are then no large dynamical factors in perturbation theory and the expansion appears to be convergent through low orders. It must then be assumed that very high order threshold singularities or even non-perturbative contributions don't affect the final result. Physically, this means that the transition from quarks and gluons into physical hadrons takes place with essentially unit probability. I am not going to explore this interesting possibility further now. Instead

I will explain a more reliable technique but one with less predictive power, the finite energy sum rule of Shankar.

Consider the contour in the complex s-plane shown in Fig. 5. Dividing the contour into two parts as shown,

$$\int_{C_1} \Pi(s)ds = - \int_{C_2} \Pi(s)ds . \tag{5.4}$$

The left hand side is given by the experimental data $R^{exp}(s)$. Along C_2, a circle of radius $|s| > 1\ \text{GeV}^2$, $\Pi(s)$ can be reliably computed using perturbation theory except for a range $\Delta\theta$ near the real axis. There one encounters the sort of uncertainties mentioned in the last paragraph. With $\Delta\theta$ large enough so that $|\text{Im}\ s| = |s|\ |\tan \Delta\theta|$ is bigger than say $1\ \text{GeV}^2$, perturbation theory should be just as reliable as along the negative real axis. The angle $\Delta\theta$ will decrease as s increases making this region a smaller and smaller part of C_2. It should thus be a good approximation to assume that $\Pi(s)$ is given reliably by $\Pi^{theor}(s)$ everywhere on C_2 for $|s| >> 1\ \text{GeV}^2$ and away from charm or other thresholds.

Now

$$\int_{C_2} \Pi^{theor}(s)ds$$

can be written in terms of the discontinuity R^{theor} along C_1 and

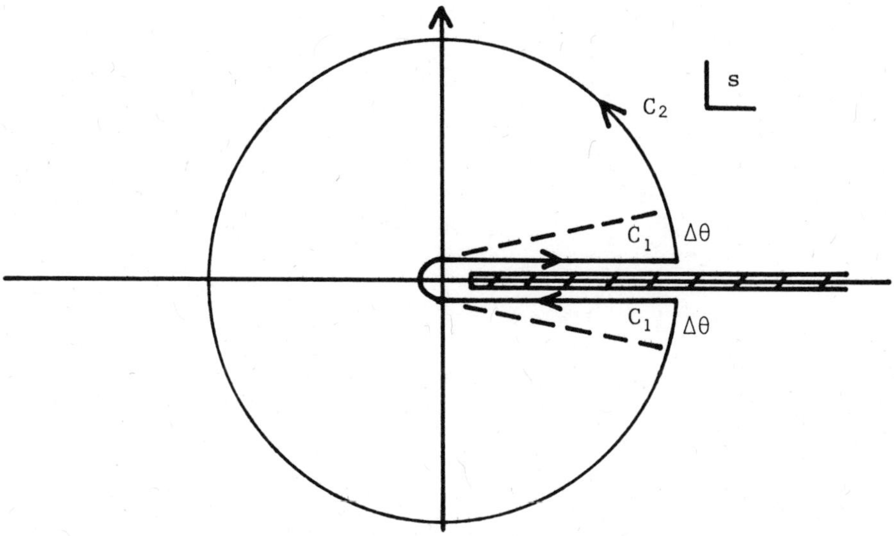

Fig. 5. The integral of $\Pi(s)$ on C_1 must cancel that on C_2.

so we have the finite energy sum rule

$$\int_{4m_\pi^2}^{s} R^{exp}(s)ds = \int_0^s R^0(s)ds + \frac{4}{3}\alpha_s(-|s|)\int_0^s R^1(s)ds \equiv \Omega(s)\ . \tag{5.5}$$

Below charm threshold and neglecting the light quark masses,

$$\Omega^{theor}(s) = 2\left\{1 + \frac{\alpha_s(-|s|)}{\pi}\right\}s\ . \tag{5.6}$$

The sum rule (5.5) is probably the best tool for confronting QCD with the total cross section data. Some comparison with experiment was given by Shankar[3] but a definitive test will have to wait for improvements in the amount and quality of the total cross section data.

Clearly the finite energy sum rule method can be applied to a variety of problems involving weak and electromagnetic currents. One simple example where it is ideally suited is the semi-hadronic weak decay of a sequential heavy lepton[17]. The recently discovered τ lepton[18] may be just that and so the possibility of experimental comparison already exists.

Consider the decay of a heavy lepton L into its own neutrino ν_L plus hadrons. The hadronic weak current can have a rather general form in terms of quark fields

$$J^\mu(x) = \sum_{\substack{ij\\ \text{flavors}}} \bar{\psi}_i C_{ij}\gamma^\mu(1-\gamma_5)\psi_j \tag{5.7}$$

where C_{ij} is some matrix in flavor. In the SU(4) model,

$$\begin{aligned} J^\mu(x) &= \bar{u}\gamma^\mu(1-\gamma_5)(d\cos\theta + s\sin\theta) \\ &+ \bar{c}\gamma^\mu(1-\gamma_5)(s\cos\theta - d\sin\theta)\ . \end{aligned} \tag{5.8}$$

We take ν_L to be massless and the charged weak current (L,ν_L) can have right handed and left handed parts. With strong interactions neglected,

$$R_L = \frac{\Gamma(L\to\nu_L + \text{hadrons})}{\Gamma(L\to\nu_L + e + \nu)} = 3\sum_{ij}(C_{ij})^2 \tag{5.9}$$

and if M_L is below charm threshold,

$$\sum_{ij} (C_{ij})^2 = \cos^2\theta + \sin^2\theta = 1 \,. \tag{5.10}$$

We expect R_L to be increased by strong interaction final state corrections and the increase can be computed in an asymptotically free theory. Let q be the momentum transferred to the hadrons. Then $0 < s \equiv q^2 < M_L^2$ and a finite energy sum rule with maximum energy M_L is ideally suited to the problem. We define

$$\begin{aligned} C_{\mu\nu}(q) &\equiv \int d^4x\, e^{-iq\cdot x} \langle 0| J^\mu(x) J^{\nu\dagger}(0)|0\rangle \\ &= -(g_{\mu\nu}q^2 - q_\mu q_\nu)\rho_1(q^2) + q_\mu q_\nu \rho_2(q^2) \,. \end{aligned} \tag{5.11}$$

$C_{\mu\nu}(q)$ is the absorptive part of a tensor $\Pi_{\mu\nu}(q)$ and its two pieces ρ_1 and ρ_2 are given in perturbation theory by

$$\begin{aligned} \rho_1^{theor}(s) &= 3\sum_{ij} (C_{ij})^2 \frac{1}{3\pi} \{1 + \frac{\alpha_s(-|s|)}{\pi}\} \\ \rho_2^{theor} &= O(m^2/s) \end{aligned} \tag{5.12}$$

where I have taken the quark masses to zero in ρ_1^{theor}.

The finite energy sum rule we need is derived as before by considering the function $\Pi(s)$ whose absorptive part is ρ_1. The integral of $s^n\Pi(s)$ about the contour of Fig. 5 can be evaluated theoretically on C_2 and with experimental data on C_1. The result is

$$\int_0^{M_L^2} ds\, s^n \rho_1^{exp}(s) = 3\sum_{ij} \frac{1}{3\pi}\left\{1 + \frac{\alpha_s(-M_L^2)}{\pi} \int_0^{M_L^2} s^n ds\right\} \,. \tag{5.13}$$

The decay width for $L \to \nu_L$ + hadrons is given by

$$\begin{aligned} \Gamma(L \to \nu_L + \text{hadrons}) &= \frac{G^2}{32\pi^2 M_L^3} \int_0^{M_L^2} ds\, (M_L^2 - s)^2 \\ &\times \{\rho_1^{exp}(s)(M_L^2 + 2s) + \rho_2^{exp}(s) M_L^2\} \end{aligned} \tag{5.14}$$

while

$$\Gamma(L \to \nu_L + e + \bar{\nu}_e) = \frac{G^2 M_L^5}{3\pi^3 2^6} \quad . \tag{5.15}$$

It then follows from the sum rule that

$$R_L = 3 \sum_{ij} (C_{ij})^2 \left\{1 + \frac{\alpha_s(-M_L^2)}{\pi}\right\} \quad . \tag{5.16}$$

For M_L below charm threshold, we then have a prediction for the leptonic branching ratio:

$$\frac{\Gamma(L \to \nu_L + e + \bar{\nu}_e)}{\Gamma(L \to \nu_L + \text{all})} = \frac{1}{5 + 3\,\frac{\alpha_s(-M_L^2)}{\pi}} \quad . \tag{5.17}$$

The factor $\alpha_s(-M_L^2)/\pi$ is probably somewhere between 0.1 and 0.2 for the τ lepton whose mass is 1.8 GeV. The value 0.2 is suggested by the e^+e^- total cross section below charm threshold and that leads to a value of 0.18 for the branching ratio (5.17). This is quite consistent with the experimental values which range between 0.15 and 0.20.[18] If yet heavier heavy leptons are discovered, their leptonic branching ratio (5.17) will also be calculable unless the lepton mass coincides with some prominent hadronic threshold.

Let me finish this topic by emphasizing, especially for the experimenters present, the theoretical status of the sum rule (5.17). Although it looks like the parton model prediction for $R(e^+e^- \to \text{hadrons})$ along with asymptotic freedom corrections, it is in fact on much more solid footing. The integration over s from 0 to M_L^2 in the definition of Γ has the effect of smoothing out source of the possible rapid variations in the cross section.

6. THE OZI RULE FOR HEAVY QUARKS

One of the earliest theoretical suggestions about charmonium was that the narrow widths of the states below charm threshold could be simply explained by quantum chromodynamics[4]. The idea is basically that since $c\bar{c}$ annihilation into light hadrons must proceed through gluons and since the effective gluon coupling constant should be small at high energies, the decay will be inhibited. The dominant contribution will come from the minimum number of intermediate gluons which depends upon the quantum numbers of the charmonium state. Some rather striking experimental predictions can be made and I will briefly describe them and review the experimental situation.

Consider the decay of the ψ/J $^3S_1$ state. Its dominant electromagnetic decay is shown in Fig.6. The $c\bar{c}$ pair must come together to annihilate into the virtual photon and if the bound state is non-relativistic then, to first approximation, the decay width will be given by

$$\Gamma(\psi/J \to \ell^+\ell^-) = \frac{4\alpha^2(2/3)^2}{M^2}|\psi(0)|^2 \quad . \tag{6.1}$$

The charge of the charmed quark is $\frac{2}{3}|e|$ and M is the charmonium mass. $\psi(0)$ is the non-relativistic wave function at the origin and one cannot expect to be able to compute it in perturbation theory. The reason for this is that the mean radius of charmonium is on the order of one Fermi, a distance scale at which the effective coupling strength for the binding has become large. Thus $\psi(0)$ will be determined primarily by the non-perturbative, long range part of the potential.

The hadronic decay of the ψ/J must proceed through a minimum of three gluons. If this is indeed the dominant contribution, that is to say, if perturbation theory is truly relevant to this problem, the decay will proceed as shown in Fig. 7. The $c\bar{c}$ annihilation will be essentially local (on the order of $1/M_c << \langle R\rangle$). The

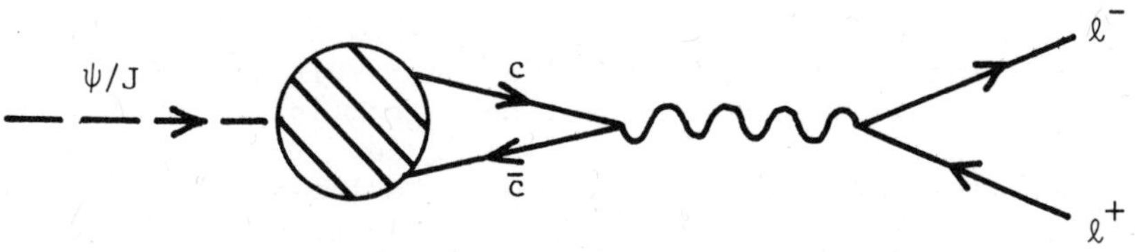

Fig. 6. The electromagnetic decay of the ψ/J. ℓ can be either μ or e.

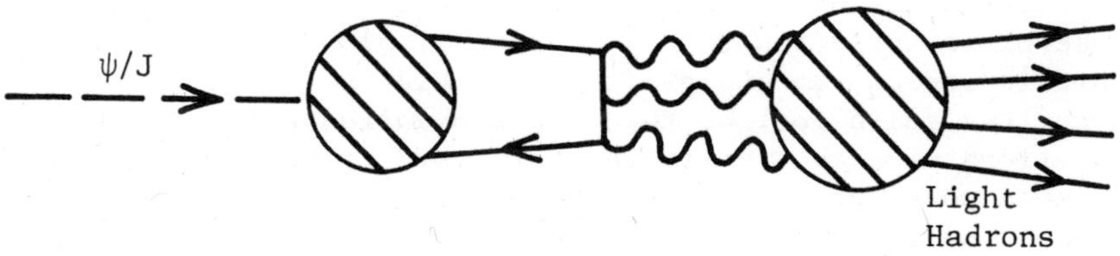

Fig. 7. The three gluon mechanism for the hadronic decay of the ψ/J.

computation of the decay matrix element is then done in analogy to the parton model computation of $\sigma_{tot}(e^+e^- \to \text{hadrons})$ as if the final state consisted of three on-mass-shell gluons. This amounts to the statement that the transition from the three gluon state into physical hadrons takes place with unit probability. A more satisfying theoretical justification of the three gluon mechanism can be given and I will turn to it shortly. If the mechanism just described is correct, then the total hadronic width of the ψ/J is given by

$$\Gamma(\psi/J \to \text{hadrons}) = \frac{40}{81\pi}(\pi^2 - 9)\frac{\alpha_s^3(M)}{M^2}|\psi(0)|^2 \quad . \tag{6.2}$$

The strong coupling constant is defined at the ψ/J mass and, as before, $\psi(0)$ is the non-relativistic wave function at the origin.

Before proceeding to the comparison of these expressions with experiment, let me review briefly the analysis that underlies Eq. (6.2). A necessary condition for the applicability of lowest order perturbation theory is that no large dynamical factors enter in higher orders to make the expansion break down. One must analyze the quantity $\sum_n |M_{c\bar{c}\to n}|^2$ where n is some quark-gluon final state. M is the decay matrix element and is defined to be two particle ($c\bar{c}$) irreducible in the decay channel. Two particle reducible contributions are absorbed into the definition of the wave function. If it can be shown the $\sum_n |M_{c\bar{c}\to n}|^2$ is free of infrared singularities for the $c\bar{c}$ pair at rest, then there can be no large dynamical factors. This is because the result will involve no small energy or momentum factors, only the (large) charm quark mass and the (large) renormalization scale. I have checked the infrared finiteness through order α_s^4 (the lowest order is α_s^3) for ψ/J decay and I conjecture that it is true to all orders. It is technically simpler to use the Coulomb rather than covariant gauges in this analysis.

I want to make one last point before proceeding to the comparison with experiment. The infrared analysis is necessary but not sufficient. It could well be that threshold singularities in high orders or even completely non-perturbative effects prevent the use of perturbation theory in this simple way. The use of asymptotic freedom to explain the OZI rule is speculative. It is on much less solid footing than the conventional deep Euclidean applications or even the finite energy sum rule I described earlier. It is even more speculative than the direct computation of $\sigma_{tot}(e^+e^- \to \text{hadrons})$ since there the production is truly local, coming from the off-shell photon.

The ratio $\Gamma(\psi/J \to \text{hadrons})/\Gamma(\psi/J \to e^+e^-)$ is experimentally known to be about 10 (48 keV/4.8 keV). From this, a value of

$\alpha_s(M_\psi)$ can be extracted and one finds that

$$\alpha_s(M_\psi) \approx 0.2 \quad . \tag{6.3}$$

This is a very small value of $\alpha_s(M_\psi)$ and whether it is consistent with $\sigma_{tot}(e^+e^- \to \text{hadrons})$ is not yet clear. The $\eta_c(^1S_0)$ decay can proceed via two gluons and with the above value of α_s, its width can be predicted. One finds

$$\Gamma(\eta_c \to \gamma\gamma) = \frac{4}{3}\Gamma(\psi \to e^+e^-) = 6\text{–}8 \text{ MeV}$$
$$\Gamma(\eta_c \to \text{hadrons}) = \frac{9}{8}\left(\frac{\alpha_s}{\alpha}\right)^2 \Gamma(\eta_c \to \gamma\gamma) = 738\,\Gamma(\eta_c \to \gamma\gamma) = 4\text{–}5 \text{ MeV}. \tag{6.4}$$

Thus the $\gamma\gamma$ branching ratio is predicted to be

$$BR(\eta_c \to \gamma\gamma) = 1.3 \times 10^{-3} \quad . \tag{6.5}$$

The state at 2800 MeV reported by DESY is the only candidate for the η_c. Let me assume this to be the case. The DESY experiment reports a product of branching ratios

$$BR(\psi/J \to \eta_c\gamma) \times BR(\eta_c \to \gamma\gamma) = (1.2 \pm 0.5)\, 10^{-4} \tag{6.6}$$

while from the inability to detect monochromatic photons in ψ/J decay, it is known that[18]

$$BR(\psi \to \eta_c\gamma) < 1.7\% \quad . \tag{6.7}$$

Thus, experimentally,

$$BR(\eta_c \to \gamma\gamma) \gtrsim 7 \times 10^{-3} \tag{6.8}$$

and the discrepancy is a factor of four or five. Perhaps this means that the total width of the η_c is smaller (by at least this factor) than the theoretical prediction of a few MeV. It could still be much broader than the ψ/J but it is probably too soon to decide whether or not the minimal gluon mechanism is verified or disproved by experiment. It is, of course, possible the η_c hasn't even been discovered yet.

The situation with respect to the P states of charmonium has been reviewed several times recently[19,20]. Since I have nothing new to add at this time, let me just briefly summarize what I know.

1. The computation is now more difficult since the wave function vanishes at the origin. The decay rates depend to leading order on the derivative of the wave function at the origin[21].

2. The prediction is in good agreement with experiment with respect to both the relative and absolute widths of the three even-C P states.

3. In the case of the $^3P_1$ state, there is a logarithmic sensitivity to the binding energy of the state[21]. On the one hand, this "large logarithm" helps to produce agreement with experiment. On the other, the logarithm might signal the breakdown of the perturbation expansion all together. This question has not really been studied very carefully yet and it is important to do so, especially in view of the fact that agreement with experiment seems best here.

Finally, if the state at 3455 MeV is identified as the η_c', there seems to be another problem for the minimal gluon mechanism. This state appears to be narrower by at least an order of magnitude[20] than the theoretical prediction of about 2 MeV. It is possible that it is not the η_c' at all but then the η_c' is as yet undiscovered.

The status of the asymptotic freedom explanation of the OZI rule remains unclear. The triplet states seem to be working rather well but the singlet states (η_c and η_c') seem to be too narrow if indeed they have been discovered at all.

As far as the charmonium states are concerned, nothing seems more important to me than finding out whether or not the 2800 and 3455 states are the η_c and η_c'. The coming of the upsilon is, of course, very exciting from the point of view of the OZI rule. The increased quark mass should make the theoretical predictions cleaner since the bound state is correspondingly smaller in size. The next generation of e^+e^- colliding beam machines should allow the upsilon states to be copiously produced and studied. Perhaps by the time of the next Banff Summer Institute on Particles and Fields, we will know a great deal more about the status of the OZI rule.

FOOTNOTES AND REFERENCES

*) Alfred P. Sloan Foundation Fellow.

1) For one recent review, see W. Marciano and H. Pagels, Rockefeller University Preprint COO-2232B-130.

2) O. Nachtman, Invited Talk at the International Conference on Leptons and Photons, Hamburg, August 1977.

3) R. Shankar, Phys. Rev. D15, 755 (1977).

4) T. Appelquist and H.D. Politzer, Phys. Rev. Lett. 34, 43 (1975) and Phys. Rev. D12, 1404 (1975).

5) S. Weinberg, Phys. Rev. D7, 2887 (1973).

6) V.N. Gribov, Lectures at the 12th Winter School of the Leningrad Nuclear Physics Institute, 1977, SLAC-TRANS-176.

7) S. Mandelstam, Invited Talk given at the Washington Meeting of the American Physical Society, April 1977.

8) E.S. Fradkin and I.B. Tyutin, Phys. Rev. D2, 2841 (1970); A. Ali and J. Bernstein, Phys. Rev. D12, 503 (1975); R.N. Mohapatra, Phys. Rev. D4, 378 (1971); D4, 1007 (1971); D.P. Sidhu, Phys. Rev. D6, 565 (1972); V.N. Popov and L.D. Fadeev "Perturbation Theory for Gauge Invariant Fields", Kiev I.T.P. report, unpublished.

9) This Ward identity may be derived by appropriately modifying the corresponding proof for covariant gauges given by E. Eichten and F. Feinberg, Phys. Rev. D10, 3254 (1974).

10) T. Appelquist and J. Carazzone, Phys. Rev. D11, 2856 (1975).

11) A. Duncan, Phys. Rev. D10, 2866 (1976); J. Frenkel and J.C. Taylor, University of Oxford Preprint 21/76 (1976).

12) T. Appelquist, M. Dine and I.J. Muzinich, Phys. Lett. 69B, 231 (1977).

13) W. Fischler, Nucl. Phys. B129, 157 (1977).

14) A.L. Fetter and J.D. Walecka, Quantum Theory of Many-Particle Systems (McGraw-Hill, N.Y., 1971). The methods of C. Bloch and collaborators are very close in spirit to our discussion. See, for example, C. Bloch, Nucl. Phys. 7, 451 (1958), and C. Bloch and C. DeDominicus, Nucl. Phys. 7, 459 (1958).

15) In examining the structure of the perturbation expansion, I am implicitly working in the zero instanton sector. Thus any A_μ giving $F_{\mu\nu} = 0$ is gauge equivalent to $A_\mu = 0$.

16) The static potential provides a natural framework for the study of QCD perturbation theory in high orders. Clearly the first order of business in such a study is to establish the general form of the expansion, perhaps by verifying our conjecture.

17) C.S. Lam and T.M. Yan, Cornell University Preprint, CLNS-357, January 1977.

18) For a review of the latest data, see the lectures of G. Feldman at this summer school.

19) T. Appelquist, Invited Talk at the XVIII International Conference on High Energy Physics, Tiblisi, Georgia, Soviet Union, August 1976.

20) J.D. Jackson, Invited Talk at the XIIIth Rencontre de Moriond, Flaine, Haute-Savoie, France, March 1977.

21) R. Barbieri, R. Gatto and R. Kögerler, Phys. Letts. 60B, 183 (1976); R. Barbieri, R. Gatto and E. Remiddi, Phys. Letts. 61B, 465 (1976).

NON-PERTURBATIVE METHODS IN FIELD THEORY

E.R. Caianiello, M. Marinaro and G. Scarpetta

Universita di Salerno

Salerno, Italy

1. INTRODUCTION

We consider a Euclidean neutral scalar field theory[1] described by the Lagrangian density

$$\mathcal{L}(x) = \mathcal{L}^{(o)}(x) + \varepsilon\mathcal{L}^{(1)}(x) \tag{1}$$

where

$$\mathcal{L}^{(o)}(x) = L[\phi(x)] \tag{2}$$

depends only on a field operator $\phi(x)$, while

$$\mathcal{L}^{(1)}(x) = -\frac{1}{2}\int dx' A(xx')\phi(x)\phi(x') \tag{3}$$

connects field operators at different points. $A(xx')$ may contain derivatives, $L[\phi(x)]$ in (2) contains no derivative terms, but is otherwise arbitrary except for qualitative requirements that secure the solvability of the (regularized) theory $\mathcal{L}^{(o)}(x)$. Eq. (3) may (but need not) be the kinetic term $-\frac{1}{2}(\partial_\mu\phi)(\partial_\mu\phi)$ in continuous space, or any spin-coupling term in a lattice space. When $\varepsilon = 0$, all Green functions of the theory can be computed. This fact, well known in lattice space, holds true also in ordinary space and the solutions can be obtained directly from the Green function equations[2] by imposing the regularizing condition

$$\delta(0) = M^d \tag{4}$$

(d = dimension of Euclidean space), or asymptotically from lattice space through functional integration[3]. The two solutions coincide when M in (4) is taken to be the inverse lattice spacing[4]. This fact can be easily ascertained by verifying that the standard expressions for the Green functions derived from the generating functional

$$\frac{W[h]}{W(0)} = \exp\left\{M^d\int dx \log \frac{\int_{-\infty}^{\infty} du\, e^{\frac{1}{M^d} L\left(\frac{M^{d/2}}{m}u\right) + \frac{uh(x)}{mM^{d/2}}}}{\int_{-\infty}^{\infty} du\, e^{\frac{1}{M^d} L\left(\frac{M^{d/2}}{m}u\right)}}\right\} \tag{5}$$

satisfy exactly[5] the equations of refs. 2 and 4, provided (4) holds.

In a previous work we have studied a general expansion for a Lagrangian decomposed as in (1) and (3). It was then shown that under the assumptions made here for $\mathcal{L}^{(o)}(x)$, which define a static ultralocal (s.u.l.) model, this is equivalent, in lattice space, to the classic cluster expansion[6]. This result, fairly obvious for lattice theories, was far from evident in continuous space. In both cases the expansion presented in ref. 1 differs from the conventional form of the cluster expansion because it exhibits a degree of formal compactness that permits additional manipulations. Our aim here is to take advantage of this feature in order both to introduce a new and possibly useful representation of the generic term of the cluster expansion, and then to sum the whole cluster expansion again into the functional integral. The steps involved in this process will be seen to be more significant than the overall proof of consistency thereby attained, because they offer new means of formulating and of approximating the functional integral itself, or the terms of its expanded forms.

2. LATTICE SPACE

2.1 General formalism

We showed in ref. 1 that, in a lattice space of volume L^d, the Nakano-Schwinger functions are given by

$$S_n(q_1 \dots q_n) = \frac{K_n(q_1 \dots q_n)}{K_o} \quad , \quad q_1, q_2 \dots q_n \in L^d \tag{6}$$

where

$$K_n(q_1 \cdots q_n) = \sum_{s=1}^{\infty} \frac{1}{s!} \sum_{P_{n_1 \cdots n_s}} \sum_{n_1 + \cdots + n_s = n} \left(\frac{M^{d/2}}{m}\right)^n \sum_{\ell_1 \cdots \ell_s = 1}^{N}$$

$$\exp\left(-\sum_{i,j=1}^{s} \frac{\varepsilon}{2m^2 M^d} A(\ell_i, \ell_j)\partial_{\rho_i}\partial_{\rho_j}\right) \prod_{j=1}^{s} \partial_{\rho_j}^{n_j} Z(\rho_j)\cdot\delta(q_{i_1} \cdots q_{i_{n_j}}, \ell_j)\Big|_{\underline{\rho}=0} \tag{7}$$

$$K_o = 1 + \sum_{s=1}^{\infty} \frac{1}{s!} \sum_{\ell_1 \cdots \ell_s = 1}^{N} \exp\left(-\sum_{i,j=1}^{s} \frac{\varepsilon}{2m^2 M^d} A(\ell_i, \ell_j)\partial_{\rho_i}\partial_{\rho_j}\right) \prod_{j=1}^{s} Z(\rho_j)\Big|_{\underline{\rho}=0} \tag{8}$$

where M^{-1} is the lattice spacing, $N = L^d M^d$, $\underline{\rho} \equiv \{\rho_1 \cdots \rho_s\}$,

$$Z(\rho) = \log \frac{\int_{-\infty}^{\infty} du\, e^{\frac{1}{M^d} L\left(\frac{M^{d/2}}{m} u\right) + \rho u}}{\int_{-\infty}^{\infty} du\, e^{\frac{1}{M^d} L\left(\frac{M^{d/2}}{m} u\right)}} \tag{9}$$

and $\sum_{P_{n_1 \cdots n_s}}$ = sum over all partitions,

$$\delta(q_1 \cdots q_n) = \begin{cases} \prod_{i=1}^{n} \delta_{q_1 q_i} & n > 1 \\ 1 & n = 1 \end{cases} .$$

It is evident, though (8) is not the standard form, that it coincides term by term with the cluster expansion of the partition function of a continuous spin system described by the reduced Hamiltonian

$$H = \frac{1}{M^d} \sum_i L(u_i) + \frac{\varepsilon}{2M^{2d}} \sum_{ij} A(i,j) u_i u_j \quad . \tag{10}$$

We introduce at this point the decompositions:

$$e^{AB} = \frac{1}{4\pi} \int d\tau \, d\sigma \, e^{-\frac{\tau^2}{4} - \frac{\sigma^2}{4}} e^{\tau(\frac{A+B}{2})} e^{i\sigma(\frac{A-B}{2})} \tag{11}$$

and write using for short the notation $n \equiv (n_1 \ldots n_d) \ldots$, $\tau \equiv (\tau_1 \ldots \tau_d) \ldots$, sums being understood accordingly

$$A(\ell_i, \ell_j) = \sum_n^Q f_n(\ell_i) g_n(\ell_j) \quad . \tag{12}$$

Substitution into (8) (it will clearly suffice to study here only K_0) yields

$$K_0 = 1 + \frac{1}{(4\pi)^{Qd}} \sum_{s=1}^{\infty} \frac{1}{s!} \sum_{\ell_1 \ldots \ell_s = 1}^{N} \int d\tau_1 \ldots d\tau_Q d\sigma_1 \ldots d\sigma_Q \, e^{-\sum_i^Q \frac{\tau_i^2 + \sigma_i^2}{4}}$$

$$\exp\left[\frac{i\sqrt{\varepsilon}}{\sqrt{2}\, M^{d/2} m} \sum_{j=1}^{N} \sum_{n=1}^{Q} \left\{\frac{\tau_n}{2} (f_n(\ell_i) + g_n(\ell_i)) + \frac{i\sigma_n}{2} (f_n(\ell_i) - g_n(\ell_i))\right\} \partial_{\rho_j}\right] \times$$

$$\times \prod_{\alpha=1}^{s} Z(\rho_\alpha) \Big|_{\underline{\rho}=0} \quad . \tag{13}$$

Hence

$$K_0 = \sum_{s=0}^{\infty} \frac{1}{s!} \langle [U_2(\underline{\tau}|\underline{\sigma})]^s \rangle \tag{14}$$

where

$$U_2(\underline{\tau}|\underline{\sigma}) = \sum_{\ell_i=1}^{N} Z\left[i\left(\frac{\varepsilon}{2m^2 M^d}\right)^{\frac{1}{2}} \sum_n^Q \left\{\frac{\tau_n}{2}(f_n(\ell_i) + g_n(\ell_i)) + \frac{i\sigma_n}{2}(f_n(\ell_i) - g_n(\ell_i))\right\}\right] \tag{15}$$

$$<F(\tau_1 \ldots \tau_Q | \sigma_1 \ldots \sigma_Q)> \equiv \frac{1}{(4\pi)^{Qd}} \int d\tau_1 \ldots d\tau_Q d\sigma_1 \ldots d\sigma_Q \, e^{-\sum_{i=1}^{Q} \frac{\tau_i^2 + \sigma_i^2}{4}} \times$$

$$\times \, F(\tau_1 \ldots \tau_Q | \sigma_1 \ldots \sigma_Q) \quad . \tag{16}$$

If in particular $f_n = g_n$

$$K_o = \sum_{s=0}^{\infty} \frac{1}{s!} <[U_1(\underline{\tau}|\underline{\sigma})]^s> \quad . \tag{17}$$

Explicitly, from (9) and (15),

$$U_{1(2)} = \sum_{\ell_j=1}^{N} \log \frac{\int_{-\infty}^{\infty} du \, e^{\frac{1}{M^d} L(\frac{M^{d/2}}{m} u) + uH_{1(2)}(\ell_j)}}{\int_{-\infty}^{\infty} du \, e^{\frac{1}{M^d} L(\frac{M^{d/2}}{m} u)}} \tag{18}$$

where

$$H_1(\ell_j) = i(\frac{\varepsilon}{2m^2 M^d})^{\frac{1}{2}} \sum_n^Q \tau_n f_n(\ell_j) \tag{19}$$

$$H_2(\ell_i) = i(\frac{\varepsilon}{2m^2 M^d})^{\frac{1}{2}} \sum_n^Q [\tau_n F_n^+(\ell_j) + \sigma_n F_n^-(\ell_j)] \tag{20}$$

in (20) we have set

$$F_n^+ = \frac{1}{2}(f_n + g_n) \quad , \quad F_n^- = \frac{i}{2}(f_n - g_n) \quad . \tag{21}$$

From (18) it is seen that $U_{1(2)}$ is the free energy of a system of N independent spins in the "external" field $H_{1(2)}$. The operation of (11) and (12) consists therefore in linearizing the quadratic expression $A(\ell_i, \ell_j)u_i u_j$ and breaking thereby the direct couplings between u_i and u_j, i.e. each "spin" u_j is coupled to a stochastic field $H_{1(2)}(\ell_j)$; the gaussian average over τ and σ restores the direct couplings. The sth term of the expression (14) is the gaussian average of the sth power of the free energy of a system of N independent spins within a stochastic external field $H_1(2)$. We have in this way introduced an alternative and compact expression for the sth term of the standard cluster expansion.

Clearly, formula (14) can be summed into

$$K_o = \langle e^{U_2(\underline{\tau}|\underline{\sigma})} \rangle . \tag{22}$$

Thus, we see that before performing the gaussian average[7], the study of the coupled theory is reduced to that of a s.u.l. model interacting with an "external" field - a problem which is solved by quadrature. The number of modes Q may, but need not be dN in an N-cell lattice; each mode is now a wave that pervades the whole lattice. Q=1 gives, as will be seen, the mean field approximation while taking Q > 1 should give closer approximations. We also notice that the decomposition (12) is not unique; its arbitrariness is evidenced if we resort to matrix notation: if

$$A(\ell_i, \ell_j) = (\mathcal{A})_{ij}$$

then

$$\mathcal{A} = F^T G$$

where F and G are Q × N matrices. Thus for any orthogonol Q × Q matrix O we can set

$$F' = OF$$

$$G' = OG$$

and write

$$\mathcal{A} = (F')^T G' \quad .$$

If $\mathcal{A}$ is Hermitian, then $\mathcal{A} = F^+F$ and the arbitrariness is that of unitary matrix. Such considerations are useful in the study of the renormalization group[8].

2.2 Mean field approximation

Let us consider, as a first approximation, Q = 1 and impose translational invariance. Eq. (12) becomes $A(\ell_i, \ell_j) = c = \text{const}$ and (22) can be written, keeping in mind (9) and choosing

$$c = -\frac{v}{N}\frac{m^2 M^d}{\varepsilon}$$

$$K_o = \frac{1}{\sqrt{4\pi}} \int d\tau\, e^{-\frac{\tau^2}{4}}\, e^{N\, Z(\sqrt{\frac{v}{2N}}\,\tau)} = \frac{\sqrt{N}}{\sqrt{2\pi v}} \int d\rho\, \exp\{-N[\frac{\rho^2}{2v} - Z(\rho)]\} \,.$$

For large N we obtain the free energy density in the mean field approximation[1,9], namely

$$F = \frac{1}{N} \log K_o = \frac{\mu^2 v}{2} - \log Z(\mu v)$$

$$= \frac{\mu^2 v}{2} - \log \frac{\int_{-\infty}^{\infty} du\, e^{\frac{1}{M^d} L(\frac{M^{d/2}}{m} u) + \mu v u}}{\int_{-\infty}^{\infty} du\, e^{\frac{1}{M^d} L(\frac{M^{d/2}}{m} u)}}$$

where μ is the magnetization per spin. In particular, by choosing (Ising model)

$$e^{+\frac{1}{M^d} L(\frac{M^{d/2}}{m} u)} = \delta(u^2 - 1)e^{hu}$$

we have:

$$F = \frac{\mu^2 v}{2} - \log \cosh(\mu v + h)$$

and the Weiss equation

$$\mu = \tanh(\mu v + h) \quad ;$$

h plays the role of an external magnetic field.

3. CONNECTION WITH FUNCTIONAL INTEGRATION

3.1 The formalism introduced thus far enables us to approach the subject of functional integration in a rather general way, the standard definition through lattice space appearing as a special case. The discussions of section 2 have been made for the sake of clarity with reference to a lattice space. In the first section, however, we have made it clear that the transition from lattice to continuous space, or vice versa, is quite trivial formally, provided (4) is respected. We wish to examine here this situation in greater detail. We recall first that the two key points of our approach are:

1) The possibility of obtaining the exact solution of any s.u.l. model in continuous space, provided only $\delta(x)$ is regularized so that (4) holds.

2) The decomposition (12), necessary for the gaussian decoupling of quadratic terms. Discrete sums occur naturally if the whole system is enclosed in a finite volume L^d as we assume throughout. We introduce at this point the approximation of truncating any such sum at some fixed integer Q, the limit $Q \to \infty$ being left as a later step in the overall computation. Then all infinite or ambiguous quantities, such as powers of $\delta(x-y)$ or of $\Delta\delta(x-y)$ that originate from the kinetic term, etc., are <u>ipso facto</u> regularized. Eq. (12) becomes, in the case of the kinetic expansion

$$A(x-y) \equiv \Delta\delta_Q(x-y) = \sum_n^Q f_n(x) g_n(y)$$

$\delta_Q(x-y)$ indicates a regularized δ function obtained by keeping Q finite. The value of M in (4) will be obviously related to Q. We recall that the choice of the functions $f_n(x)$ is only subject to the requirement that they form a complete set in the finite volume L^d. Each different choice yields a different approximation to the functional integral, though in the limit $Q \to \infty$ they shall all formally coincide.

We define now the regularized generating functional of the Nakano-Schwinger functions in L^d as:

$$S_Q(x_1 \ldots x_n) = \frac{\delta}{\delta h(x_1)} \cdots \frac{\delta}{\delta h(x_n)} \frac{W_Q[h]}{W_Q(0)}$$

where

$$W_Q[h] = \left\langle \exp\left\{ M^d \int dx \log \frac{\int_{-\infty}^{\infty} du\, e^{\frac{1}{M^d} L(\frac{M^{d/2}}{m} u) + u(H_2(x) + \frac{h(x)}{mM^{d/2}})}}{\int_{-\infty}^{\infty} du\, e^{\frac{1}{M^d} L(\frac{M^{d/2}}{m} u)}} \right\} \right\rangle_Q . \tag{23}$$

Thus,

$$W_Q(0) = W_Q[h]\Big|_{h=0} = (K_o)_Q .$$

Clearly, (23) is the gaussian average of the generating functional (5) of the s.u.l. model in the "external" field $H_2(x)$. Exact physical quantities are obtained from (23) in the limits $Q \to \infty$, $L \to \infty$; these limits are of course connected with the well known problems of renormalization.

We show now that the definition of W[h] in lattice space is contained as a particular case in our definition (23). As was stipulated, we identify M with the inverse spacing of the lattice which is obtained by approximating $\int dy$ in (23) with a summation over cubes of volume M^{-d}. Eq. (23) then reduces to

$$W_Q(h) = \left\langle \exp\left\{ \sum_{i=1}^{N} \log \frac{\int_{-\infty}^{\infty} du\, e^{\frac{1}{M^d} L(\frac{M^{d/2}}{m} u) + u(H_2(i) + \frac{h(i)}{mM^{d/2}})}}{\int_{-\infty}^{\infty} du\, e^{\frac{1}{M^d} L(\frac{M^{d/2}}{m} u)}} \right\} \right\rangle_Q . \tag{24}$$

Then, after standard manipulations, taking the limit $Q \to \infty$ and considering M as an additional regularizing parameter which is kept fixed, we arrive at

$$W_Q[h] = \int d\phi_1 \ldots d\phi_N \; e^{\frac{1}{M^d}\sum_i^N (L(\phi_i) + \phi_i h(x)) + \frac{1}{M^{2d}}\sum_{ij}^N \frac{A(ij)}{2}\phi_i\phi_j} \tag{25}$$

which is just the ordinary lattice approximation to the functional integral.

3.2 We demonstrate now how our definition works in the standard case of a free field. This is obtained by setting

$$\mathcal{L}^{(o)}(x) = -\frac{m^2\phi^2}{2} \quad \text{in eq. (2)} \quad ,$$

$$A(x-y) = \Delta\delta(x-y) \quad \text{in eq. (3)}$$

and $\varepsilon = 1$ throughout.

The regularized functional (23) reduces to

$$W_Q^o[h] = \langle \exp M^d \int dx \log \frac{\int_{-\infty}^{\infty} e^{-\frac{u^2}{2} + u(H_2(x) + \frac{h(x)}{mM^{d/2}})}}{\int_{-\infty}^{\infty} du\, e^{-\frac{u^2}{2}}} \rangle_Q \; . \tag{26}$$

It follows then

$$S_Q^o(xy) = \frac{1}{W_Q^o(0)} \frac{\delta}{\delta h(x)} \frac{\delta}{\delta h(y)} W_Q^o[h] \Big|_{h=0}$$

$$= \frac{1}{W_Q^o(0)} \langle \exp[M^d \int dz \frac{[H_2(z)]^2}{2}] [H_2(x)H_2(y) + \frac{\delta(x-y)}{M^d}] \frac{M^d}{m^2} \rangle_Q \; . \tag{27}$$

To speed up the computation we choose $F^{\pm}$ in H_2 to be such that:

$$\int dy\, F^{\pm}_m(y) F^{\pm}_n(y) = b^{\pm}_m \delta_{mn}$$

$$\int dy\, F^{\pm}_m(y) F^{\mp}_n(y) = 0 \quad ; \tag{28}$$

substitution in (27) then yields

$$S^{o}_{Q}(xy) = \frac{\delta(x-y)}{m^2} - \frac{1}{m^2} \sum_n^Q \left\{ \frac{F^{+}_n(x)F^{+}_n(y)}{m^2 + b^{+}_n} + \frac{F^{-}_n(x)F^{-}_n(y)}{m^2 + b^{-}_n} \right\} . \tag{29}$$

The special choice

$$F^{+}_n(x) = L^{-d/2} K_n \cos(K_n x)$$

$$F^{-}_n(x) = - L^{-d/2} K_n \sin(K_n x) \tag{30}$$

where $K_n = 2\pi n/L$ with $n \equiv (n_1 \dots n_d)$ finally yields

$$S^{o}_{Q}(xy) = \frac{1}{m^2} \left\{ \delta(x-y) - \frac{1}{L^d} \sum_n^Q e^{K_n(x-y)} \right\} + \frac{1}{L^d} \sum_n^Q \frac{e^{iK_n(x-y)}}{m^2 + K_n^2} \tag{31}$$

so that

$$\lim_{Q\to\infty} S^{o}_{Q}(xy) = \frac{1}{L^d} \sum_n \frac{e^{iK_n(x-y)}}{m^2 + K_n^2}$$

which is the Euclidean free propagator in the box L^d.

3.3 $g\phi^4$ theory: asymptotic for large g. Here

$$\mathcal{L}^{(o)}(x) = - \frac{m^2}{2} \phi^2 - g\phi^4 \quad ; \quad A(xy) = \Delta\delta(x-y)$$

$$\varepsilon = 1 \quad , \quad g \stackrel{\rightarrow}{\sim} \infty \quad .$$

To obtain the leading term when $g \to \infty$, we may expand the 4th order polynomial in the inner exponent of (23) written for short

$$-\frac{u^2}{2} - g\frac{M^d}{m^4} n^4 + \frac{u}{mM^{d/2}} \rho(x)$$

where $\rho(x) = h(x) + mM^{d/2}H(x)$. The latter given by (20), around its minimum value

$$u_o \underset{g\to\infty}{\simeq} \frac{m}{M^{d/2}} [\frac{\rho(x)}{4g}]^{1/3} .$$

An estimate of the leading term gives immediately

$$(W[h])_Q \underset{g\to\infty}{\lessapprox} \langle \exp\int dx \frac{3}{4} (\frac{1}{4g})^{1/3} (h(x) + \frac{i}{\sqrt{2}} \sum_n^Q [\tau_n F_n^+(x) + \sigma_n F_n^-(x)])^{4/3} \rangle_Q .$$

ACKNOWLEDGEMENTS

E.R.C. and M.M. wish to express their sincere thanks to the members of the Theoretical Physics Institute and of the Center for Quantum Field Theory and Complex Systems of the University of Alberta, Edmonton, for their kind hospitality, which has greatly and in many ways helped in the completion of this work.

REFERENCES

1. E.R. Caianiello, M. Marinaro and G. Scarpetta, "Some unorthodox expansion in Quantum Field Theory and Statistical Mechanics", Nuovo Cimento (in press).

2. E.R. Caianiello and G. Scarpetta, Nuovo Cimento 22A, 448 (1974).

3. W. Kainz, Lett. Nuovo Cimento 12, 217 (1975);
H.G. Dosch, Nucl. Phys. B96, 525 (1975).

4. M. Marinaro, Nuovo Cimento 32A, 355 (1976).

5. We note, as a non-trivial point, that (5) yields one particular solution of the complete set of Green's functions equations of ref. (2). The latter equations admit also solutions that, unlike those derived from (5), do not reduce to the s.u.l.

free model when $g \to 0$, and exist only when $g >$ some fixed positive value. Solutions of this type (symmetry breaking?) can be derived also from a functional of type (5), provided the $\int du$ is taken over some range not symmetrical around $u = 0$.

6. M. Wortis, "Linked cluster expansion", in Phase Transitions and Critical Phenomena, Vol. 3, Ed. by C. Domb and S. Green (Academic Press, New York, 1974).

7. A.C. Siegert, Physica 26, 830 (1960); S.F. Edwards, Phil. Mag. 4, 1171 (1959).

8. E.R. Caianiello, M. Fusco-Girard and M. Marinaro, "Renormalization group transformation by moving bands approach", preprint, Universita di Salerno.

9. R.H. Brout, Phase transition (Benjamin, New York, 1965).

APPENDIX

We show here explicitly, by choosing $L(M^{d/2} u/m)$ in (5) as follows:

$$L(\frac{M^{d/2}}{m} u) = - \lambda u^4 - \frac{u^2}{2} \quad . \tag{A.1}$$

That the Nakano-Schwinger (N.S.) functions derived from the generating functional

$$W[h(x)] = \exp\{M^d \int dx \log \frac{\int_{-\infty}^{\infty} du \exp - [\lambda u^4 + \frac{u^2}{2} - \frac{uh(x)}{mM^{d/2}}]}{\int_{-\infty}^{\infty} du \exp - [\lambda u^4 + \frac{u^2}{2}]}\} \tag{A.2}$$

satisfy the set of branching equations of I and II type reported in ref. 2. We note first that the following identity is satisfied

$$\int_{-\infty}^{\infty} du[4\lambda u^3 + u - \frac{h(x)}{mM^{d/2}}]\exp - [\lambda u^4 + \frac{u^2}{2} - \frac{uh(x)}{mM^{d/2}}] \equiv 0 \quad . \tag{A.3}$$

Eq. (A.3), keeping in mind (A.2), can be written in terms of the functional derivative of W[h]: we find

$$\frac{\delta W[h]}{\delta h} + 4\lambda \frac{m^2}{M^d} \frac{\delta^3 W[h]}{\delta h^3} - \frac{h(x)}{m^2} W[h] = 0 \quad . \tag{A.4}$$

In deriving (A.4) the regularization prescription

$$\delta(0) = M^d \tag{A.5}$$

has been adopted. From (A.4), by performing functional differentiations with respect to L(x), it is straightforward to verify that the N.S. functions generated by (A.2) satisfy all the set of branching equations of I type. Then, by differentiating W[h] with respect to λ and keeping in mind (A.5), one readily obtains the following eq. for W[h]:

$$\frac{d}{d\lambda} W[h(x)] = - \frac{m^4}{M^d} \frac{\delta^4 W[h]}{\delta h^4} + \frac{m^4}{M^d} W[h] \left(\frac{\delta^4 W[h]}{\delta h^4}\right)_{h=0} \tag{A.6}$$

From (A.6), by using the same procedure as before, one verifies that also the set of branching equation of II type are satisfied. It may be convenient, in order to facilitate the comparison with the results of ref. 2, to write the equations satisfied by the generator of the truncated N.S. functions setting

$$Z[h(x)] = \log W[h] = M^d \int dx \log \frac{\int_{-\infty}^{\infty} du \exp - [\lambda u^4 + \frac{u^2}{2} - \frac{uh(x)}{mM^{d/2}}]}{\int_{-\infty}^{\infty} du \exp - [\lambda u^4 + \frac{u^2}{2}]} \tag{A.7}$$

and substituting in (A.4) one has

$$\frac{\delta Z[h(x)]}{\delta h(x)} + 4\lambda \frac{m^2}{M^d} [(\frac{\delta Z}{\delta h})^3 + 3 \frac{\delta Z}{\delta h} \frac{\delta^2 Z}{\delta h^2} + \frac{\delta^3 Z}{\delta h^3}] - \frac{h(x)}{m^2} = 0 \tag{A.8}$$

from which one generates the set of eq. (15) of ref. 2. Besides, by combining (A.4) and (A.6), it is easy to see that

$$4\lambda \frac{d}{d\lambda} W[h(x)] = m^2 \frac{\delta^2 W[h]}{\delta h^2} - h(x) \frac{\delta W[h(x)]}{\delta h} - m^2 W[h] \left(\frac{\delta^2 W[h]}{\delta h^2}\right)_{h=0} \tag{A.9}$$

(A.9) in terms of Z[h] becomes

$$4\lambda \frac{d}{d\lambda} Z[h(x)] = m^2\left(\frac{\delta Z[h]}{\delta h}\right)^2 + m^2 \frac{\delta^2 Z[h]}{\delta h^2} - h(x) \frac{\delta Z[h]}{\delta h} - m^2\left\{\left(\frac{\delta Z[h]}{\delta h}\right)^2 + \frac{\delta^2 Z[h]}{\delta h^2}\right\}_{h=0} \qquad (A.10)$$

from which the set of eqs. (16) of ref. 2 are generated. Summarizing, we have shown that the truncated N.S. functions obtained from the relations

$$\eta_n(x_1 \dots x_n) = \frac{\delta}{\delta h(x_1)} \cdots \frac{\delta}{\delta h(x_n)} Z[h(x)] \Big|_{\{\underline{h}=0\}}$$

where Z[h] is given in (A.7), satisfy the set of equations (15) and (16) of ref. 2 and therefore are a solution of the s.u.l. model described there.

CHARMED PARTICLE SPECTROSCOPY*

G. J. Feldman

Stanford Linear Accelerator Center

Stanford, California 94305, U.S.A.

1. INTRODUCTION

These lectures[1] will attempt to review what we have learned about charmed particles in the little over a year since they were definitively identified.[2,3] The recent discovery of the ψ(3772)[4,5] has given us a powerful tool for the detailed study of D mesons. By studying its decays we have been able to measure D meson masses and absolute branching ratios with an accuracy that only a few months earlier had been thought to be beyond our reach. Accordingly, the emphasis of these talks will be on this recent work.

Figure 1 shows the ratio R of the total hadronic cross section to the muon pair production cross section in the threshold regions for charmed meson production.[4,6] Most of the work on charmed mesons has been done at the prominent peaks at 3.77, 4.03, and 4.41 GeV. The ψ(3772) (or ψ") is just above $D\bar{D}$ threshold and below $D\bar{D}^*$ threshold. The peak at 4.03 is just above $D^*\bar{D}^*$ threshold.

Section 2 will discuss the accurate determination of D meson masses and their consequences. Section 3 will cover the determination of D meson branching fractions for hadronic decay modes and their use in measuring the amount of charm production in e^+e^- annihilation at various energies. Semi-leptonic D meson branching ratios will be the subject of section 4. Section 5 will explore some of the uses of tagged decays at the ψ(3772). Sections 6 and 7 will cover D meson spin and parity determinations and the study of D^0-$\bar{D}^0$ mixing. Finally sections 8 and 9 will discuss what little is known about F mesons and charmed baryons.

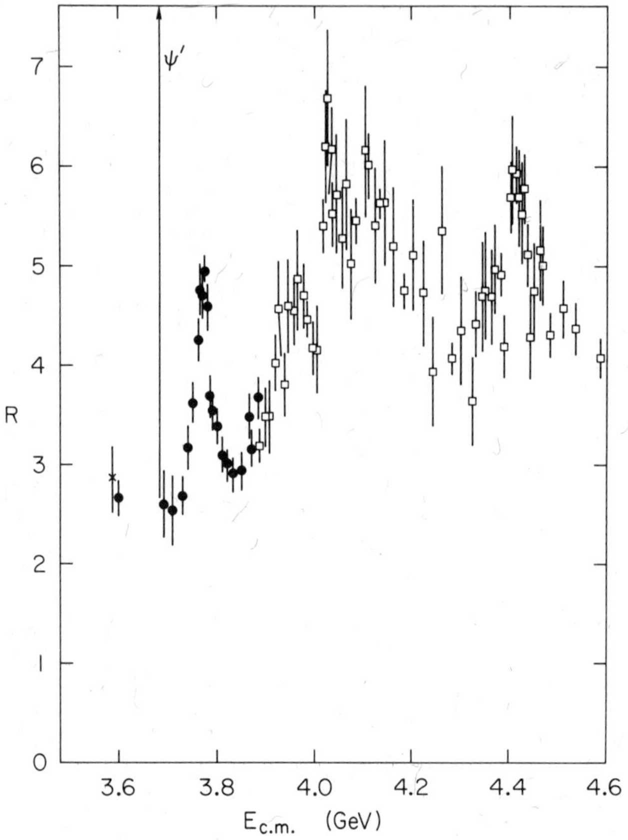

Figure 1: R in the threshold region for charmed meson production.

2. MASSES

2.1 D^0 and D^+ Masses

To calculate a mass one uses the formula

$$m = (E^2 - p^2)^{\frac{1}{2}} \tag{1}$$

The advantage of studying $e^+e^- \rightarrow D\bar{D}$ is that the energy E must equal E_b, the energy of one of the incident beams. E_b has an rms spread, due to quantum fluctuations in synchrotron radiation, of only about 1 MeV,[7] and its central value can be monitored to high precision.[8] For $D\bar{D}$ production near threshold, as in ψ' decays, we have the additional advantage that p^2 is small, about 0.08 $(GeV/c)^2$. Thus any error in p is demagnified in its effect on the determination of the mass. The final result is that we measure masses in ψ'' decays with an rms resolution of about 3 MeV/c^2, which is a factor of 5 to 10 better than they can be measured at higher energies.

In the SPEAR Magnetic Detector charged kaons are identified by time-of-flight measurements[2,3] and neutral kaons are identified by measurement of the dipion mass and the consistency of the dipion vertex with the assumed kaon decay.[9] For each particle combination we first require that the measured energy agree with E_b to within 50 MeV and then calculate the mass from Eq. 1 with $E = E_b$. The results,[10] given in Figure 2 in 4 MeV/c^2 wide bins, show clear signals in five modes including the previously unreported mode $D^{\pm} \to K_s \pi^{\pm}$. Figure 3 shows the D^+ and D^0 mass spectra for the sum of all observed modes in 2 MeV/c^2 bins. The mass difference of about 5 MeV/c^2 between the D^+ and D^0 is clearly visible. Fits to the mass spectra give

$$M_{D^0} = 1863.3 \pm 0.9 \text{ MeV/c}^2 \qquad (2)$$

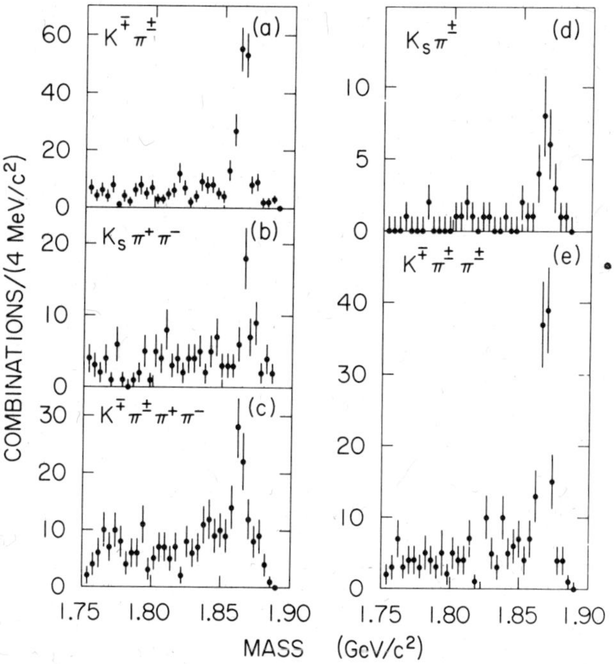

Figure 2: Invariant mass spectra for various D decay modes at the ψ''.

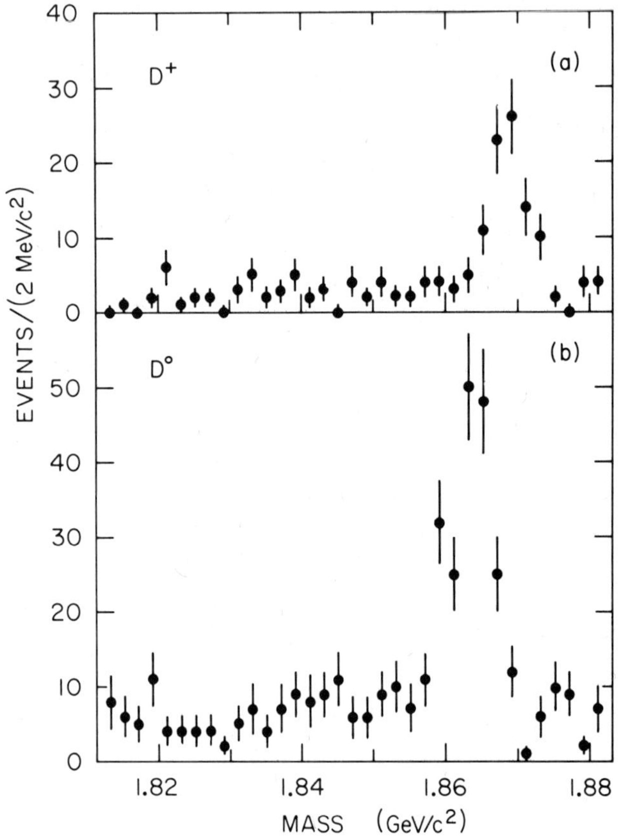

Figure 3: Invariant mass spectra for the sum of all observed (a) D^+ and (b) D^0 decay modes decaying into all charged particles.

and

$$M_{D^+} = 1868.3 \pm 0.9 \text{ MeV}/c^2 \quad . \tag{3}$$

The errors are dominated by systematic uncertainties such as the absolute momentum calibration and the stability of E_b monitoring. The D^+-D^0 mass difference is determined to be 5.0 ± 0.8 MeV/c^2; it is known more precisely than either D mass because several systematic errors cancel in the mass difference. The theoretical estimate of this mass difference has been widely discussed with estimates ranging from 2 to 15 MeV/c^2.[11]

2.2 D^{*0} Mass

To obtain the D^{*0} mass we employ the same trick with $D^{*0}\bar{D}^{*0}$ production at 4.028 GeV with the following differences:

a) We observe the D^0 from $D^{*0} \to D^0\pi^0$ decay. Since the Q value of the reaction is small, the D^0 carries off most of the D^{*0} momentum. Thus the detection of the D^0 rather than the D^{*0} causes no real problem.

b) There is contamination from $D^{*+} \to D^0\pi^+$ and $D^{*0} \to D^0\gamma$ decays.

Figure 4a shows the contributions to the D^0 momentum spectrum. The problem here is to determine the center of peak $B[D^{*0} \to D^0\pi^0]$ in the presence of peaks $A[D^{*+} \to D^0\pi^+]$ and $C[D^{*0} \to D^0\gamma]$.

The data and a fit to the data are shown in Figure 4b.[12] The D^{*0} mass is determined to be 2006 ± 1.5 MeV/c^2. The uncertainty is larger here than it was for the D^0 or D^+ because of the difficulty of extracting the signal and because the fit is not perfect.

2.3 D^{*+} Mass

There are insufficient statistics to enable us to observe $D^{*+}D^{*-}$ production at 4.028 GeV (see Figure 4c), so another method is used to obtain the D^{*+} mass: We observe the $D^{*+} \to D^0\pi^+$ decay directly. Since the Q value is small the π^+ momentum will be only m_π/m_{D^*} ($\approx$7%) of the D^* momentum. It is thus necessary to use high momentum D^*'s from high energy data ($E_{c.m.}$ = 6.8 GeV) to obtain pions with enough momentum to be visible in the magnetic detector.

The kinematics in this case are not as transparent as they were in the previous cases, but the essential point is that the Q value determines the kinematics and even a crude measurement of the Q value translates into a very precise measurement of the D^{*+} mass. Figure 5 shows the D^{*+}-D^0 mass difference in 1 MeV/c^2 bins.[13] The Q value is determined to be 5.7 ± 0.5 MeV which, when combined with the D^0 mass, yields a D^{*+} mass of 2008.6 ± 1.0 MeV/c^2.

2.4 Mass Difference and Q Values

We previously gave the mass difference

$$\delta \equiv m_{D^+} - m_{D^0} = 5.0 \pm 0.8 \text{ MeV}/c^2 \quad . \qquad (4)$$

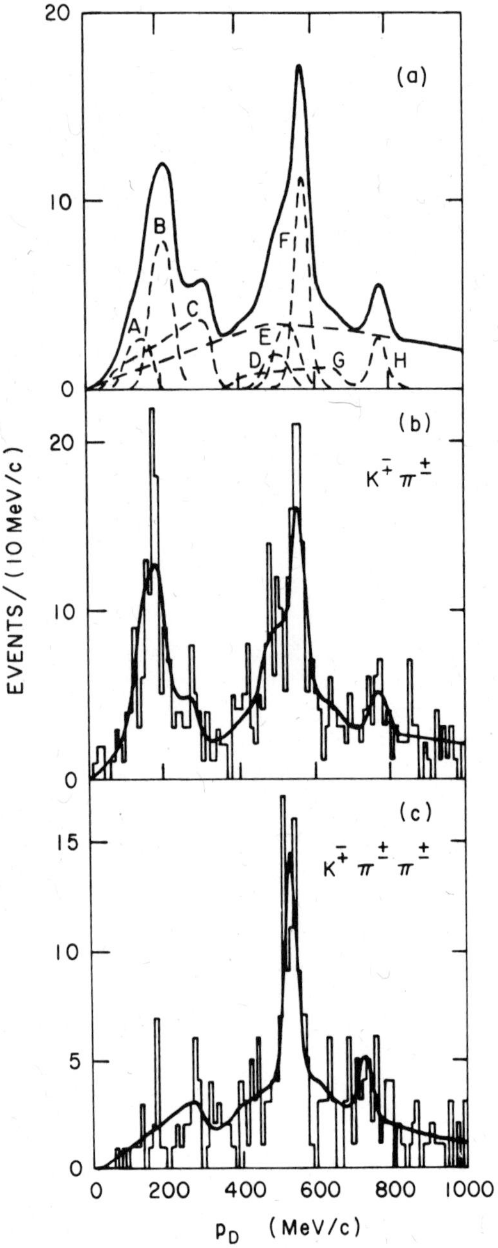

Figure 4:(a) Fit to D momentum spectrum for $D^0 \to K^{\mp}\pi^{\pm}$ at 4.028 GeV; A, B, and C are contributions from $D^*\bar{D}^*$ production with A: $D^{*+} \to D^0\pi^+$, B: $D^{*0} \to D^0\pi^0$ and C: $D^{*0} \to D^0\gamma$. D, E, F and G are contributions from $D^*\bar{D}+D\bar{D}^*$ with D: $D^{*+} \to D^0\pi^+$, E: $D^{*0} \to D^0\pi^0$, F: direct D^0 and G: $D^{*0} \to D^0\gamma$. H is the contribution from $D^0\bar{D}^0$ production.

(b) D momentum spectrum for $D^0 \to K^{\mp}\pi^{\pm}$ at 4.028 GeV.

(c) D momentum spectrum for $D^{\pm} \to K^{\mp}\pi^{\pm}\pi^{\pm}$ at 4.028 GeV.

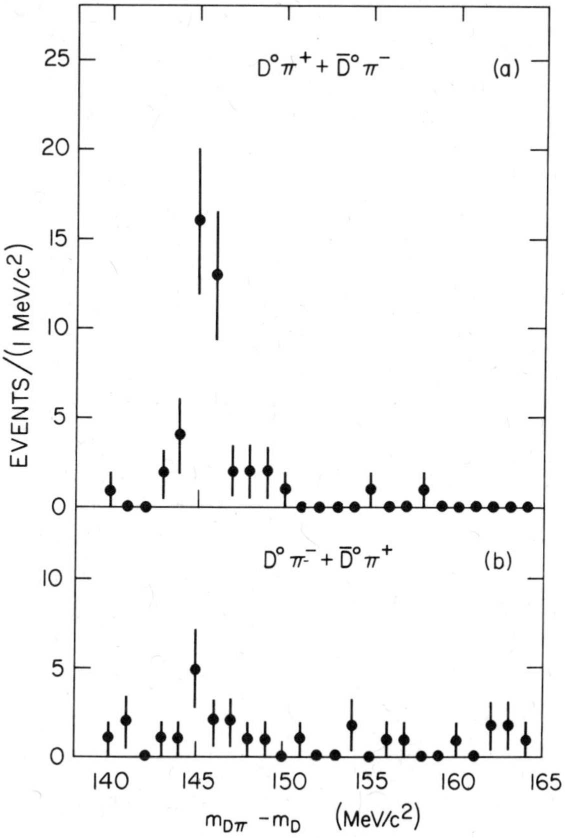

Figure 5: $D\pi$ - D mass difference spectra for (a) $D^0\pi^+$ and $\bar{D}^0\pi^-$ combinations and (b) $\bar{D}^0\pi^+$ and $D^0\pi^-$ combinations.

We can now add

$$\delta^* \equiv m_{D^{*+}} - m_{D^{*0}} = 2.6 \pm 1.8 \text{ MeV}/c^2 \tag{5}$$

and $\delta-\delta^* = 2.4 \pm 2.4$ MeV/c^2 (6)

The quantity $\delta-\delta^*$ is an electromagnetic hyperfine splitting for which theoretical estimates vary between 0 and 3 MeV/c^2. The error given in Eq. 6 is somewhat larger than would be naively expected from Eqs. 4 and 5 due to correlations in the errors.

The Q values for $D^* \to D\pi$ and $D^* \to D\gamma$ are given in Figure 6. The decay $D^{*0} \to D^+\pi^-$ appears to be kinematically forbidden in the limit of zero D^{*0} width. Even allowing for finite D^* width, it cannot be an important decay mode.

2.5 D^* Branching Fractions

The D^{*0} branching fractions have been determined from the D^0 momentum spectrum at 4.028 GeV by fitting the relative contributions of curves B and C in Figure 4a.[12] The result is $B(D^{*0} \to D^0\gamma) = 0.45 \pm 0.15$.[14] The D^{*+} branching fractions were not well determined from the 4.028 GeV data due to insufficient statistics, but we can now calculate them using the $D^* \to D\pi$ Q values and a few reasonable assumptions. The inputs are:

a) D and D^* masses, and

b) $B(D^{*0} \to D^0\gamma)$.

The assumptions are:

a) Isospin conservation in $D^* \to D\pi$ decays,

b) $\Gamma(D^* \to D\pi)$ is proportional to p^3 where p is the D momentum in the rest frame, and

c) the quark model prediction for $\Gamma(D^* \to D\gamma)$:[15]

$$\frac{\Gamma(D^{*+} \to D^+\gamma)}{\Gamma(D^{*0} \to D^0\gamma)} = \frac{(\mu_c - \mu_{\bar{d}})^2}{(\mu_c - \mu_{\bar{u}})^2} , \tag{7}$$

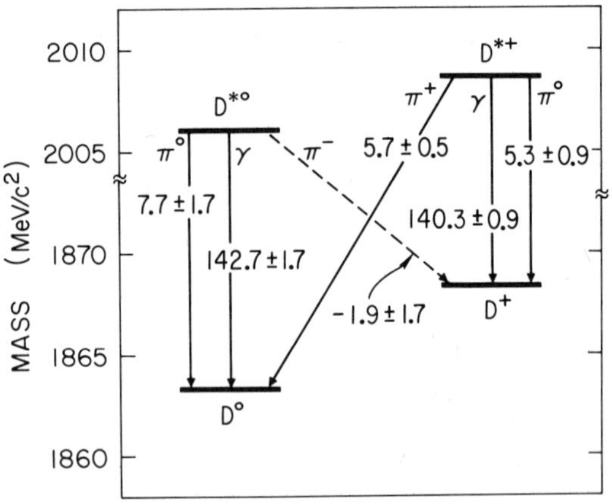

Figure 6: Q values for $D^* \to D$ transitions.

where μ is a quark magnetic moment which we assume is inversely proportional to the quark mass. Thus,

$$\frac{\Gamma(D^{*+} \to D^{+}\gamma)}{\Gamma(D^{*0} \to D^{0}\gamma)} = \left(\frac{2\frac{m_u}{m_c} - 1}{2\frac{m_u}{m_c} + 2} \right)^2 \tag{8}$$

taking $m_d = m_u$. The quark masses are not real masses and cannot be determined with any real accuracy. We will thus take two extreme cases to test the sensitivity of this assumption: $m_u/m_c = m_\rho/m_\psi$ and $m_u/m_c = 0$. We

$$\frac{\Gamma(D^{*+} \to D^{+}\gamma)}{\Gamma(D^{*0} \to D^{0}\gamma)} = \begin{cases} 1/25 \text{ for } \frac{m_u}{m_c} = \frac{m_\rho}{m_\psi} \\ \\ 1/4 \text{ for } \frac{m_u}{m_c} = 0 \quad . \end{cases} \tag{9}$$

The results are given in Table 1. Independent of the details of assumption (c), $B(D^{*+} \to D^{+}\gamma)$ is small, and $B(D^{*+} \to D^{0}\pi^{+})$ is about twice as large as $B(D^{*+} \to D^{+}\pi^{0})$. By accident, the total D^{*0} width is about equal to the D^{*+} width. The best experimental information on D^{*} widths comes from Figure 5.from which we can deduce that $\Gamma_{D^{*+}} < 2.0$ MeV/c^2 at the 90% confidence level.[13]

3. BRANCHING FRACTIONS INTO HADRONIC DECAY MODES

3.1 Modes With All Charged Particles

The cross section times branching fractions ($\sigma \cdot B$) for inclusive D production in various modes involving only charged particles have been determined at $E_{c.m.}$ = 3.774, 4.028, and 4.414 GeV. The data from the ψ'' (i.e. 3.774 GeV) are shown in Figure 2 and the data from 4.028 and 4.414 GeV are shown in Figure 7.[16] The results are given in Table 2.[17] The relative D^{0} branching fractions from 3.774 GeV and 4.028 GeV are in good agreement. The data from 4.414 GeV are not in as good agreement, but as can be seen from Figure 7, there are higher backgrounds at 4.414 GeV than at lower energies and it is more difficult to extract the signal. No conclusive evidence for Cabbibo suppressed decays has been seen at any energy. There are upper limits given in Table 2 for $E_{c.m.}$ = 4.028 GeV which are consistent with the expected degree of suppression.

TABLE 1. D* branching fractions and widths. See the text for a discussion of the input data and the assumptions which were used.

Decay Mode	D^{*0}	D^{*+} $\frac{m_u}{m_c} = \frac{m_\rho}{m_\psi}$	D^{*+} $\frac{m_u}{m_c} = 0$
$B(D^* \to D\pi^{\pm})$	0	0.68 ± 0.08	0.63 ± 0.09
$B(D^* \to D\pi^{0})$	0.55 ± 0.15	0.30 ± 0.08	0.27 ± 0.07
$B(D^* \to D\gamma)$	0.45 ± 0.15	0.02 ± 0.01	0.10 ± 0.05
$\Gamma(D^{*0})/\Gamma(D^{*+})$	-	1.0 ± 0.3	0.9 ± 0.3

TABLE 2. $\sigma \cdot B$ in nb for various D decay modes at three values of $E_{c.m.}$. See footnote 17.

Decay Mode	$E_{c.m.}$ (GeV)		
	3.774	4.028	4.414
$D^0 \to K^{\mp}\pi^{\pm}$	0.25 ± 0.05	0.53 ± 0.10	0.28 ± 0.08
$\to \bar{K}^0\pi^+\pi^-$+c.c	0.46 ± 0.12	1.01 ± 0.28	0.85 ± 0.32
$\to K^{\mp}\pi^{\pm}\pi^+\pi^-$	0.36 ± 0.10	0.77 ± 0.25	0.85 ± 0.36
$\to \pi^+\pi^-$	-	< 0.04	-
$\to K^+K^-$	-	< 0.04	-
Total D^0 observed modes	1.07 ± 0.21	2.31 ± 0.39	1.97 ± 0.49
$D^+ \to \bar{K}^0\pi^+$+c.c	0.14 ± 0.05	< 0.17	-
$\to K^{\mp}\pi^{\pm}\pi^{\pm}$	0.36 ± 0.05	0.37 ± 0.09	0.31 ± 0.11
$\to \pi^{\pm}\pi^+\pi^-$	-	< 0.03	-

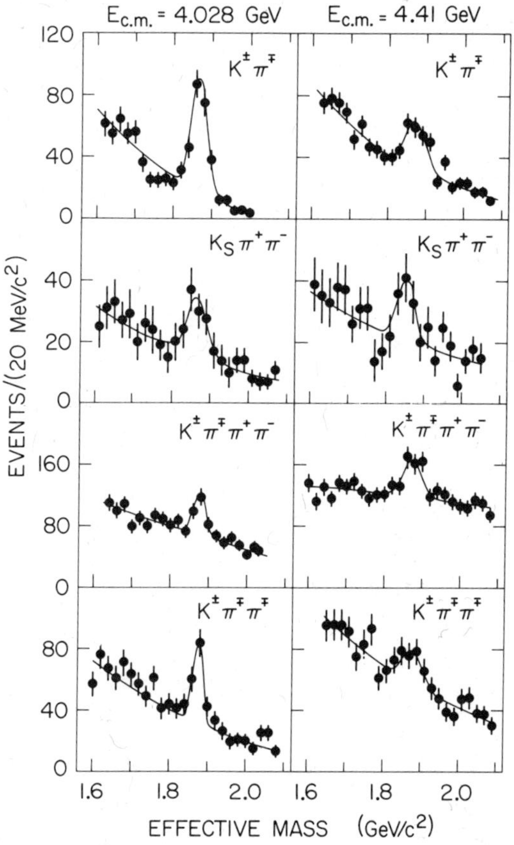

Figure 7: Invariant mass spectra for various D decay modes at 4.028 and 4.414 GeV.

3.2 Modes With a π^o

The addition of a wall of lead-glass blocks (LGW)[18] to the SPEAR Magnetic Detector allows us to search for D hadronic decay modes containing a π^o.[19] The LGW subtends about 10% of the solid angle covered by the detector's spark chambers and counters, as is shown in Figure 8.

Figure 9a shows the reconstructed $\gamma\gamma$ mass spectrum for minimum γ momentum of 100 and 150 MeV/c. The π^o signal is clearly visible in both cases, although the mass resolution is poor for low energy photons which can lose a large fraction of their energy in the aluminum solenoid coil which preceeds the LGW. The acceptance for π^o's in the LGW is shown in Figure 9b. At 600 MeV/c, the region of interest for D decays, the acceptance is about 1%.

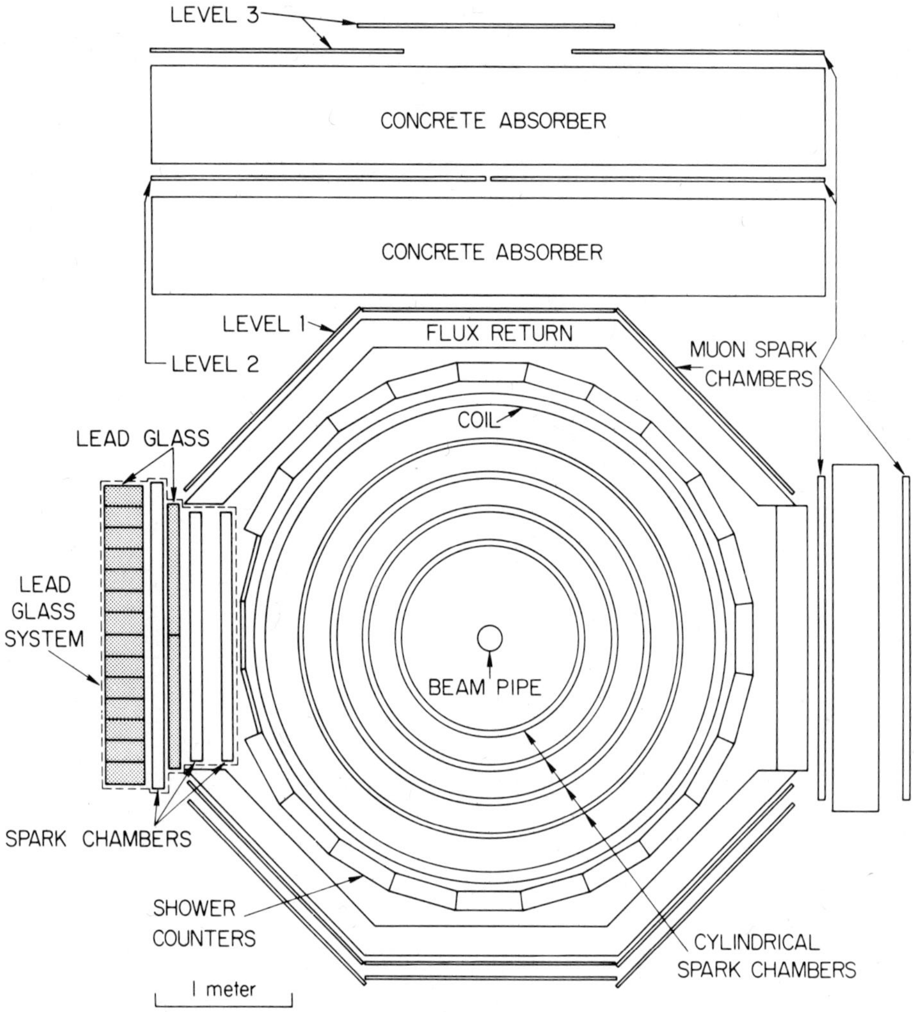

Figure 8: The SPEAR magnetic detector as seen looking along the incident beams. The proportional chambers around the beam pipe and the trigger counters are not shown. The LGW is shown on the left of the detector.

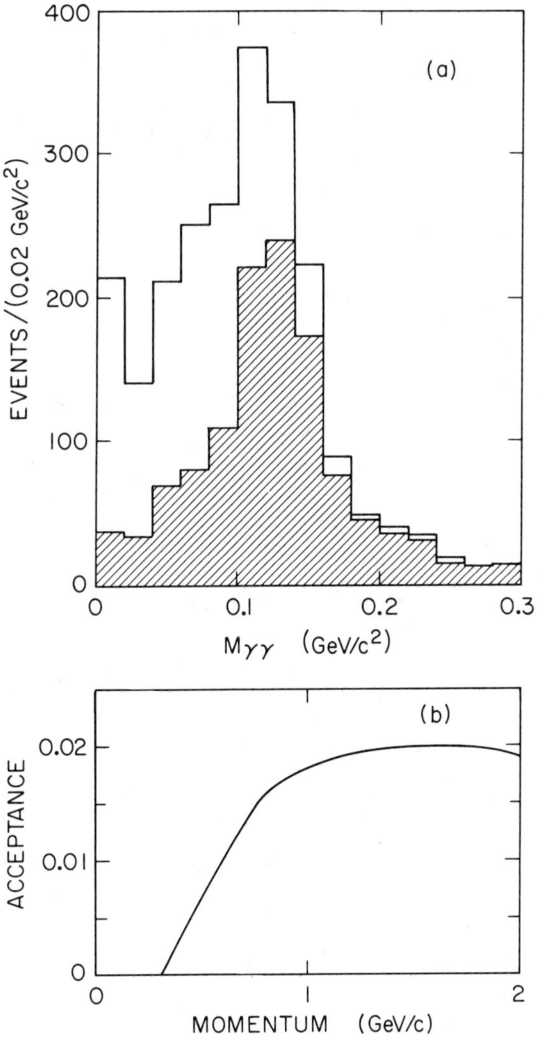

Figure 9: (a) Two photon invariant mass distribution for photon pairs detected in the LGW. A photon energy cutoff of 150 MeV is required for events in the shaded region. (b) π^o acceptance in the LGW as a function of π^o momentum.

Photon pairs detected in the LGW at the ψ'' are combined with charged particles found in the detector and the combination is fit with two constraints: that the total reconstructed energy equal the beam energy and that the $\gamma\gamma$ invariant mass equal the π^o mass. The invariant mass spectrum for $K^{\mp}\pi^{\pm}\pi^o$ combinations which gives acceptable fits is shown in Figure 10. A signal of 7.3 events over an estimated background of 1.7 events is seen at the D^o mass. This corresponds to a $\sigma \cdot B$ of 1.4 ± 0.6 nb.

No significant signals are seen in other possible decay modes. The upper limit on $\sigma \cdot B$ for $D^o \to \bar{K}^o\pi^o$ plus $\bar{D}^o \to K^o\pi^o$ is 0.7 nb at the 90% confidence level.

3.3 Absolute Branching Fractions

In the ψ'' we have for the first time a situation in which charm production is sufficiently simple that we can use measurements of the total cross section and of $\sigma \cdot B$ for D decay modes to calculate absolute branching fractions.

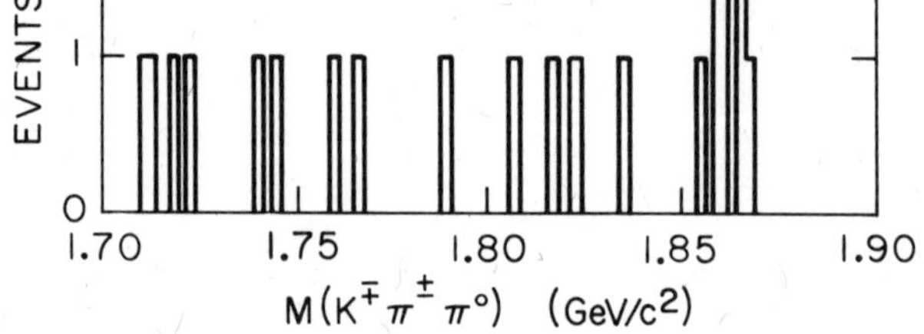

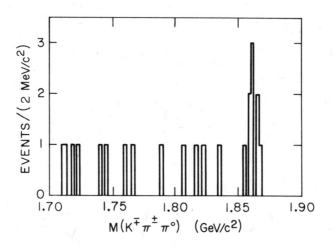

Figure 10: $K^{\mp}\pi^{\pm}\pi^o$ invariant mass distribution for constrained events.

The inputs are

a) $\sigma \cdot B$ measurements at the ψ'' which we have just discussed, and

b) the total cross section measurements in the vicinity of the ψ''.[4]

The assumptions are

a) The ψ'' is a state of definite isospin, either 0 or 1. This assumption gives equal partial widths to $D^0\bar{D}^0$ and D^+D^- except for phase space factors.

b) The phase space factors are given by

$$\frac{p^3}{1 + (rp)^2} \tag{10}$$

where p is the D momentum and r is an interaction radius.[20] The value of r is not known,[21] but as r varies from 0 to infinity, the fraction of $D^0\bar{D}^0$ changes from 0.59 to 0.53. We can thus take this fraction to be 0.56 ± 0.03. The error due to the uncertainty in r is small compared to other systematic errors.

c) $D\bar{D}$ is the only substantial decay mode of the ψ''. The rationale for this assumption is that the ψ' and ψ'' differ in mass by only 88 MeV/c^2 and thus should have similar decay modes to channels which are open to both states. However, the total ψ'' width is two orders of magnitude larger than the ψ' width. The most reasonable explanation for the difference in widths is to attribute most of the ψ'' width to the $D\bar{D}$ channel, which is accessible to it, but not to the ψ'. This is a dramatic example of the Okubo-Zweig-Iizuka rule.

The results are given in Table 3 . The $\bar{K}^0\pi^+$ decay mode of the D^+ is comparable in size to the $D^0 \to K^-\pi^+$ decay mode. This decay does not appear to be suppressed as was suggested from the analogue of octet enhancement in strangeness changing decays.[22]

TABLE 3. D branching fractions for hadronic decay modes. See the text for a discussion of the input data and the assumptions which were used.

Mode	Branching fraction (%)
$D^0 \to K^-\pi^+$	2.2 ± 0.6
$\to \bar{K}^0\pi^+\pi^-$	4.0 ± 1.3
$\to K^-\pi^+\pi^0$	12 ± 6
$\to K^-\pi^+\pi^-\pi^+$	3.2 ± 1.1
$D^+ \to \bar{K}^0\pi^+$	1.5 ± 0.6
$\to K^-\pi^+\pi^+$	3.9 ± 1.0

3.4 Comparison to the Statistical Model

It is instructive to compare the absolute branching fractions given in Table 3 to the predictions of a statistical model. This model, due to Rosner,[23] uses a Poisson multiplicity distribution and, within each multiplicity, equal contributions from each isospin amplitude. There is no real theoretical justification for this model and one should probably view its predictions with a certain degree of skepticism. Nevertheless, it can serve as a crude guide to the reasonableness of the measurements.

The statistical model predictions are given in terms of the ratio of a given state to the sum of all states of the form $K + n\pi$. Therefore we will define $f_{Kn\pi}$ to be the ratio

$$f_{Kn\pi} = \frac{\Sigma B(D \to K + n\pi)}{B(D \to \text{all})} \tag{11}$$

In addition to $K + n\pi$, "all" will include $K\eta + n\pi$, semi-leptonic decays, and Cabbibo suppressed decays. In Table 4 we list the prediction times $f_{Kn\pi}$ divided by the measurement. We would expect $f_{Kn\pi}$ to have a value of around 0.6 to 0.7, if, as we will soon see, the semi-leptonic branching fractions are of the order of 0.2 for the sum of electronic and muonic modes. The value of $f_{Kn\pi}$ seems smaller than this if deduced from modes with all charged particles, but, with sizeable errors, larger than this if deduced from the $K^-\pi^+\pi^0$ mode.

TABLE 4. Comparison of the D branching fractions from Table 3 with the statistical model of Ref. 23. See text for the definition of $f_{Kn\pi}$.

Mode	Prediction/measurement
$D^0 \to K^-\pi^+$	$(2.7 \pm 0.7)\ f_{Kn\pi}$
$\to \bar{K}^0\pi^+\pi^-$	$(3.0 \pm 1.0)\ f_{Kn\pi}$
$\to K^-\pi^+\pi^0$	$(0.8 \pm 0.4)\ f_{Kn\pi}$
$\to K^-\pi^+\pi^-\pi^+$	$(2.2 \pm 0.8)\ f_{Kn\pi}$
$D^+ \to \bar{K}^0\pi^+$	$(8.7 \pm 3.5)\ f_{Kn\pi}$
$\to K^-\pi^+\pi^+$	$(2.6 \pm 0.7)\ f_{Kn\pi}$

3.5 Charm Production at 4.028 and 4.414 GeV

With the absolute branching fractions from Table 3 we are now in a position to calculate the amount of charm production at two of the prominent peaks in the 4 GeV region.

The inputs are

a) $\sigma \cdot B$ from Table 2, and

b) B from Table 3.

There are no additional assumptions to those already used in constructing Table 3.

We define $R_D = \sigma_D/2\sigma_{\mu\mu}$. The factor of 2 in the denominator takes into account the fact that charmed particles are produced in pairs, so that R_D can be directly compared to the total hadronic R. In particular, we compare it to

$$R_{new} \equiv R - R_{old} - R_\tau \qquad (12)$$

where R is taken from measurements of the total hadronic cross section,[6,17] R_{old}(2.4) is taken from measurements of the total hadronic cross section below charm threshold,[4] and R is the theoretical expression for the production of a 1.9 GeV/c^2 mass lepton.

$$R_\tau = \tfrac{1}{2}\beta(3 - \beta^2) \quad . \qquad (13)$$

If D's and τ's are the only new particles being produced then R_{new} should equal R_D. If the production of F's, charmed baryons, or even other new particles are sizeable, then R_{new} will be larger than R_D.

The results are given in Table 5. R_{D^0} is calculated from all observed D^0 modes and also from just the better-measured $K^\pm\pi^\mp$ mode. At 4.028 GeV these two measurements are consistent and R_{new} is consistent with being equal to R_D. At 4.414 GeV the two measurements differ somewhat, but nevertheless it is clear that whatever else may be happening at 4.414 GeV, most of the excess cross section is going into D production. Note that the fraction of charged D production at 4.028 GeV is significantly smaller than that at 3.774 GeV where we have assumed $R_{D^\pm}/R_D = 0.44 + 0.03$. In the near future one should be able to combine this result with inclusive lepton production at 4.028 and 3.774 GeV to calculate the ratio of D^+ to D^0 semi-leptonic branching fractions. Since semi-leptonic decays are I = 0 transitions, this rate is the inverse of the ratio of D^+ to D^0 lifetimes.[24]

4. BRANCHING FRACTIONS INTO SEMI-LEPTONIC DECAY MODES

The general technique which has been used to measure the semi-electronic branching fraction of D mesons is to count the number of electrons in events in which three or more charged particles are detected, correct for backgrounds and losses, and divide by the total cross section for charm production. There are several problems inherent in this approach:

a) It is necessary to subtract garden variety backgrounds such as γ conversions, Dalitz pairs, and ψ and ψ' decays. This is straightforward but these backgrounds can become large at low electron momentum and limit the statistical accuracy in this region.

TABLE 5. Charm production at 4.028 and 4.414 GeV. See text for definitions and a discussion of the input data and assumptions which were used.

	4.028 GeV		4.414 GeV	
	All D^0 modes	$K^{\mp}\pi^{\pm}$ only	All D^0 modes	$K^{\mp}\pi^{\pm}$ only
R_{D^0}	2.3 ± 0.6	2.2 ± 0.8	2.3 ± 0.8	1.4 ± 0.6
$R_{D^\pm}$	0.9 ± 0.4	0.9 ± 0.4	0.9 ± 0.4	0.9 ± 0.4
R_D	3.2 ± 0.7	3.1 ± 0.9	3.2 ± 0.9	2.3 ± 0.7
R_{new}	3.1 ± 1.0	3.1 ± 1.0	2.5 ± 1.0	2.5 ± 1.0
$R_{D^\pm}/R_D$	0.28 ± 0.09	0.29± 0.10	0.38 ± 0.11	0.39 ± 0.13

b) Around 15% of τ decays involving an electron are expected to occur in events with more than two charged particles, and these decays must be subtracted. This is not a serious problem in practice since this τ contribution is relatively small and relatively well known.[25]

c) Some D decays involving an electron will not be counted because they occur in events with only two charged particles. This correction is relatively small, but somewhat model dependent.

d) Each experiment has some momentum below which the apparatus cannot detect electrons. Accordingly, a correction must be made for that part of the electron spectrum which is below this threshold. This correction is also model dependent, but fortunately this model dependence partially cancels that of the correction for two-prong D decays.[26]

e) One must determine the total cross section for charm production in order to divide by it. Curiously, at present it appears that discrepancies between different experiments are due more to differences in the total cross sections they measure than to the number of electrons they count.

f) Finally, what is measured is a natural average of charm particle semi-leptonic decays. As we have seen, at 3.772 GeV there are roughly equal numbers of neutral and charged D's, while at 4.028 GeV, there are probably slightly over twice as many neutral D's as charged D's. At higher energies F's and charmed baryons will contribute to the average. As we have already mentioned, in the future we will be able to use these differences to unfold the various particles' decays. For the present, however, they only contribute some additional uncertainty to the measurements. In order to concentrate on the D meson branching ratios and to reduce as much as possible the model dependence of the results, we will study only the measurements at the lowest possible c.m. energies.

Three experiments have now measured the average D meson semi-electronic branching fractions. Table 6 lists these experiments, the electron detection technique, the solid angle, the lowest c.m. energy, and the average semi-electronic branching fraction measured at that energy.[26-29] The DELCO experiment at SPEAR, shown in Figure 11, is exceptional in having excellent electron detection and an order of magnitude more solid angle than the other two experiments. Its statistical power is shown in Figure 12 where the ψ'' line shape is clearly seen in the cross section for electron production as a function of c.m. energy.

The three experiments are in good agreement given the sizeable statistical and systematic errors. The world average of about 9% is about a factor of two smaller than the 20% which would be expected from naive quark and lepton counting.

TABLE 6. Semi-electronic D meson branching fractions.

Experiment	Technique	Ω	Lowest $E_{c.m.}$	$\langle B_e \rangle$(%)
DASP[27]	Cerenkov and shower counters	7%	3.99 to 4.08	10±3
LGW[26]	Lead glass	6%	3.774	7.2±2.8
DELCO[28]	Cerenkov and shower counters	60%	3.774	11±3
		World Average		9.3±1.7

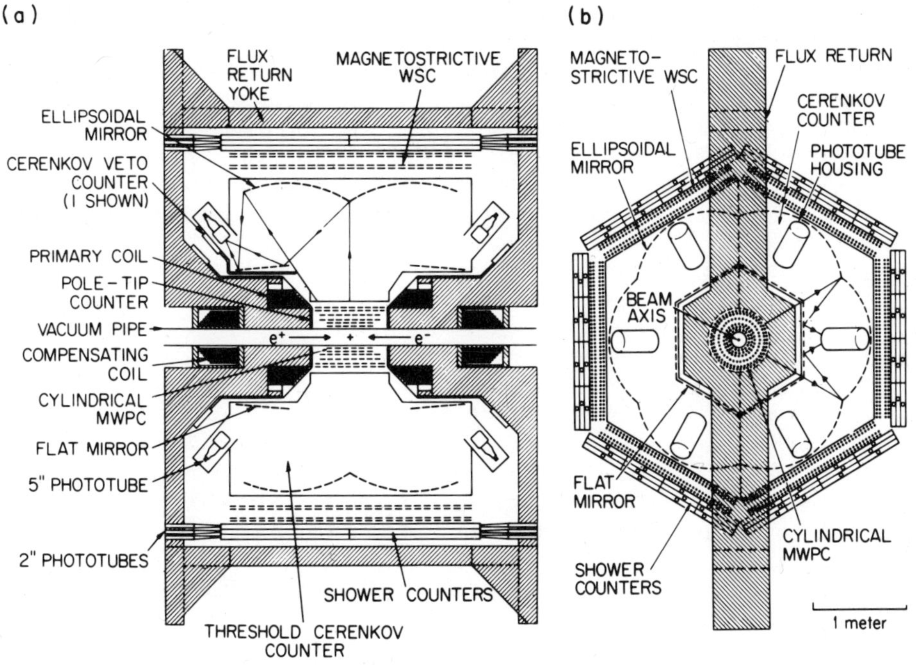

Figure 11: The DELCO apparatus at SPEAR as seen (a) perpendicular to the incident beams and (b) along the incident beams.

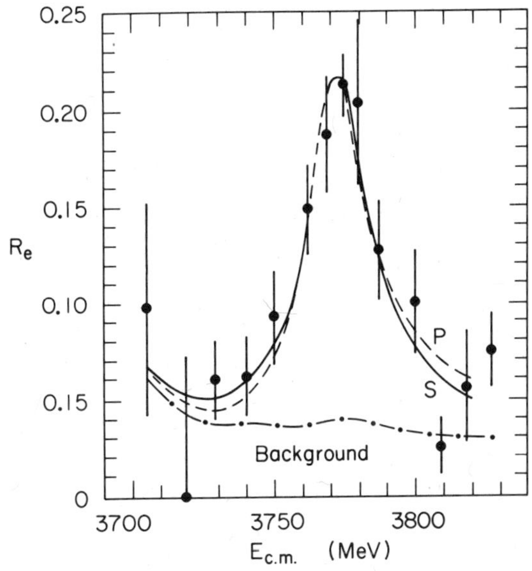

Figure 12: R_e, the ratio of the cross section for inclusive electrons in multiparticle events to the theoretical cross section for μ pair production, in the vicinity of the ψ''.

Each of the three experiments has measured the electron momentum spectrum, shown in Figures 13-15. All of the spectra agree with the relatively soft spectra expected from $K e\nu$ and $K^* e\nu$ decay modes, and all of the experiments can obtain good fits to their spectra if they include sizeable fractions of these two modes. None of the experiments has had the sensitivity to study the V,A structure of the $K^* e\nu$ decay mode. This remains probably the single most interesting outstanding question in the study of D meson decays.

5. TAGGED EVENTS

With the discovery of the ψ'', it becomes possible for the first time to "tag" charmed particle decays. For example, if we detect a D^0 decay into $K^-\pi^+$ in a ψ'' decay, then we are also looking, essentially without bias, at a $\bar{D}^0$ decay where the $\bar{D}^0$ has a momentum of 300 MeV/c in a known direction. In this section, we will give some examples, based on a preliminary analysis, of the type of studies which are possible with tagged events. In the future this technique will probably become the prime method of studying D mesons.

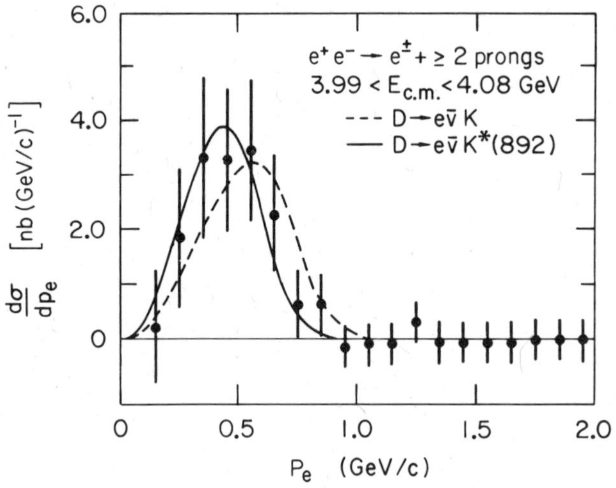

Figure 13: The momentum spectrum of inclusive electrons in multi-particle events in the $E_{c.m.}$ range 3.99 to 4.08 GeV as measured by the DASP collaboration. Curves show theoretical spectra expected for $K^*e\nu$ and $Ke\nu$ decay modes of the D.

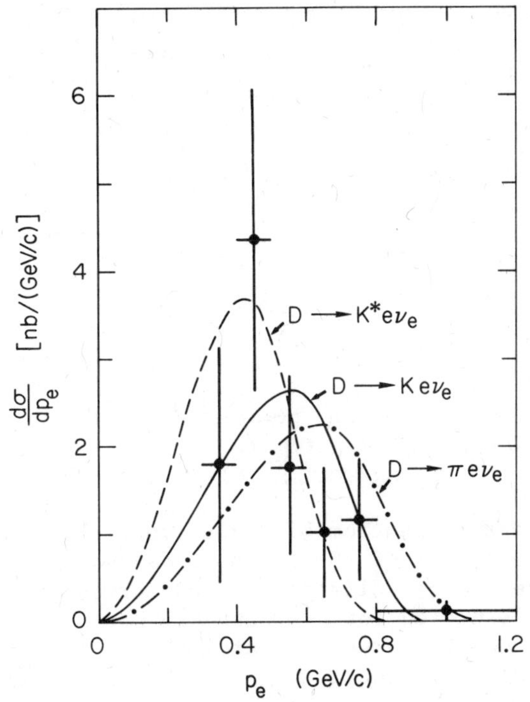

Figure 14: The momentum spectrum of inclusive electrons in multiparticle decays of the ψ'' as measured by the LGW experiment on the SPEAR magnetic detector. Curves show theoretical spectra expected for $K^*e\nu$, $Ke\nu$, and $\pi e\nu$ decay modes of the D.

5.1 Decays With All Charged Particles

In section 3.3 we determined absolute D branching fractions by employing some very reasonable assumptions about the nature of ψ''. We can now check these assumptions by using tagged events to measure D branching fractions.

We use the five decay modes shown in Figure 2 as the tagging decays and look for events in which all or all but one of the other D are detected. There are 194 tagging D^0 decays and 82 tagging $D^\pm$ decays.

We find eight cases of $K^\mp\pi^\pm$ or $K^\mp\pi^\pm\pi^+\pi^-$ decay opposite the tagging decays. These eight cases come from six events because in two cases both halves of the event tag each other, and such events must be counted twice. Correcting for detection and triggering efficiencies, these events give

$$B(D^0 \to K^-\pi^+ \text{ or } K^-\pi^+\pi^-\pi^+) = (6.2 \pm 2.7)\% \qquad (14)$$

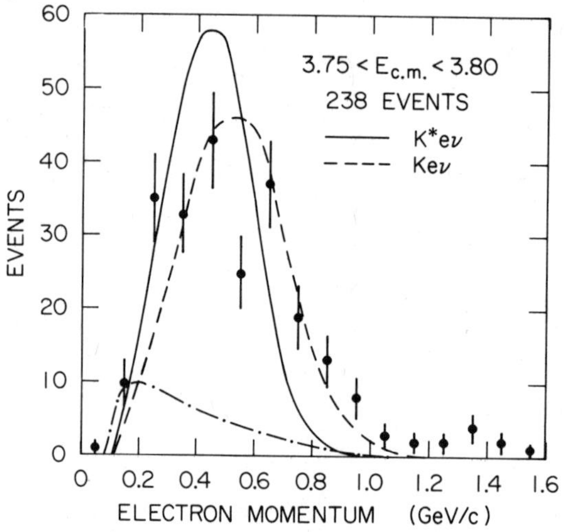

Figure 15: The momentum spectrum of inclusive electrons in multi-particle decays of the ψ'' as measured by the DELCO experiment. The solid and dashed curves show theoretical spectra expected for $K^*e\nu$ and $Ke\nu$ decay modes of the D. The dot-dashed curve indicates the estimated background remaining in the data.

which is consistent with the value of (5.4 ± 1.5)% from Table 3.

There are two cases of a $K^{\mp}\pi^{\pm}\pi^{\pm}$ decay opposite a tagging $D^{\pm}$ decay, each from a separate event. These two events give

$$B(D^+ \to K^-\pi^+\pi^+) = (3.4 \pm 2.4)\% \quad , \qquad (15)$$

which is in good agreement with the value of (3.9 ± 1.0)% from Table 3.

We can now turn these results around and use them to calculate ψ'' branching fractions without the aid of any assumptions.

The inputs are

a) $\sigma \cdot B$ for D decays at the ψ'' (Table 2),

b) B for D decays from the tagged events (Eqs. 14 and 15), and

c) ψ'' total cross section measurements.[4,17]

The results are given in Table 7 along with the values which were assumed in section 3.3. The agreement is excellent, but given the enormous errors, clearly fortuitous.

TABLE 7. ψ'' branching fractions in per cent. See the text for a discussion of the input data.

Decay Mode	B measured	B (assumed in Sec. 3.3)
$\psi'' \to D^0\bar{D}^0$	49 ± 25	56 ± 3
$\psi'' \to D^+D^-$	50 ± 38	44 ± 3
$\psi'' \to D\bar{D}$	99 ± 48	100

5.2 Decays With One Missing Neutral

We can also look for events with a charged kaon, another charged particle, and a missing, near zero-mass, neutral particle opposite a tagging D^0 decay. These decays could be $D^0 \to K^{\mp}\pi^{\pm}\pi^0$, $D^0 \to K^{\mp}e^{\pm}\nu$, or $D^0 \to K^{\mp}\mu^{\pm}\nu$ modes, which we shall designate as $D_{\pi 3}$, D_{e3}, and $D_{\mu 3}$ for short. There are ten cases of these events, which leads to

$$B(D_{\pi 3}) + B(D_{e3}) + B(D_{\mu 3}) = (11.7 \pm 4.1)\% \quad . \tag{16}$$

Unfortunately it is quite difficult to distinguish between $D_{\pi 3}$, D_{e3}, and $D_{\mu 3}$ decays in the magnetic detector because

a) the leptons often have low momentum and are not discriminated from pions, and

b) low energy π^0's are difficult to detect.

However, we can obtain some information on D_{e3} decays by subtracting the measured branching fraction for $D_{\pi 3}$ decays, $(12 \pm 6)\%$[19] and setting $B(D_{e3}) = B(D_{\mu 3})$. This gives

$$B(D^o \to K^- e^+ \nu) < 3.6\% \text{ at } 1\sigma \text{ c.l.} \tag{17}$$

Dividing by the world average given in Table 6,

$$\frac{B(D^o \to K^- e^+ \nu)}{B(D^o \to e^+ + \text{anything})} < 39\% \text{ at } 1\sigma \text{ c.l.} \tag{18}$$

By fitting $Ke\nu$ and $K^* e\nu$ spectra to their measured electron momentum spectrum (see Figure 13), the DASP group obtained $(35 \pm 30)\%$ for the ratio given in Eq. 18.[27]

5.3 Charged Multiplicity in D Decays

To determine the charged multiplicities in D decays, we count the charged particles opposite a tagging decay and use a Monte Carlo calculation of efficiencies to unfold the true distributions from the observed distributions. In this preliminary analysis we have used only the $K^\mp \pi^\pm$ and $K^\mp \pi^\pm \pi^\pm$ modes as tagging decays. Backgrounds, which are typically about 10% have been explicitly subtracted from the data. No attempt has been made to identify neutral kaons so that a K_s decaying to two charged pions will count as two charged particles.

The raw data are displayed in the top portion of Figure 16 and the unfolded data are displayed in the bottom portion. D^o's decay primarily to two charged particles, while D^+'s decay to roughly equal numbers of one and three charged particles. The mean charged multiplicities are

$$\langle n_c \rangle_{D^o} = 2.3 \pm 0.2 \tag{19}$$

$$\langle n_c \rangle_{D^+} = 2.3 \pm 0.3 \; . \tag{20}$$

The statistical model assumed somewhat higher charged multiplicities, typically about 2.7.[23]

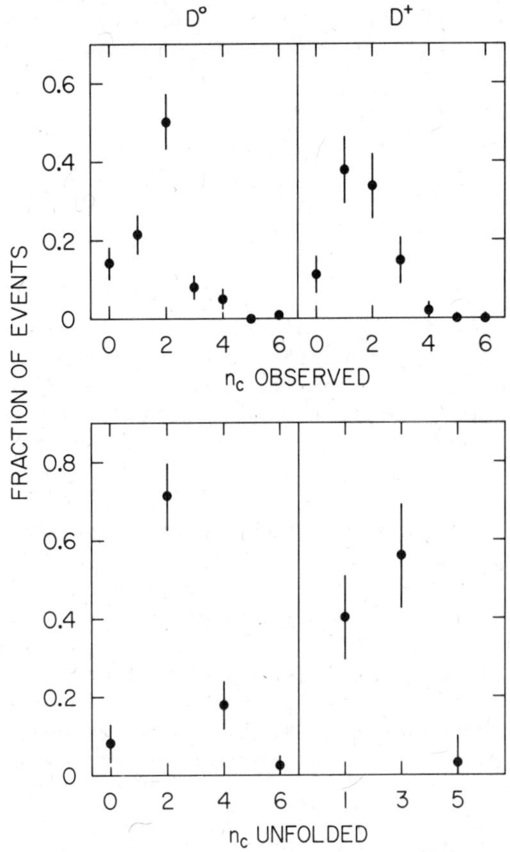

Figure 16: Observed and true (unfolded) charged multiplicities for D decays.

5.4 Two-prong $D\bar{D}$ Decays

Events in which only two charged particles are produced are of special interest experimentally because

a) there is background in two-prong events from QED processes,

b) they have a much lower detection efficiency in the magnetic detector and most other detectors, and

c) $\tau^+\tau^-$ decays occur primarily in two-prong events.

We can calculate the fraction of $D\bar{D}$ decays that go into two charged particles directly from the data in Figure 16:

$$D^0\bar{D}^0 \; : \; 2f_0 f_2 = (11 \pm 7)\% \qquad (21)$$

$$D^+D^- \; : \; f_1^2 = (17 \pm 8)\% \quad . \qquad (22)$$

Here f_n represents the fraction of decays to n charged particles. The fractions of two charged particle events given by Eqs. 21 and 22 are not vastly different from the overall fraction of two prong events. We thus expect to see the same type of variation with energy in the two prong cross section as in the multiprong cross section.

Semi-leptonic decays of D's and leptonic decays of τ's are often separated experimentally by multiplicity: the D's are presumed to decay primarily into events with four or more charged particles while τ's are presumed to decay primarily into events with two charged particles. It is thus important to measure the extent to which semi-leptonic D decays can occur in two-charged-prong events. We do not presently have enough information to determine this but we can set upper limits by assuming that semi-leptonic decays always occur in the lowest possible charged multiplicities.

For $D^0\bar{D}^0$ decays the lowest charged multiplicity is clearly two, therefore

$$D^0\bar{D}^0 \text{ 2-prong lepton fraction} < f_0$$

$$< 13\% \text{ at } 1\sigma \text{ c.l.} \qquad (23)$$

For D^+D^- decays one might expect that the upper limit is just f_1. However we can obtain a better limit if we assume that Cabbibo suppressed decays are unimportant. Then the simplest semi-leptonic decay is $D^+ \rightarrow K^0 \ell^+ \pi$. The $\bar{K}^0$ looks like zero prongs two-thirds of the time and like two prongs one-third of the time. Therefore

$$D^+D^- \text{ 2-prong lepton fraction} < 0.66 \, f_1$$

$$< 34\% \text{ at } 1\sigma \text{ c.l.} \qquad (24)$$

6. SPINS AND PARITIES

If the D meson family is similar to other mesons, then we would expect the lowest lying state to be a pseudoscalar ($J^P = 0^-$) and the second lowest lying state to be a vector particle ($J^P = 1^-$). Although a combination of the experimental data and our theoretical prejudice that low lying states should have low spins strongly favor this choice, it is worthwhile to state precisely what we have determined about D meson spins and parities from experimental measurements alone:

a) The absolute D parity cannot be measured. This is simply because in reactions in which charm is not conserved, parity is also not conserved.[30] We are thus free to define the D parity,

$$P_D = -1 \quad . \tag{25}$$

If the D is a bound s-wave state of two quarks, then this definition sets the parity of the charmed quark equal to the parity of the other quarks,

b) If either the D or D^* has spin 0, then from the observation of $D^* \to D\pi$,[13]

$$P_{D*} = (-1)^{J_D + J_{D*}} \quad . \tag{26}$$

c) Both the D and D^* cannot have spin 0. If they did then from Eq. (26), the D^* would be a scalar particle. But the reaction

$$e^+e^- \to D\bar{D}^* \tag{27}$$

has clearly been observed[12] and e^+e^- annihilation through a single virtual photon cannot couple to a scalar plus a pseudoscalar by parity conservation.

d) The next simplest hypothesis is that the sum of the absolute value of the two spins is one. There are two possibilities:

$$J_D^P = 0^- \text{ and } J_{D^*}^P = 1^- \tag{28a}$$

or $$J_D^P = 1^- \text{ and } J_{D^*}^P = 0^- \quad . \tag{28b}$$

Three independent measurements strongly favor the former of these possibilities:

i) The angular distributions for

$$e^+e^- \rightarrow D\bar{D}^* \quad \text{with} \quad D \rightarrow K\pi \tag{29}$$

are of the form

$$P(\Theta,\theta) \propto 1 + \cos^2\Theta \tag{30}$$

for hypothesis (28a) and of the form

$$P(\Theta,\theta) \propto \sin^2\theta\ (1 + \cos^2\Theta) \tag{31a}$$

for hypothesis (28b), where Θ is the angle between the DD^* production direction and the incident beams, and θ is the angle between the $K\pi$ decay direction and the D momentum in the D center of mass. It is clear from Eqs. (30) and (31a) that one should look at the θ dependence. This is done in Figure 17b, for 4.028 GeV data.[31] Equation (30) has a confidence level of 51% while Eq. (31a) has a confidence level of 0.6%.

Another way of studying these same data with different systematic errors has also been used. First Eq. (31a) is expanded to include the dependence on the azimuthal angle between $D\bar{D}^*$ production plane and the D decay plane:

$$P(\Theta,\theta,\phi) \propto \sin^2\theta\ (\cos^2\phi + \cos^2\Theta\ \sin^2\phi) \quad . \tag{31b}$$

Then the events are divided into two groups depending on whether the right side of Eq. (31b) is greater or less than 0.32, and a $K\pi$ invariant mass spectrum is constructed for each set of events. The number of events in each group is extracted by fitting the spectra, as shown in Figure 18. For hypothesis (28a) there should be roughly equal numbers of events in the two groups, while for hypothesis (28b) there should be more than twice as many events in Figure 18b as in Figure 18a. The data are consistent with hypothesis (28a) with a confidence level of nearly 100% and inconsistent with hypothesis (28b) with a confidence level of only 0.35%.

ii) Additional evidence favoring hypothesis (28a) over (28b) comes from measurements of the production angle for $D^*\bar{D}^*$ and $D\bar{D}$ final states. Here the dependence on Θ must be

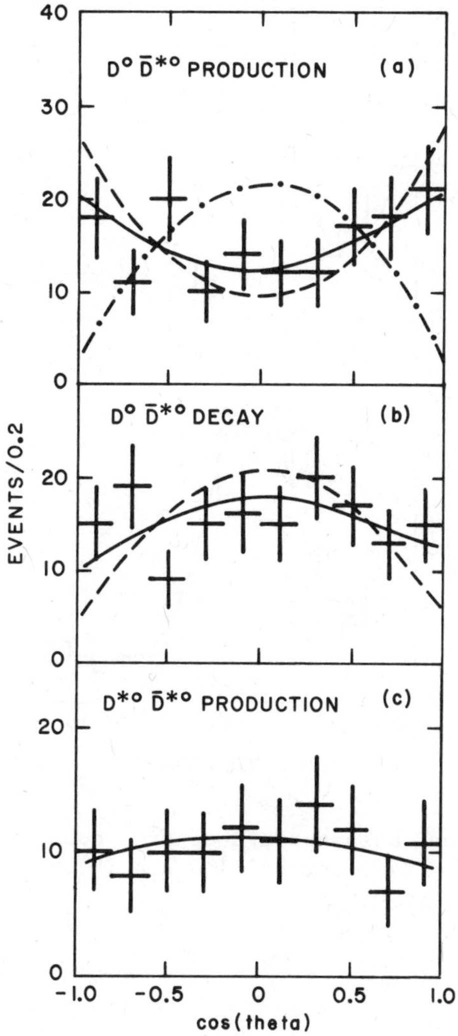

Figure 17: (a) Θ distribution of D^o in reaction (27). Solid curve corresponds to hypothesis (28a) and dashed curve corresponds to hypothesis (28b). The dot-dashed curve corresponds to spinless D's and D*'s. (b) θ distribution in reaction (27). Curves as in part (a). (c) Θ distribution for D^o's from $D^*\bar{D}^*$ production. Curve is a fit to Eq. (32).

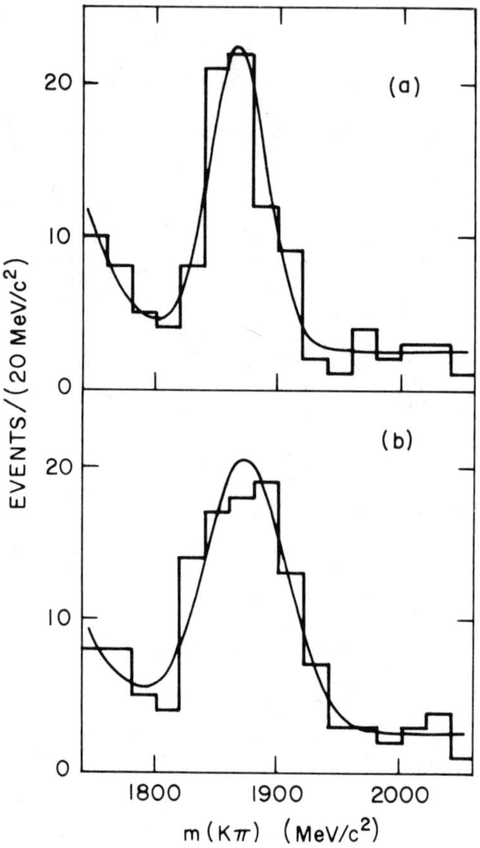

Figure 18: Invariant mass spectra of $K^{\mp}\pi^{\pm}$ decay modes in reaction (27) for (a) low and (b) high values of the right side of Eq. (31b).

$$P(\Theta) \propto 1 + \alpha\cos^2\Theta \quad , \tag{32}$$

where $|\alpha| < 1$ in general and $\alpha = -1$ for spin 0. For $D^*\bar{D}^*$ production at 4.028 GeV[31] the Θ distribution is shown in Figure 17c and

$$\alpha_{D^{*0}} = -0.30 \pm 0.33 \quad . \tag{33}$$

This represents a 2% confidence level for the D^* spin to be zero.

In contrast the Θ distribution for $D\bar{D}$ production at 3.774 GeV[10] is shown in Figure 19. The fits to Eq. (32) give

$$\alpha_{D^0} = -1.00 \pm 0.09 \quad \text{and} \quad \alpha_{D^+} = -1.04 \pm 0.10 \quad , \qquad (34)$$

in complete agreement with a spin zero assignment for the D. While these data do not prove the D spin is zero, it seems unlikely that a particle of higher spin would give $\alpha \approx -1$ by accident.

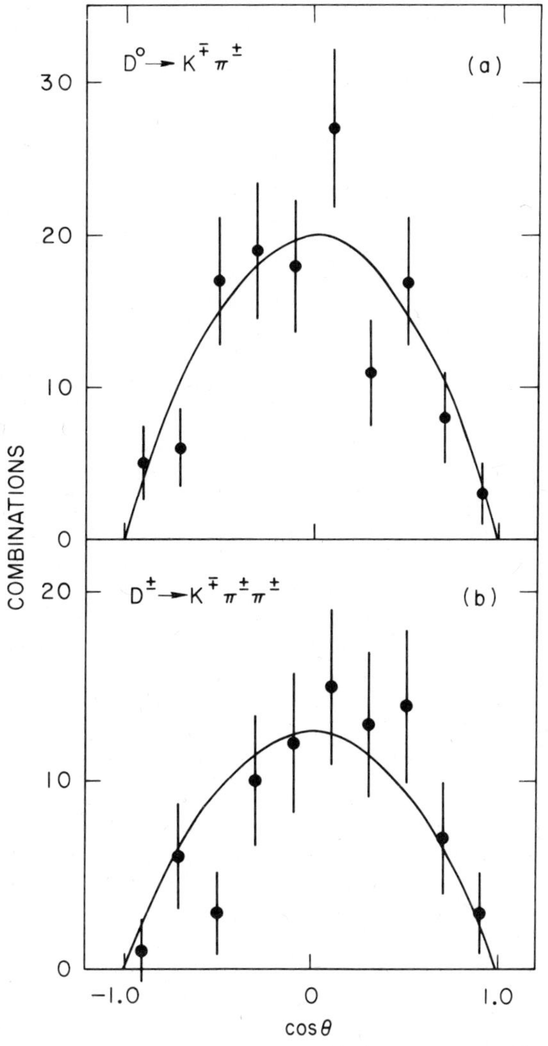

Figure 19: θ distributions for (a) $K^{\mp}\pi^{\pm}$ and (b) $K^{\mp}\pi^{\pm}\pi^{\pm}$ decay modes from $D\bar{D}$ production. The curves represent Eq. (32) with $\alpha = -1$.

Thus, to conclude, if $|J_D| + |J_{D*}| < 2$, the data allow only the assignment given in (28a). In addition, there is circumstantial evidence that independent of the D* spin, the D has spin zero.

7. $D^0 - \bar{D}^0$ MIXING

In principle, the D^0 and $\bar{D}^0$ could mix in much the same way as the K^0 and $\bar{K}^0$ mix to form the K_s and K_L. If the GIM mechanism works for charm then the time scale for mixing should be long compared to the D^0 lifetime and there should be a negligible amount of mixing, around 0.2%. On the other hand, if there is even a small charm-changing neutral current, then the time scale for mixing should be small compared to the D^0 lifetime and there should be complete D^0-$\bar{D}^0$ mixing - that is, D^0's should decay equally to states of positive and negative strangeness.[32]

Two independent measurements find no evidence for D^0-$\bar{D}^0$ mixing in agreement with the standard model. One technique has been to find $D^0 \to K^-\pi^+$ decays in the 4 GeV region in which the kaon is well identified by time of flight and search for additional well identified charged kaons.[12] Of 77 events, in 62 the kaons appear to have opposite charge and in 15 events they appear to have the same charge. However 16 events with kaons of apparently the same charge are expected in the absence of mixing from π-K misidentifications. Thus, there is no evidence for mixing and the fraction of events containing a D^0 which exhibit strangeness violation is less than 18% at the 90% confidence level.

The second technique has been to study $D^{*+} \to D^0\pi^+$ decays. In these decays the sign of the cascade pion tags the charm content of the D^0. One can then measure the sign of the strangeness, that is, the fraction of the time the D^0 decays to $K^+\pi^-$ instead of $K^-\pi^+$. A glance at Figure 5 shows that $D^{*+} \to (K^-\pi^+)\pi^+$ decays (Figure 5a) are much more common than $D^{*+} \to (K^+\pi^-)\pi^-$ decays. To obtain a more sensitive measure, we can eliminate events in which the kaons are not well identified. There remain 26 events in the peak in Figure 5a and only three events in the same region in Figure 5b, two of which are expected from backgrounds and misidentifications. Thus again, there is no evidence for D^0-$\bar{D}^0$ mixing and the fraction of D^0's which decay to states of positive strangeness is less than 16% at the 90% confidence level.

8. F^+ MESONS

Compared to the detailed information which has been compiled on D mesons, we have only a few hints concerning the F^+ mesons,

$c\bar{s}$ bound states. The DASP group has reported four events at 4.4 GeV which can be fit to the hypothesis of either an $F\bar{F}^*$ or $F^*\bar{F}^*$ final state.[33] In these events they observe the decay $F^* \to F\gamma$. The mass of the F is determined to be 2039 $\pm$ 60 MeV/c^2. The large uncertainty in mass comes partially from the ambiguity of whether the final state was $F\bar{F}^*$ or $F^*\bar{F}^*$.

The second hint comes from SPEAR magnetic detector data at 4.16 GeV. There is a peak in the invariant mass spectrum for the sum of $K^+K^-\pi^\pm$, $K_sK^\pm$, and $K^+K^-\pi^+\pi^-\pi^\pm$ final states fit to the hypothesis of $F\bar{F}$ production.[34] The mass is accurately determined (by the same technique which was used for D mesons) to be 2039.5 $\pm$ 1.0 MeV/c^2. The significance of this result is still being investigated.

Either or both of these results may turn out to be correct, but in both cases the level of significance is certainly less than that of the original data which showed the existence of D mesons.[2,3] Second generation detectors such as the Mark II at SPEAR should be able to easily confirm or deny these results in the near future.

9. CHARMED BARYONS

Our experimental information on charmed baryons is also somewhat meager. There have been two reports of observation of the Λ_c^+, the isosinglet cud bound state. The first is a single neutrino event in the Brookhaven 7-ft bubble chamber of the form[35]

$$\nu p \to \mu^- \Lambda\pi^+\pi^+\pi^+\pi^- \quad . \tag{35}$$

This event violates the $\Delta S = \Delta Q$ rule and thus is likely to have a charm baryon decay in the final state. The mass of one $\Lambda\pi^+\pi^+\pi^-$ combination is 2.26 GeV/c^2.

The second report is of a peak in the inclusive $\bar{\Lambda}\pi^-\pi^-\pi^+$ mass spectrum at 2.26 GeV/c^2 from high energy photoproduction at Fermilab.[36] Although the authors do not give a production cross section corresponding to this peak, it is clear that it is sizeable and that it implies the exciting possibility of a rather large photoproduction cross section for charmed particles, perhaps close to 1 μb. An improved version of this experiment is currently running at Fermilab and results should be available soon.

No peaks in invariant mass spectra corresponding to any charmed baryons have been seen in e^+e^- annihilation. Upper limits have been set which are about an order of magnitude lower than typical $\sigma \cdot B$ measurements for D meson production at the same energies.[34] The only positive indication in e^+e^- annihilation is a sharp rise in inclusive baryon production from around 4.4 GeV to

5.0 GeV,[37] the threshold region for charmed baryon production. The proton and Λ data are shown in Figures 20a and 20b. The statistical accuracy of the Λ data is too low to establish the precise region in which the rise occurs. However, the most significant feature is that the Λ cross sections are consistent with being between 10 and 15% of the proton cross sections at all energies. This low value tends to indicate that charmed baryons may preferentially decay into final states containing protons and kaons rather than states containing strange baryons.

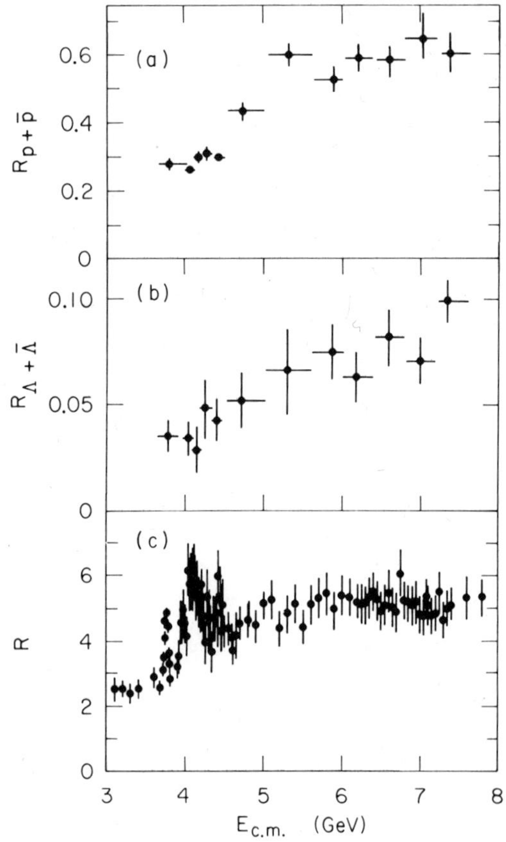

Figure 20: The ratios of cross sections for (a) inclusive p plus $\bar{p}$ production, (b) inclusive Λ plus $\bar{\Lambda}$ production, and (c) hadron production to the theoretical cross section for μ pair production.

REFERENCES AND FOOTNOTES

* Work supported by the Department of Energy.

1) In addition to charmed particle spectroscopy, the topics of the charmonium spectroscopy and the τ lepton were covered in the oral version of these lectures. These latter two topics have been omitted from the written version because they are adequately reviewed elsewhere. For a discussion of both these topics see G. J. Feldman and M. L. Perl, Phys. Rep. 33C, 285 (1977), also issued as SLAC report SLAC-PUB-1972 (1977). A few more recent results can be found in G. J. Feldman in the Proceedings of the V International Conference on Experimental Meson Spectroscopy, Northeastern University, Boston, Mass., April 29-30, 1977, also issued as SLAC report number SLAC-PUB-1977(1977) and M. L. Perl in the Proceedings of the 1977 International Symposium on Lepton and Photon Interactions at High Energies, Hamburg, Germany, August 25-31, 1977, also issued as SLAC report SLAC-PUB-2022 (1977). The lecture notes presented here are based in part on G. J. Feldman in the Proceedings of SLAC Summer Institute on Particle Physics, July 11-22, 1977, Stanford, California, also issued as SLAC report number SLAC-PUB-2000 (1977).

2) G. Goldhaber et al., Phys. Rev. Lett. 37, 255 (1976).

3) I. Peruzzi et al., Phys. Rev. Lett. 37, 569 (1976).

4) P. A. Rapidis et al., Phys. Rev. Lett. 39, 526 (1977).

5) W. Bacino et al., SLAC report number SLAC-PUB-2030 (1977).

6) J. Siegrist et al., Phys. Rev. Lett. 36, 700 (1976).

7) P. B. Wilson et al., SLAC report number SLAC-PUB-1894 (1977).

8) In our error analysis, we have assigned the error due to the long-term drift in E_b determination to be 0.5 MeV. This is probably conservative.

9) V. Lüth et al., Phys. Lett. 70B, 120 (1977).

10) I. Peruzzi et al., Phys. Rev. Lett. 39, 1301 (1977).

11) D. B. Lichtenberg, Phys. Rev. D12, 3760 (1975); A. De Rujula, H. Georgi, and S. L. Glashow, Phys. Rev. Lett. 37, 398 (1976); K. Lane and S. Weinberg, Phys. Rev. Lett. 37, 717 (1976); S. Ono, Phys. Rev. Lett. 37, 655 (1976); W. Celmaster, Phys. Rev. Lett. 37, 1042 (1976); H. Fritzsch, Phys. Lett. 63B, 419 (1976); J. Dabaul and M. Krammer, Phys. Lett. 64B, 341 (1976); N. G. Deshande et al., Phys. Rev. Lett. 37, 1305 (1976); J. S. Guillen, Nuovo Cimento Lett. 18, 218 (1977), D. H. Boal and A. C. D. Wright, Phys. Rev. D16, 1505 (1977); L. H. Chan, Lousiana State University

report (1976); D. C. Peaslee, Phys. Rev. D15, 3495 (1977); and N. F. Nasrallah and K. Schilcher, Universitat Mainz report NZ-TH 76/15 (1976).

12) G. Goldhaber et al., Phys. Lett. 69B,503 (1977).

13) G. J. Feldman et al., Phys. Rev. Lett. 38, 1313 (1977).

14) The value is from the "normal fit" of Reference 12. The "isospin constrained fit" give an erroneous value for $B(D^{*0} \to D^0\gamma)$ because the D^+ mass was fit to be 1873 MeV/c^2. This forbade the decay $D^{*+} \to D^+\pi^0$ and forced all $D^{*+} \to D^+$ decays into the mode $D^{*+} \to D^+\gamma$, which in turn forced too high a value of $B(D^{*0} \to D^0\gamma)$ through the constraint conditions.

15) See F. J. Gilman in the Proceedings of the SLAC Summer Institute on Particle Physics, July 11-22, 1977, Stanford, California for a discussion of the validity of these models for ordinary quarks.

16) M. Piccolo et al., Phys. Lett. 70B, 260 (1977).

17) The values of References 4, 6 and 16 have been reduced by 7.5% to account for a revised calculation of the external radiative correction for Bhabha scattering. See footnote 11 of Reference 10.

18) A. Barbaro-Galtieri et al., Phys. Rev. Lett. 39, 1058 (1977).

19) D. L. Scharre et al., SLAC report number SLAC-PUB-2019 (1977).

20) J. D. Jackson, Nuovo Cimento 34, 1645 (1964); A. Barbaro-Galtieri in Advances in Particle Physics, Vol. II, ed. R. L. Cool and R. L. Marshak (Wiley, New York, 1968).

21) In Reference 4, acceptable fits for the ψ'' line shape were obtained for all values of r greater than one fm.

22) G. Altarelli, N. Cabibbo, L. Maiani, Nucl. Phys. B88, 285 (1975); R. L. Kingsley, S. B. Treiman, F. Wilczek and A. Zee, Phys. Rev. D11, 1919 (1975); M. B. Einhorn and C. Quigg, Phys. Rev. D12, 2015 (1975).

23) J. L. Rosner, Invited talk at Orbis Scientiae, Jan. 17-21, 1977, Coral Gables, Fla., (Institute for Advanced Study report COO-2220-120).

24) A. Pais and S. B. Treiman, Phys. Rev. D15, 2529 (1977).

25) See references given in footnote 1.

26) J. M. Feller et al., SLAC report number SLAC-PUB-2028 (1977).

27) R. Brandelik et al.,Phys. Lett. 70B, 387 (1977).

28) J. Krikby in the Proceedings of the 1977 International Symposium on Lepton and Photon Interactions at High Energies, Hamburg, Germany, August 25-31, 1977, also issued as SLAC report number SLAC-PUB-2040 (1977).

29) The experimental results given here are the best known values as of the writing of this report and thus differ slightly from those given in the oral version.

30) J. E. Wiss et al., Phys. Rev. Lett. 37, 1531 (1976).

31) H. K. Nguyen et al., Phys. Rev. Lett. 39, 262 (1977).

32) See Reference 13 for an extensive list of references on this topic.

33) R. Brandelik et al., Phys. Lett. 70B, 132 (1977).

34) D. Lüke, Proceedings of the 1977 Meeting of the Division of Particles and Fields of the American Physical Society, Argonne National Laboratory, Argonne, Ill., Oct. 6-8, 1977.

35) E. G. Cazzoli et al., Phys. Rev. Lett. 34, 1125 (1975).

36) B. Knapp et al., Phys. Rev. Lett. 37, 882 (1976).

37) M. Piccolo et al., Phys. Rev. Lett. 39, 1503 (1977).

DIMENSIONAL REGULARIZATION AND HYPERFUNCTIONS*

Yasunori Fujii

University of Tokyo-Komaba

Meguro-ku, Tokyo 153

1. INTRODUCTION

The method of dimensional regularization has proved itself very useful especially when it is applied to the gauge theories.[3] The applications have been limited mainly to the calculations of Feynman integrals. We want to apply the method also to the calculation of spectral representations. This was motivated by our previous attempt to remove the difficulty of the critical dimension in the dual string model.[4]

What we needed was regularizing the integral of a spectral function of the massless field. It turned out that to this kind of integral the usual argument of analytic continuation no longer applies. We found that the best way is to appeal to the concept of hyperfunctions.[5]

We also recognized that the same argument can be applied to many other models of the relativistic quantum field theory. In order to illustrate the points we choose to discuss the Schwinger term in quantum electrodynamics. We calculate the Schwinger term by the method of dimensional regularization interpreted in the sense of hyperfunctions. It turns out that the Schwinger term vanishes contrary to the general belief. This is the consequence of the fact that the naive positivity argument is no longer justified once the divergent integral is regularized.

In section 2 we review briefly the most important part of the dimensional regularization method. We analyze how the positivity condition is violated. Section 3 is a short course on the theory of hyperfunctions. It is shown that the method of dimensional regularization can be understood through the concept of hyperfunctions without explicit reference to the argument of analytic continuation. In section 4 we apply our method to the Schwinger term. The consequence is discussed from other points of view. The final section 5 is devoted to the discussions of other related topics.

2. DIMENSIONAL REGULARIZATION

Consider an example of Feynman integrals;

$$I = \int d^N k \frac{1}{(k^2+a^2)^2} \qquad (a^2 \neq 0) \tag{2.1}$$

where N is allowed to be continuous or even complex. After the Wick rotation and the angular integration we are left with the one-dimensional integral,

$$I = i \frac{2\pi^\nu}{\Gamma(\nu)} \int_0^\infty dk \frac{k^{N-1}}{(k^2+a^2)^2} \qquad (\nu \equiv N/2) \quad . \tag{2.2}$$

This is convergent if $\nu<2$ (or $N<4$). To evaluate the integral we introduce the variable

$$x = \frac{a^2}{k^2+a^2} \quad . \tag{2.3}$$

Notice that the ultra-violet limit $k^2 \to \infty$ is mapped to the point $x=0$. We then obtain

$$I = i \frac{\pi^\nu}{\Gamma(\nu)} \int_0^1 x^{1-\nu}(1-x)^{\nu-1} dx \quad . \tag{2.4}$$

The x integration is given by the beta function $B(2-\nu, \nu)$, which can be analytically continued to the whole ν plane except at the poles, $2-\nu = 0, -1, -2, \ldots; \nu = 0, -1, -2, \ldots$.

If $\mathrm{Re}\nu>2$, the integral is divergent and the original expression loses its meaning. But analytic continuation allows us to give a meaning even for $\mathrm{Re}\nu>2$. The infinity at $N=4$ is separated in the form of the pole, and is subtracted by the renormalization technique. Obviously the naive positivity argument no longer holds in general if $\mathrm{Re}\nu>2$ (the beta function assumes both signs whereas the intergrand is still positive as long as ν is real).

We want to study more closely how the ultra-violet divergence, i.e. the divergence at $x=0$, has been removed with the violation of the positivity condition. For this purpose it is more convenient to analyze the gamma function which is closely connected with the beta function.

Consider Euler's gamma function

$$\Gamma(\alpha) = \int_0^\infty e^{-x} x^{\alpha-1} dx \quad , \tag{2.5}$$

which is well-defined for $\mathrm{Re}\alpha>0$; the result can be analytically continued to the whole α plane except for the poles at $\alpha = 0, -1, -2, \ldots$. The positivity condition is also violated in general if $\mathrm{Re}\alpha<0$.

Another expression is given by the Hankel's integral representation;

$$\Gamma(\alpha) = \frac{1}{2i \sin\pi(\alpha-1)} \int_C e^{-z}(-z)^{\alpha-1} dz \ , \quad (\alpha \neq \text{integer}) \tag{2.6}$$

where the contour C is shown in Fig. 1(a). In the integrand, $(-z)^{\alpha-1}$ is defined by

$$(-z)^{\alpha-1} = \exp[(\alpha-1)\mathrm{Log}(-z)] \quad , \tag{2.7}$$

where $\mathrm{Log}(-z)$ is the principal valued function having the cut along the real axis from 0 to ∞. Hence the function $e^{-z}(-z)^{\alpha-1}$ has the cut along the same part of the real axis. The gap (discontinuity) is given by

$$-2ie^{-x}x^{\alpha-1} \sin\pi(\alpha-1)$$

for $x>0$ and 0 for $x<0$.

It is now easy to show that (2.6) is equivalent to (2.5) if $\mathrm{Re}\alpha>0$ (deform the contour as shown in Fig. 1(b)). If $\mathrm{Re}\alpha<0$, on the other hand, (2.5) is meaningless whereas (2.6) is defined for any $\alpha \neq$ integer. The result is the same as that obtained by analytic continuation.

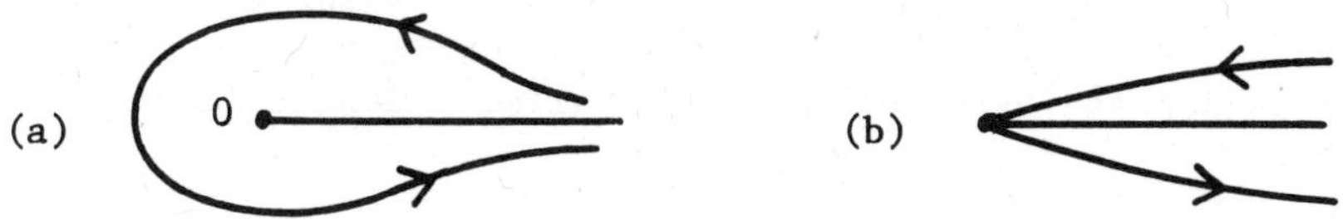

Fig. 1

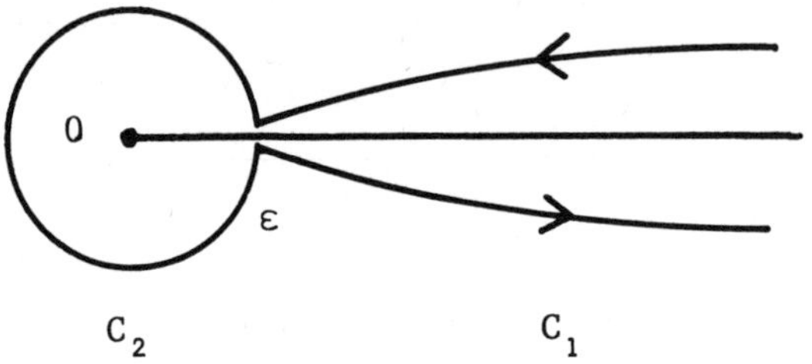

Fig. 2

It is convenient to divide the contour C into two parts, C_1 and C_2, as in Fig. 2. We can easily check that there is no contribution from C_2 if $\mathrm{Re}\alpha>0$. If $\mathrm{Re}\alpha<0$, the contribution from C_1 diverges as $-\varepsilon^{\alpha}/\alpha$ as $\varepsilon\to 0$, while the contribution from C_2 gives $\varepsilon^{\alpha}/\alpha$ which cancels exactly the divergent part from C_1.

The cutoff calculation of Feynman integrals with $k^2<\Lambda$ is equivalent to ignoring the small circle C_2 ($\varepsilon\sim\Lambda^{-1}$), whereas the regularized calculation corresponds to integrating around the point $k^2 = \infty$. As a price, however, the positivity condition is lost in general; the above contribution from C_2 is negative if $\alpha = \text{real} < 0$.

We emphasize that, although the result of the Hankel's formula agrees with that of analytic continuation, the calculation itself is independent of the analytic continuation argument. The contour integration as in Hankel's representation has been elevated by Sato to a general principle to define integrals of hyperfunctions.[5]

3. HYPERFUNCTIONS

In the elementary level of applications, the hyperfunctions are just Schwartz's distributions.[6,7] In the more sophisticated level, the concept of hyperfunctions is more general than that of distributions.

Definition: Suppose there is a function $F_+(z)$ which is analytic in a certain domain above the real axis. Suppose also there is a function $F_-(z)$ which is analytic below the real axis. The set of these is denoted by $F(z)$, which is called a defining function. A hyperfunction $f(x)$ is defined by

$$f(x) \stackrel{\text{def}}{=} F_+(x+i\varepsilon) - F_-(x-i\varepsilon) \equiv [F(z)] \quad . \tag{3.1}$$

Example 1

$$\delta(x) = -\frac{1}{2\pi i}\left(\frac{1}{x+i\varepsilon} - \frac{1}{x-i\varepsilon}\right) = \left[-\frac{1}{2\pi i}\frac{1}{z}\right] \quad . \tag{3.2}$$

Example 2

$$\theta(x) = \left[-\frac{1}{2\pi i}\operatorname{Log}(-z)\right] \quad . \tag{3.3}$$

Example 3

$$x_+^\mu = \left[-\frac{1}{2i\sin\pi\mu}(-z)^\mu\right] \qquad (\mu \neq \text{integer}) \tag{3.4}$$

$$= \begin{cases} x^\mu & (x>0) \\ 0 & (x<0) \end{cases} \quad .$$

Example 4

$$\lim_{\mu\to 0} \mu x_+^{\mu-1} = \left[-\frac{1}{2\pi i}\frac{1}{z}\right] = \delta(x) \quad . \tag{3.5}$$

A hyperfunction can be multiplied by a usual analytic function. But a product of two or more hyperfunctions is not well-defined, in general. A usual function $\varphi(z)$ can also be regarded as a hyperfunction ($F_+(z) = \varphi(z)/2$, $F_-(z) = -\varphi(z)/2$).

Derivatives: The derivative of a hyperfunction is defined by the derivative of its defining function.

$$\frac{df}{dx} \overset{\text{def}}{=} \left[\frac{dF(z)}{dz}\right] \quad . \tag{3.6}$$

Example 5

$$\frac{d\theta}{dx} = \left[-\frac{1}{2\pi i}\frac{1}{z}\right] = \delta(x) \quad . \tag{3.7}$$

Example 6

$$\frac{d\delta}{dx} = \left[\frac{1}{2\pi i}\frac{1}{z^2}\right] \quad . \tag{3.8}$$

Example 7

$$\frac{d}{dx}x_+^\mu = \mu x_+^{\mu-1} \quad . \tag{3.9}$$

Integrations: The definite integral of f(x) is defined by

$$\int_a^b f(x)dx \overset{\text{def}}{=} -\int_C F(z)dz \quad , \tag{3.10}$$

where F(z) is the defining function which gives a hyperfunction f(x) for $a \le x \le b$, whereas it gives zero for $x<a$, $x>b$. The contour C, shown in Fig. 3(a), can be deformed as long as it stays within the region in which F(z) is analytic. If f(x) is regular for $a \le x \le b$, we find immediately, by deforming the contour to that in Fig. 3(b) and recalling (3.1), that the right-hand side of (3.10) agrees with the usual definition. If f(x) has some singularities in the region $a \le x \le b$, then the right-hand side of (3.10) gives the regularization of the original divergent integral. (This way of defining integrations can be viewed as an outgrowth of Hadamard's idea of finite part of divergent integrals. Further developments in the same direction may be found in the works by Pham, Hua and Teplitz.[8])

Example 8

$$\int_0^\infty e^{-x} x^{\alpha-1} dx = \frac{1}{2i \sin \pi(\alpha-1)} \int_C e^{-z}(-z)^{\alpha-1} dz \quad . \tag{3.11}$$

This is the Hankel's formula. Combining this with the explanation before, we now have the statement: dimensional regularization is achieved by interpreting the integration in the sense of hyperfunctions on the Riemann sphere that includes the point at infinity.

Example 9

$$I(\alpha) = \int_0^1 x^{\alpha-1} dx = \frac{1}{\alpha} \quad . \tag{3.12}$$

This integral is convergent in the usual sense only if $\mathrm{Re}\alpha > 0$. We interpret (3.12) as

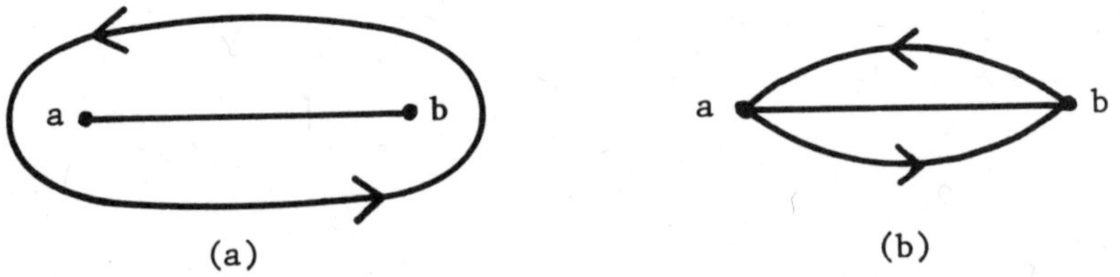

Fig. 3

$$\int_{a<0}^{1} x_+^{\alpha-1}dx = \frac{1}{2i\sin\pi(\alpha-1)}\int_C (-z)^{\alpha-1}dz \quad . \tag{3.13}$$

The contour C may be chosen as the unit circle centered at the origin $z = 0$. The elementary integration gives $1/\alpha$. We have <u>not</u> appealed to the argument of analytic continuation.

<u>Example 10</u>

$$\mathcal{J}(\alpha) = \int_0^\infty x^{\alpha-1}dx = 0 \qquad (\alpha \neq \text{integer}) \quad . \tag{3.14}$$

The divergence from $x=0$ may be avoided if $\mathrm{Re}\,\alpha>0$, whereas the divergence from $x=\infty$ may be avoided if $\mathrm{Re}\,\alpha<0$. These two regions do not overlap each other. This makes it difficult to apply the method of analytic continuation. (See, however, example 1, p. 70 of ref. 7.) But we can interpret (3.14) in the sense of hyperfunctions.

In order to go around the point $x=\infty$ in an explicit way we introduce the variable

$$t = \frac{x}{1+x} \quad . \tag{3.15}$$

The point $x=\infty$ is now mapped to $t=1$. In terms of t (3.14) is put into the form

$$\mathcal{J}(\alpha) = \int_0^1 t^{\alpha-1}(1-t)^{-\alpha-1}dt$$

$$= B(\alpha, -\alpha) = 0 \quad . \tag{3.16}$$

<u>Example 11</u>

If we combine (3.14) with (3.5) we obtain

$$\int_{a<0}^{\infty} \delta(x)dx = 0 \quad . \tag{3.17}$$

This shows that our delta function has a support not only at $x=0$ but also at $x=\infty$. In terms of the "usual" delta function, our delta function may be given by $\delta(x) - x^{-2}\delta(x^{-1})$. Of course our delta function behaves in the same way as the usual one as long as it is multiplied by a function which falls off rapidly for $x\to\infty$.

4. SCHWINGER TERM IN SPINOR QUANTUM ELECTRODYNAMICS

The Schwinger term S is defined by

$$<0|[j_i(x),\ j_0(0)]|0>_{t=0} = iS\partial_i\delta(\vec{x}) \quad , \tag{4.1}$$

$$S = \int_{4m^2}^{\infty} \frac{\rho(s)}{s}\, ds \quad . \tag{4.2}$$

In the lowest order calculation, the intermediate state of the electron and the positron gives the spectral function

$$\rho(s) = \frac{2^{2-3\nu}\pi^{-\nu+1/2}}{\Gamma(\nu+\frac{1}{2})} s^{-1/2}(s-4m^2)^{\nu-3/2}[(\nu-1)s + 2m^2] \quad . \tag{4.3}$$

Substituting (4.3) into (4.2) we obtain

$$S = \frac{2^{2-3\nu}\pi^{-\nu+1/2}}{\Gamma(\nu+\frac{1}{2})} [(\nu-1)J(1/2,\ \nu-3/2) + 2m^2J(3/2,\ \nu-3/2)], \tag{4.4}$$

where

$$J(\alpha,\ \beta) = \int_{4m^2}^{\infty} s^{-\alpha}(s-4m^2)^{\beta}ds \quad . \tag{4.5}$$

If we choose N=4 (ν=2), the integrand of the first integral in (4.4) approaches constant as $s\to\infty$, so that we have a "quadratic" divergence. However, let us interpret the integrals in the sense of hyperfunctions on the Riemann sphere that includes the point $s=\infty$. Then $J(\alpha,\ \beta)$ in (4.5) is given by the beta function;

$$J(\alpha,\ \beta) = (4m^2)^{-\alpha+\beta+1} B(\alpha-\beta-1,\ \beta+1) \quad . \tag{4.6}$$

We apply the usual recursion formula for the gamma function and obtain the result

$$S = 0 \quad , \tag{4.7}$$

for any N and any m.

This is in sharp contradiction to the long-held belief that the Schwinger term is non-zero. This belief is based on the positivity argument. Our result (4.7) demonstrates in the most dramatic way that the naive positivity argument is no longer justified once the divergent integral is regularized. From the studies in the preceding sections, however, it is evident that the negative contribution comes only from $s=\infty$. In this sense no contradiction occurs with the unitarity condition, for example, in the finite energy region. (In fact our conclusion is not in contradiction to the usual result; the Schwinger term calculated by the cutoff method diverges in accordance with the situation that the e^+e^- cross section tends to a constant asymptotically. Our regularized calculation is certainly better suited, however, in discussing the commutator of the current densities, not the integrated charges.)

To reinforce our conclusion (4.7) we add the following arguments:

(i) Canonical Result for $1<\mathrm{Re}N<2$

The spectral function (4.3) is no longer positive definite for $\nu<1$. The integration (4.2) is convergent in the usual sense if $1<\mathrm{Re}N<2$. Therefore there is no reason why the canonical value $S=0$ follows for $1<\mathrm{Re}N<2$. The result (4.7) may be considered to be a smooth extrapolation from this range of N.

(ii) Pauli-Villars Method

For N=4 we need two regulators. By adding (subtracting) the regulator contribution to (4.2) we again obtain the result (4.7). In the Pauli-Villars method the regulators contribute negative terms from $s = 4M^2 \to \infty$. In fact the dimensional regularization method may be regarded as a very convenient mathematical method which gives automatically, by including the negative contribution at infinity, the final result obtained by the Pauli-Villars method after a series of complicated steps.

(iii) An Alternative Method with the Equal-Time Limit Taken at the End

In deriving (4.2) we took the limit $t\to 0$ before we integrated over s. These two processes are not commutable in general. If we go to the limit $t\to 0$ after the s integration, we obtain the result which is indefinite in general. The calculation is well-defined only if $\mathrm{Re}N<2$. The result is again $S=0$. The same remark applies also to the point-splitting technique and the method of BJL limit.

5. RELATED DISCUSSIONS

(i) One of the advantages of the method of hyperfunctions is that it gives the simplest proof of

$$\int d^N k \frac{1}{k^2} = 0 \quad , \tag{5.1}$$

which was first obtained by 't Hooft and Veltman in the calculation of the massless Yang-Mills theory.[9] A simpler example of (5.1) is found in the tadpole term in the photon self-energy part of the massless scalar quantum electrodynamics. Leibbrandt and Capper gave a proof of (5.1) for N = 3, 4, ...[10] Our proof is simply based on (3.14) and is valid for any N.

(ii) The result (4.7) holds true for any m. But this may not guarantee that the same result follows if we start with m=0 at the outset. For m=0, (4.3) becomes

$$\rho(s) = \frac{2^{2-3\nu}\pi^{-\nu+1/2}}{\Gamma(\nu+\frac{1}{2})} (\nu-1) s^{\nu-1} \quad . \tag{5.2}$$

Then the integral is

$$\int_0^\infty s^{\nu-2} ds \tag{5.3}$$

which vanishes according to (3.14). In fact (3.14) is the formula which guarantees the smoothness of the massless limit.

(iii) It has been pointed out that the free string theory is, to some extent, equivalent to the massless scalar field theory in the (hypothetical) two-dimensional space-time.[11] In the light-like gauge, the commutator of the energy-momentum tensor (the Virasoro algebra) has the anomaly which is the analogue of the Schwinger term in quantum electrodynamics. This anomaly is the origin of the difficulty of the critical dimension. By applying the method of dimensional regularization, we can show that as the anomaly disappears so does the critical dimension.

REFERENCES

*) Talk based on references 1 and 2.

1) Y. Fujii and K. Mima, "Current Commutator Anomaly Regularized Dimensionally by Means of Hyperfunctions", Prog. Theor. Phys., (in press).

2) Y. Chikashige, Y. Fujii and K. Mima, "Lorentz Invariant String Regularized Dimensionally by Means of Hyperfunctions, I, II", UT-Komaba preprints 77-8, 9.

3) G. 't Hooft and M. Veltman, Nucl. Phys. B44 (1972), 189; C.G. Bollini and J.J. Giambiagi, Phys. Letters 40B (1972), 566; Nuovo Cim. 12B (1972), 20.

4) Y. Chikashige and Y. Fujii, Prog. Theor. Phys. 57 (1977), 632; 1038.

5) M. Sato, Sūgaku 10 (1958), 1; M. Sato, J. Fac. Sci. (Univ. of Tokyo); Sec. I. Vol. 8 (1959-60), 139.

6) L. Schwartz, Théorie des Distributions, Hermann Press, Paris (1966).

7) I.M. Gel'fand and G.E. Shilov, Generalized Functions, Vol. I, Academic Press, N.Y. (1964).

8) F. Pham, Introduction à l'étude topologique des singularités de Landau, Gauthiers Villars (1976).
R.C. Hua and V.L. Teplitz, Homology and Feynmann Integrals, Benjamin, N.Y., Amsterdam (1966).

9) G. 't Hooft and M. Veltman, "Diagrammar" in Particle Interactions at Very High Energies, Part B, Edited by D. Speiser, F. Halzen and J. Weyers, Plenum Press, N.Y. (1974).

10) G. Leibbrandt and D.M. Capper, J. Math. Phys. 15 (1974), 82; 15 (1974), 86.

11) S. Ferrara, A.F. Grillo and R. Gatto, Nuovo Cimento 12A (1972), 959.
S. Fubini, A.J. Hanson and R. Jackiw, Phys. Rev. D7 (1973), 1732.

HADRON SPECTROSCOPY AND THE NEW PARTICLES*

Frederick J. Gilman

Stanford Linear Accelerator Center

Stanford, California 94503, U.S.A.

1. INTRODUCTION: THE QUARK MODEL

Hadron spectroscopy would be of interest to physicists if only because the manner of organization of matter falls within the definition of the subject of physics. More specifically, with hundreds of hadronic states it behooves us to find some order and degree of understanding of what is known to exist and to be able to predict particles yet to be found.

However, present interest in hadron spectroscopy stems largely from a different source: the large body of evidence, spectroscopic and otherwise, for hadron substructure and in particular for a quark basis of hadronic matter. In many ways the focus of particle physics has shifted down to the quarks and leptons as the primary components of matter and hence of spectroscopy. Even though hadrons are then a secondary spectroscopy, it is only through a study of them that we learn the properties of the quarks (at least in the absence of free quarks). Moreover, the dynamics between quarks can only be studied by the spectroscopy and interactions of hadrons. We still have much to learn about these quark-quark forces: how quark confinement comes about, the nature of the force law, the spin dependent forces, etc., etc.

1.1 Quarks

We have already cited the fact that much points to quarks as the building blocks of hadrons. Let us review briefly some of the evidence:

1. Deep Inelastic eN, μN, νN, and $\bar{\nu}$N Scattering

The magnitude of the cross section, scaling behavior, and the relationship of structure functions observed in deep inelastic scattering indicate that the nucleon has point, spin 1/2 constituents with which the weak or electromagnetic current interacts.[1] Further, the amount of scattering depends on whether the target is a neutron or proton and on the spin orientation of the proton,[2] so that the constituents which are related to the nucleons isospin or spin are also what is "seen" by the weak or electromagnetic currents.

2. Electron-Positron Annihilation

The ratio R of the cross section for $e^+e^- \rightarrow$ hadrons to that for $e^+e^- \rightarrow \mu^+\mu^-$ is a (different) constant both below and above charm threshold, as it should be if the basic process were production of a pair of point particles, followed by their eventual materialization as hadrons.[3] In fact the part of R due to charmed meson production at SPEAR agrees with what is expected from the basic process of production of a pair of charmed quarks.[4] Furthermore, the observation of back to back jets at SPEAR yields the additional information that their angular distribution is that characteristic of production of a pair of spin 1/2 particles.[3]

3. Hadron Spectroscopy

With a few possible exceptions,[5] the hundreds of hadrons we now know are understood as quark-antiquark bound states (mesons) or three quark bound states (baryons). An enormous simplification has taken place and is part of the standard "lore". Now one often forgets that something as basic as the ordering of spins and parities of states (0^-,1^- as lowest mass mesons and $1/2^+$, $3/2^+$ as lowest mass baryons) is trivially understood in the quark model but is otherwise quite mysterious.

4. Weak and Electromagnetic Current Matrix Elements

The quark model gives us a quantitative understanding of both the magnetic moments and magnetic transition moments between the ground state baryons, as we will see in detail later. When formulated in the general framework of the transformation from current to constituent quarks,[6] one can discuss the photon transition amplitudes from the nucleon to excited nucleon resonances. When a few reduced matrix elements are fixed in terms of known amplitudes, one gets correct predictions for the signs and magnitudes of a fair number of other amplitudes.[7] Further, if one is willing to use PCAC to relate matrix elements of the axial-vector current to pion amplitudes, then a similar theory of pionic transitions ensues. Again the signs and magnitudes of many amplitudes are correctly given.[7] It would seem very unlikely that all this is an accident.

5. High Transverse Momentum Phenomena

It seems very likely that high transverse momentum hadron production in hadron-hadron collisions has its origin in "hard scattering" of constituents of the hadrons.[8] The connection to quarks is much less direct, and certainly not unique, when compared to (1)-(4) above. But the similarities to hadron production in deep inelastic scattering and electron-positron annihilation, especially the production of jets in each case, are quite striking. Although it is much harder to get precise information on quarks in this case, this is an important area of research exactly because it may give us information on quark dynamics in a different setting.[9]

1.2 Color

Quarks are thought to carry a strong interaction "charge" called color. There are three such colors, which we take as red, yellow, and blue. Present experimental evidence for the need for color comes from three sources:

1. The rate for $\pi^0 \rightarrow \gamma\gamma$

The amplitude for $\pi^0 \rightarrow \gamma\gamma$, when related to that for $\partial_\mu A_\mu \rightarrow \gamma\gamma$ by PCAC, has a magnitude and sign given by the triangle graph (with a closed fermion loop) anomaly[10] in the coupling of two vector currents to an axial-vector current. Without color, one gets the wrong rate. With it, the amplitude is increased by a factor of three and the rate by a factor 9 and then agrees with experiment.[11]

2. The ratio $R = \sigma(e^+e^- \rightarrow \text{hadrons})/\sigma(e^+e^- \rightarrow \mu^+\mu^-)$

Color increases the predicted cross section (on the basis of the quark model) by a factor of three. This is needed to get even rough agreement with experiment both below and above charm threshold.[3]

3. The Baryon Wave Function

The wave function for fermions should be totally antisymmetric. If the three quarks in a baryon are a singlet with respect to color (see below), the color part of the wave function is antisymmetric. Thus the remainder (spin, space and quark type or flavor) must be symmetric. This is indeed the case from the experimental spectrum and in particular is true for the ground state, which has a symmetrical spatial wave function combined with one symmetrical in spin and flavor.

Each of these experimental pieces of evidence for color needs some theoretical analysis to deduce the appropriateness of the concept of color, but they only involve "counting" the color quantum numbers. There are other, non-experimental, reasons for color, which have a much less solid basis in concrete facts. They all involve using color as a non-Abelian charge in a gauge field theory context. Nevertheless, they are important and have much to do with the overwhelming acceptance of the idea that colored quarks and gluons are the basis of all strong interactions.

1. Quantum Chromodynamics (QCD)

The theory of quarks coupled via the color "charge" to gauge vector bosons (gluons) is often referred to as QCD. It is a non-trivial point that QCD is the only known field theory (and a non-Abelian gauge theory at that) which has a chance of being the correct one for strong interactions.

2. Asymptotic Freedom

A non-Abelian gauge theory like QCD has the property of asymptotic freedom:[12] the effective coupling constant vanishes logarithmically at small distances, i.e. at large four momentum squared. This allows one to "understand" the scaling behavior[1] (characteristic of free field theory) observed in deep inelastic scattering.

Even more, the theory predicts that scaling is not exact and the (small) predicted logarithmic breaking is consistent with what is seen in recent experiments.[1]

3. Infrared Slavery

The increase in the coupling constant as one goes to larger distances inspires the hope that the color forces become infinite and quarks (and other objects with color) are confined. Up to now this has not been shown rigorously, but there are some suggestive model calculations of how it might come about.[13]

4. Okubo-Zweig-Iizuka[14] Rule Violation

Certain strong interaction decays where none of the final hadrons contain the quarks of the initial hadron, are very much suppressed in rate. This is particularly well exemplified in the case of the "new" particles. A theory of these processes involving intermediate gluons leads to a systematics of mass and spin-parity dependence of the degree of suppression[15] which we will discuss later in these lectures.

5. Spin-Dependent Quark-Quark Forces

Such forces result in hadron states with the same quark content but different relative spin orientations being split in mass. From the experimental observations it seems that the force between quark and quark in a baryon must have the same sign as between quark and antiquark in a meson. Exchange of a neutral vector meson without color does not yield such a result, but exchange of gluons coupled to color does if within color singlet mesons and baryons.[16] While single gluon exchange does not have to be the origin of all such forces, it still is to be desired that the lowest order effect have at least the right sign.

6. Dynamical Gluons

Sum rules for deep inelastic scattering indicate that quarks do not carry all the momentum or energy of the nucleon.[1] If the remainder is assigned to the gluons they should manifest themselves in a variety of ways by interacting with quarks and other gluons to produce hadrons in hadron-hadron collisions, to produce gluon jets in e^+e^- collisions,[17] etc.

1.3 Confinement

As we have already indicated above in our discussion of "infrared slavery", color is central to another aspect of quarks, that of confinement. We will take as a principle, perhaps derivable at a later time from QCD, that color is confined, i.e. only color singlet states can be seen. Then both quarks and gluons are not found among the asymptotic states of the theory. Bound states which are colorless can be and are seen: they are the hadrons.

The form of the effective color confining potential is not known for sure. Some arguments[13] in QCD and the string model suggest that the effective potential is linear, $V(r) = kr$, so that the force, $-dV/dr = -k$ is a constant. It then takes infinite energy to move a quark infinitely far away, as expected for a confining potential. Estimates of the force, k, principally from fitting charmonium spectroscopy[18] suggest that $k \approx 0.2\ \text{GeV}^2$ = 17 metric tons × (the acceleration of gravity).

1.4 Flavor

In addition to carrying color, quarks are distinguished from one another by their "flavor". At present we know of four flavors for quarks: up, down, strange, and charm. A fifth flavor (at least) is strongly suspected on the basis of the recently discovered Υ enhancement[19] at ∿ 9.5 GeV in the muon pair spectrum produced in proton-nucleon collisions. A particle data group type summary of the quark flavors is given in Table 1.

The masses given in Table 1 of course cannot have the usual meaning since we do not see the quarks as free particles. They are so-called "constituent masses" and occur as parameters with the dimensions of mass in certain equations. They are different than current quark masses which occur in other equations. Any meaning to be attached to them is only within these equations, if then.

The values of Q/e are most easily obtained by noting that baryons contain three quarks. Then the Δ^{++}, Δ^{+}, Δ^{o}, Δ^{-} charge states of the 3-3 resonance yield the u and d quark charges, while the Σ^{*+}, Σ^{*o}, Σ^{*-} or Ω^{-} tell us the charge on the s quark must be -1/3. In the case of the charmed quark, the best present evidence for Q/e = 2/3 comes from the charges of the D^{o} and D^{+}, the (non-strange) mesons containing a charmed quark. That it is a charmed quark and not antiquark in the D^{o} and D^{+} follows from assuming that c → s in weak decays (so that the states with a charmed quark decay into final hadrons with strangeness -1).

TABLE 1: Quark Flavors à la Particle Data Group

Quark	J^P	Mass (MeV)	Q/e	Baryon No.	Strangeness	Charm
u	$1/2^+$	∿ 350	2/3	1/3	0	0
d	$1/2^+$	∿ 350	-1/3	1/3	0	0
s	$1/2^+$	∿ 500	-1/3	1/3	-1	0
c	$1/2^+$	∿1650	2/3	1/3	0	1
?	$1/2^+$	∿5000	?	1/3	0	0

There is other confirmatory evidence for all these charge assignments, such as the size of the change in R in e^+e^- annihilation in crossing the appropriate threshold, the size of the electromagnetic coupling of the vector mesons, etc.

1.5 Weak Couplings

We have already touched on the weak couplings of quarks. For the moment, we recall the standard model[20] where the weak-electromagnetic gauge group is SU(2)×U(1) and the left-handed quarks fall into doublet representations:

$$\begin{pmatrix} u \\ d_\theta \end{pmatrix} \qquad \begin{pmatrix} c \\ s_\theta \end{pmatrix} \quad ,$$

where

$$d_\theta = d\cos\theta + s\sin\theta \tag{1a}$$

$$s_\theta = -d\sin\theta + s\cos\theta \tag{1b}$$

and θ is the Cabibbo angle ($\sin^2\theta \approx 0.05$). Right-handed quarks are singlets. The charged weak current has V-A transitions between c and s with strength proportional to cosθ, while u and s and c and d have strength proportional to sinθ. A very important question, which we will return to later, is the addition of more quarks and the possibility of having right-handed new and/or old quarks in doublet rather than singlet representations. But we note now that experiment prohibits the right-handed u quark from being coupled with more than a few percent of full strength to the d or s quark, nor can the right-handed c quark couple to the d quark with strength comparable to the left-handed c to s quark coupling.[21]

1.6 Hadrons

We now review briefly the states composed of quarks which we do observe, the hadrons. The simplest possible hadrons are made of a quark and an antiquark forming a meson, or three quarks, forming a baryon. All other combinations of one, two, or three quarks and/or antiquarks have a net color (i.e., are not singlets under color SU(3)), and are forbidden by the principle of color confinement.

In the case of mesons it is simple to see that the color wave function

$$(\bar{q}_{1R}q_{2R} + \bar{q}_{1B}q_{2B} + \bar{q}_{1Y}q_{2Y})/\sqrt{3}$$

is a normalized color singlet for any antiquark $(\bar{q}_1)$-quark(q_2) bound state.

The quark and antiquark spin may be combined to form a total quark spin, S, which is either 0 or 1. When coupled with the relative internal orbital angular momentum L, we can form a total meson angular momentum, $\vec{J} = \vec{L} + \vec{S}$.

To complete the meson wave function we can choose any of the four (or more?) flavors (u,d,s,c) for the quark and any of the four flavors for the antiquark. Thus there are 16 possible flavor possibilities for each value of L, S, and J. A meson wave function in the quark model then can be written in factorized form as

$$\psi_{\text{color}}(\text{singlet})\times\psi_{\text{flavor}}\times[\psi_{\text{spin}}(S=0,1)\times\psi_{\text{orbital}}(L=0,1,\ldots)]_{\text{total } J}$$

For baryons the situation is a little more complicated. The normalized color singlet state with quarks $q_1q_2q_3$ is

$$\frac{1}{\sqrt{3!}}\begin{vmatrix} q_{1R} & q_{2R} & q_{3R} \\ q_{1B} & q_{2B} & q_{3B} \\ q_{1Y} & q_{2Y} & q_{3Y} \end{vmatrix}$$

as is easily seen by noting that a transformation induced by an element of the color SU(3) group multiplies the matrix above by another matrix of determinant unity.

The total quark spin, S, may be 1/2 or 3/2. This is to be combined with the net internal orbital angular momentum L, to form the total baryon angular momentum $\vec{J} = \vec{L} + \vec{S}$. The internal orbital angular momentum can be constructed in several ways, but most simply one may take $\vec{\ell}_{12}$ as the orbital angular momentum between quarks 1 and 2 and add to it $\vec{\ell}_3$, the orbital angular momentum of the third quark relative to the center of mass of the first two, to form $\vec{L} = \vec{\ell}_{12} + \vec{\ell}_3$.

The flavor wave function is also a bit more complicated than for mesons because not all flavor states are allowed for a given L and S due to Fermi statistics. With a color singlet wave function which is antisymmetric, the remainder of the baryon wave function must be symmetric. We will discuss the detailed implications of this later.

1.7 Exotics

The meson and baryon states we have discussed so far are the conventional ones of the quark model and involve the minimum number of quarks and/or antiquarks which can form a color singlet. We might well define a manifest exotic as any state which cannot be made out of a quark-antiquark pair in the case of a meson and three quarks in the case of a baryon.

Traditionally, one breaks up exotics into two categories. Exotics of the first kind, or "flavor exotics", are states in SU(2), SU(3),... representations not found when hadrons are formed as described above. Examples include doubly charged mesons, a baryon with positive strangeness, a meson with two units of charm, etc.

Exotics of the second kind are sometimes called "CP exotics". These are specifically mesons with parity $P = (-1)^J$ which have $CP = -1$ or a meson with $J^{PC} = 0^{--}$. Neither of these can be formed from a quark and an antiquark. A particular example of such an exotic is a vector meson with even charge conjugation.

In models which have a mechanism for forming exotic states, very often there are hadrons which do not have manifestly exotic quantum numbers themselves, but which have a quark content such that they have exotic relatives. These states are sometimes called "crypto-exotics". It is convenient to extend the definition of exotic to include them. Then an exotic is a meson which is not a quark-antiquark state or a baryon which is not three quarks. To use this definition we of course imply that we can tell what quarks are inside a given hadron!

There are many examples of predictions of such exotic states:

1. $q\bar{q}q\bar{q}$ mesons and qqq $q\bar{q}$ baryons as in bag model calculations;[22]

2. $(c\bar{q})(\bar{c}q)$ bound states of two charmed mesons to form "molecular charmonium;[23]

3. Baryonium;[24]

4. Mesons composed of $\bar{q}q$ in a color octet state coupled to a gluon;[25]

5. Quarkless states composed of gluons alone or "glueballs";[26]

6. States where the energy-momentum and perhaps spin are carried by fields other than the quarks, such as a neutral "soul";[27]

7. String excitations in a model of quark binding through a field theoretic string. String excitations may also be coupled to the quark orbital angular momentum to produce a fairly complicated spectroscopy.[28]

In fact, it is difficult to avoid exotic states with any real dynamics in a field theoretic framework. For no matter whether we confine quarks with gluons, with strings, or with some other fields, in a true field theory, the binding-field will have dynamical degrees of freedom of its own. Then, in addition to the quarks, there will be other fields which carry energy and momentum - which have their own spectrum of excitations and can "slosh" around inside the hadron relative to the quarks. The coupling of these excitations to the quark excitations in general gives rise to extra, sometimes manifestly exotic, states in the hadronic spectrum in addition to the ones usually expected. It is thus not a question so much of whether exotic states exist at all: almost any theory of hadrons worthy of the name predicts them at some mass. The important question is quantitative: at what mass and with what quantum numbers do they occur?

2. MESONS

We now consider in more detail the spectroscopy of mesons. Their flavor quantum numbers can be read off directly from their quark content, while their parity $P = (-1)^{L+1}$ and, for charge self-conjugate states ($\bar{u}u$, $\bar{d}d$, $\bar{s}s$, $\bar{c}c$), their charge conjugation $C = (-1)^{L+S}$.

The flavor states in the case of four quarks (u,d,s, and c) are:

$\bar{u}u$	$\bar{u}d$	$\bar{u}s$	$\bar{u}c$
$\bar{d}u$	$\bar{d}d$	$\bar{d}s$	$\bar{d}c$
$\bar{s}u$	$\bar{s}d$	$\bar{s}s$	$\bar{s}c$
$\bar{c}u$	$\bar{c}d$	$\bar{c}s$	$\bar{c}c$

To the extent that the u and d quarks are degenerate in mass and have the same strong interactions, one has a strong interaction SU(2) symmetry (called isotopic spin). Similarly, to the extent that u, d, and s are degenerate and have the same strong interactions, one has an SU(3) symmetry. The grouping of mesons composed of u, d, s, and c quarks into multiplets which are irreducible representations of these symmetries is given in Table 2.

For the L = 0 ground state mesons we have quark spin 0 or 1 and hence $J^P = 0^-$ or 1^-. The approximate assignment of observed pseudoscalar and vector mesons to the flavor states composed of u, d, s and c quarks is indicated in Table 3.

TABLE 2: SU(2) and SU(3) Multiplets for Mesons

Quark Flavor State	Isospin (SU(2) Representation	SU(3) Representation
$\bar{d}u$ $(\bar{u}u-\bar{d}d)/\sqrt{2}$ $\bar{u}d$	1	
$\bar{s}u$ $\bar{s}d$	1/2	8
$\bar{d}s$ $\bar{u}s$	1/2	
$(\bar{u}u+\bar{d}d-2\bar{s}s)/\sqrt{6}$	0	
$\bar{d}c$ $\bar{u}c$	1/2	$\bar{3}$
$\bar{s}c$	0	
$\bar{c}u$ $\bar{c}d$	1/2	3
$\bar{c}s$	0	
$(\bar{u}u+\bar{d}d+\bar{s}s)\sqrt{3}$	0	1
$\bar{c}c$	0	1

TABLE 3: Ground State Mesons[34]

Quark Flavor State			$J^P=0^-$ Observed Meson[11,29,30,31]	$J^P=1^-$ Observed Meson[11,30,31]
$\bar{d}u$	$(\bar{u}u-\bar{d}d)/\sqrt{2}$	$\bar{u}d$	$\pi^{+,o,-}(140)$	$\rho^{+,o,-}(770)$
$\bar{s}u$	$\bar{s}d$		$K^{+,o}(495)$	$K^{*+,o}(890)$
	$\bar{d}s$	$\bar{u}s$	$\bar{K}^{o,-}(495)$	$\bar{K}^{*o,-}(890)$
	$(\bar{u}u+\bar{d}d)/\sqrt{2}$		$\eta(550)$ ↖ mixture	$\omega(783)$
	$\bar{s}s$		$\eta'(958)$ ↙	$\phi(1020)$
$\bar{d}c$	$\bar{u}c$		$D^{+,o}(1865)$	$D^{*+,o}(2010)$
$\bar{s}c$			$F^{+}(2030)$	$F^{*+}(2140)$
	$\bar{c}u$	$\bar{c}d$	$\bar{D}^{o,-}(1865)$	$\bar{D}^{*o,-}(2010)$
		$\bar{c}s$	$\bar{F}^{-}(2030)$	$\bar{F}^{*-}(2140)$
	$\bar{c}c$		$X(2830)$	$\psi(3095)$

With the discovery of D^o, D^+ and D^{*o}, D^{*+} last year[29], the additional evidence[30] for the X(2.83), and the new indications[31] from the DASP group for the F and F^*, we have a known particle for every one of the 16 pseudoscalar and 16 vector mesons expected for the mesonic ground states with four quarks. If we take the $\Upsilon(9.5)$ to be the ground state vector meson composed of a fifth quark and its antiquark then we still have 25–16=9 pseudoscalar and 8 vector mesons yet to be found! But given the existence of the 32 ground state mesons composed of four quarks, we have no doubt they all will eventually be found. For in fact, we do not have 32 unaffiliated particles, but really 32 examples of the same thing – the ground state of a quark-antiquark system – obtained by putting in different flavors for the quark and antiquark and changing their spin orientation.

The ground state, by implication of its name, is not the only level found in the quark-antiquark spectrum. There are (at least) radial and orbital excitations. We define a radial excitation as a state which has all the same quantum numbers, including internal quark L and S, as another $\bar{q}_1 q_2$ state at lower mass. The idea as well as the name for such states is borrowed from non-relativistic potential theory. There, in a potential of sufficient strength, one finds a series of such levels, each successive radial excitation having another node in its radial wave function. Familiar examples of such a situation occur for the Coulomb, harmonic oscillator and linear potentials.

Suppose such a higher mass pseudoscalar or vector meson is discovered; is it necessarily a radial excitation of the ground state? For a $J^P=0^-$ state the answer is yes; one can only make a pseudoscalar out of a quark and antiquark if L=S=0. Thus all quantum numbers including L and S are the same as that for the ground state pseudoscalar. For a $J^P=1^-$ state, this is not necessarily so. Both internal L=0, S=1 and L=2, S=1 can result in $J^P=1^-$ states and only the first case meets our definition of a radial excitation of the ground state. Furthermore, the closeness in mass of L=0 radial excitations and L=2 states in linear and harmonic potentials makes mixing between the corresponding $J^P=1^-$ states very likely.

Barring such complete mixing, how can we tell the L=0 from L=2 vector mesons? First, if a pseudoscalar partner is found nearby in mass, we know it must be a radial excitation, and hence also the vector meson. Second, if we have enough confidence in our knowledge of the potential binding the quark and antiquark together, then we can calculate the mass expected for a given state and expect experiment to agree. Along the same lines, if we know experimentally the mass of expected nearby states, it may be possible to associate a new state with L=0 or L=2 depending on its mass. Third, in a nonrelativistic picture $\Gamma(V^0 \to e^+e^-) \propto |f(r=0)|^2$, the square of the spatial wave function at the origin. This vanishes for L=2 in the nonrelativistic approximation. For charmed quarks at least, even after relativistic corrections, the L=2 vector mesons should have a very much smaller leptonic width than those with L=0. Last, in a theory of pionic decays based on the quark model, the relative signs of various vector meson decay amplitudes are different depending on whether L=0 or 2. For example, the amplitudes for $\rho' \to \pi\omega$ vs. $\rho' \to \pi\pi$ have a different relative sign[32] if the ρ' is a quark-antiquark state with L=2 rather than L=0. Similar considerations led to the establishment[33] of a $J^P=3/2$, I=3/2 pion-nucleon resonance at ∿1700 MeV as a radial excitation of the Δ(1232) rather than an L=2 baryon state.

The most persuasive evidence for a sequence of mesonic radial excitations comes from charmonium. There we have[34] the $\psi \equiv \psi(3095)$

and its radial excitation $\psi' \equiv \psi(3684)$. The new state[35], $\psi(3772)$, on the basis of its leptonic width and agreement with potential model calculations is most likely an L=2 state, though with some mixture of the L=0 radial excitation, ψ'. The mass region between ~4 and ~4.2 GeV contains several bumps, with one very likely another radial excitation of the ψ. The $\psi(4414)$ fits fairly well as yet a third radial excitation. There is every reason to expect still higher mass radially excited states but they become very difficult to distinguish from background because of the increasing total width and smaller coupling to e^+e^-.

At the moment, with some recent additions to the list of known states, the evidence for radial excitations in the "old" meson spectrum is fairly convincing by itself. The only established mesonic radial excitation[11] for quite some time was the $\rho'(1600)$. In the last year or so it has been joined by a $K'(1400)$ which was found[36] in an isobar model analysis of the $K\pi\pi$ final state produced in $K^{\pm}p$ collisions at 13 GeV/c. It is a $J^P=0^-$ state decaying to $K(\pi\pi)_S$ and as noted before, must be a radial excitation of the ground state K(495). It has a possible K*(1650) vector meson partner found in some $K\pi$ phase shift analysis solutions of the same experiment.[37]

The last few months have seen a population explosion among vector mesons composed of "old" quarks. The initial result from Orsay[38] was an indication of a bump in $e^+e^- \to 5\pi$ near 1780 MeV. This has been followed by evidence for a relatively narrow bump at ~1820 MeV from Frascati.[39] Even more recent data indicates that the region from 1500 to 2000 MeV may be quite complicated with as many as half a dozen (or even more!) vector meson states found in that region.[40] Inasmuch as we do expect both L=0 radial excitations and L=2 vector mesons, all composed of $\bar{u}u$, $\bar{d}d$, and $\bar{s}s$ in that mass region, such a complicated situation is not totally unexpected. In any case, although much remains to be sorted out, both charmonium and the old mesons emphatically indicate that <u>radial excitations of mesons do exist</u>.

The other clear set of excitations in the meson spectrum is that corresponding to non-zero orbital angular momentum between the quarks. We recall from the first section that in the quark model when L=1 each quark flavor combination occurs in an S=0 state with $J^{PC}=1^{+-}$ and three S=1 states with $J^{PC}=0^{++}$, 1^{++}, and 2^{++}.

The most spectacular examples of the L=1, S=1 states are the $\chi(3414)$, $\chi(3508)$, and $\chi(3552)$ charmonium ($\bar{c}c$) levels which very likely have $J^{PC}=0^{++}$, 1^{++} and 2^{++} respectively. For u, d and s quarks, only the $J^P=2^+$ states are completely found (see Table 4).

TABLE 4 : $J^P=2^+$ Mesons Composed of u, d, and s Quarks

Quark Flavor State			Observed Meson[11]
$\bar{d}u$	$(\bar{u}u - \bar{d}d)/\sqrt{2}$	$\bar{u}d$	$A_2(1310)$
	$(\bar{u}u + \bar{d}d)/\sqrt{2}$		f(1270)
$\bar{s}u$	$\bar{s}d$		K*(1420)
	$\bar{d}s$	$\bar{u}s$	$\bar{K}$*(1420)
	$\bar{s}s$		f'(1515)

The $J^P=1^+$ states of u, d and s quarks are a traditional area of experimental confusion. However, in the last year or so the situation is beginning to clarify. The biggest single advance has been the evidence[41,42,43] for two Q mesons, Q_1($\sim$1300) and Q_2($\sim$1400), which are axial-vector states containing a strange quark and a u or d quark. The observed states are actually mixtures[42] of the S=0 and S=1 quark model states. The B(1235) meson is an established candidate for the isospin one axial-vector state composed of u and d quarks with quark spin S=0. The D(1285) (not to be confused with the charmed mesons) is the only established[11] isospin zero meson which likely had $J^P=1^+$ (and from its positive charge conjugation would correspond to S=1).

Along with the Q mesons, the traditional problem child of the axial-vector mesons is the A_1. Even here some real progress is being made. Although earlier analyses of diffractive three pion production were never able to show evidence for a real resonance at the peak mass of $\sim$1100 MeV, more recent theoretical work[44] with multichannel analyses do indicate resonance behavior, although perhaps at a higher mass (even possibly 1400 to 1500 MeV). At the same time, more direct experimental indications of a resonance decaying to $\pi\rho$ at $\sim$1100 MeV come from several different experiments performed at CERN.[43] It seems unlikely that the uncertainty with regard to the A_1 will persist very much longer. With, in addition, the new evidence[45] for the heavy lepton decay $\tau \to A_1 \nu_\tau$, the establishment of a suitable isospin one meson to match the L=1, S=1 axial-vector meson composed of u and d quarks seems finally to be within sight.

With the situation for 1^+ states composed of u, d and s quarks straightening out that for the $J^P=0^+$ states is still confusing. Several of these states, like the δ(970), which would be the I=1 scalar meson composed of u and d quarks, are established.[11] But the confusion surrounding the isoscalar $J^P=0^+$ states (of which there may be too many, and at the wrong masses) prevents one from being very optimistic at the moment.[46] There are in fact various proposals assigning some or all these scalar mesons to what we would call exotic multiplets.[22,26,47]

At the next level of orbital excitation, L=2, only a few states are pinned down for sure. We have already noted the ψ(3772), which is very likely L=2 and S=1 combined to form $J^P=1^-$. The g(1690), ω*(1675) and K*(1780) all are now established[11] to have $J^P=3^-$ and hence correspond to L=2 and S=1 combined to form $J^P=3^-$ for states with u, d, and s quarks. While many states, particularly with $J^P=2^-$, remain to be established experimentally, enough has been found to give us assurance that all the L=2 levels must exist for all possible quark flavors.

When we get to L=3, the only established state[11] is the h(2040) which fits well as the L=3, S=1 isoscalar state composed of u and d quarks with $J^P=4^+$. Although most specific quantum number assignments are unknown above ∿2 GeV for mesons, there is no indication that the sequence of orbital (or radial) excitations stops here. On the contrary, there are clear signals[11,48] for meson resonances extending above 2.5 GeV and we have every reason to expect that broad, difficult to isolate states exist at masses much higher than that.

A comparison of the known charmonium ($\bar{c}c$) spectroscopy with that expected from a linear potential is shown in Table 5. The match between the $\bar{c}c$ states expected and the experimental observations is rather convincing evidence even taken by itself, that we are dealing with the bound states of a fermion-antifermion system.

However, the knowledge of strange mesons acquired over the years, or of isospin one mesons, is fairly impressive also (see Table 6.) Even more important, where gaps exist in the established charmonium states, they are often filled in the case of strange or I=1 mesons, and _vice versa_. On the one hand, this gives us great confidence that all the J^P states corresponding to a given L level do in fact exist. On the other hand, from the ground state mesons and some of the excited levels where most of the different flavor states have been found, we also have great confidence that each level comes in _all_ possible quark-antiquark flavor states. Sooner or later they will all be found. The most important question is whether other, non-$\bar{q}q$ levels exist in the meson spectrum.

TABLE 5: J^{PC} Levels of $\bar{c}c$ in a Linear Potential (not to scale)

Quark Model Level		Observed Meson[4,11]
1D	3^{--}	?
	2^{--}	?
	2^{-+}	?
	1^{--}	$\psi(3772)$
2S	1^{--}	$\psi(3684)$
	0^{-+}	$\chi(3455)$?
1P	2^{++}	$\chi(3552)$
	1^{++}	$\chi(3508)$
	1^{+-}	?
	0^{++}	$\chi(3414)$
1S	1^{--}	$\psi(3095)$
	0^{-+}	$\chi(2830)$

TABLE 6: J^{PC} Levels of $\bar{q}_1 q_2$ in a Linear Potential

Quark Model Level		Observed Strange Meson[11]		Observed I=1 Meson[11]
1D	3^{--}	K*(1780)		g(1690)
	2^{--}	?		?
	2^{-+}	L(1770)?		A_3(1640)?
	1^{--}	?		?
2S	1^{--}	K*'(1650)		ρ'(1600)
	0^{-+}	K'(1400)		?
1P	2^{++}	K*(1420)		A_2(1310)
	1^{++}	Q_2(1400)	mix	A_1?
	1^{+-}	Q_1(1300)		B(1235)
	0^{++}	κ(1250)?		δ(970)?
1S	1^{--}	K*(890)		ρ(770)
	0^{-+}	K(495)		π(140)

3. BARYONS

The baryon ground state has all three quarks in relative s-states and therefore overall L=0. If the total quark spin S=3/2, the spin wave function is completely symmetric. Since the space wave function is also symmetric for the ground state, the quark flavor wave function must also be symmetric. Then in combination with the antisymmetric color singlet wave function, one has an overall antisymmetric three quark wave function, in accord with Fermi-Dirac statistics. With four quark flavors from which to choose, there are 20 possible symmetric three quark flavor states. These are shown in Table 7 together with the corresponding observed baryon, if known.

TABLE 7: S=3/2 Baryon Ground States

Quark Flavor States	Observed States[11,49]
uuu uud udd ddd	$\Delta^{++,+,o,-}$(1232)
uus uds dds	$\Sigma^{*+,o,-}$(1385)
uss dss	$\Xi^{*o,-}$(1530)
sss	Ω^-(1670)
uuc udc ddc	Σ_c^* or $C_1^{*++,+,o}$(2500?)[50]
usc dsc	$S^{*+,o}$ (?)
ssc	T^{*o} (?)
ucc dcc	X_u^{*++}, X_d^{*+} (?)
scc	X_s^{*+} (?)
ccc	Θ^{++} (?)

In the case of total quark spin S=1/2, it may be shown that the spin wave function is of "mixed symmetry". With a symmetric ground state spatial wave function, Fermi-Dirac statistics now demands a mixed symmetry flavor wave function. With four quarks, it turns out there are again 20 such quark flavor states. It is purely an accident that the number of flavor states is the same as for a symmetric flavor wave function, and, as we will see below, is not true when there are other than four quark flavors. The appropriate mixed symmetry states composed of u, d, s, and c quarks, together with their experimental counterparts, are shown in Table 8.

TABLE 8: S=1/2 Baryon Ground States

Quark Flavor States	Observed States[11,49]
uud udd	$N^{+,o}$(940)
uus {ud}s dds	$\Sigma^{+,o,-}$(1190)
[ud]s	Λ(1115)
uss dss	$\Xi^{o,-}$(1320)
uuc {ud}c ddc	Σ_c or $C_1^{++,+,o}$(2426?)[51]
{us}c {ds}c	$S^{+,o}$ (?)
ssc	T^o (?)
[us]c [ds]c	$A^{+,o}$ (?)
[ud]c	Λ_c or C_o^+ (2260)[50]
ucc dcc	X_u^{++}, X_d^+ (?)
scc	X_s^+ (?)

{ } ≡ symmetrized, in flavor

[ ] ≡ antisymmetrized, in flavor

Radial excitations of the baryon ground state, as for meson radial excitations, differ only in having a different radial wave function and should have the same spin and flavor states available as the ground state. For S=3/2 we then have a symmetric flavor wave function, while for S=1/2 one of mixed symmetry. The number of possible baryon (three quark) flavor states as a function of the number of different quark flavors is given in Table 9. Also shown is the number of flavor states times the number of S_z states available for the entire ground state or its radial excitation. We often refer to the set of these states by their total spin (S_z) and flavor multiplicity, e.g. for three quarks (u, d, s) it is the "56", made up of an SU(3) octet with S=1/2 and a decuplet with S=3/2 .

Besides the ground state or its radial excitations, we will of course have the same accounting of baryon spin and flavor states whenever the quark spatial wave function is symmetric. For then the flavor times spin wave function is required to be symmetric, and we have exactly the same arguments on the available spin and flavor states that led us to Table 9 for the ground state.

TABLE 9: Multiplicity of the Baryon Ground State or its Radial Excitations

N=No. of Quark Flavors	No. of Baryon Flavor States S=1/2	S=3/2	No. of Spin Times Flavor States
1	0	1	4
2	2	4	20
3	8	10	56
4	20	20	120
5	40	35	220
6	70	56	364

For baryon orbital excitations one can in principle have quark spatial wave functions which are symmetrical, antisymmetrical, or or mixed symmetry. The lowest orbital excitation, that with L=1, turns out to have a spatial wave function with mixed symmetry among the three quarks. For the case of quark spin S=3/2 (a symmetric spin wave function), this forces a mixed symmetry flavor wave function. However, when S=1/2 (mixed symmetry spin wave function) the overall Fermi-Dirac statistics can be satisfied with either a symmetrical, mixed symmetry, or antisymmetrical flavor wave function. The situation with regard to the multiplicity of baryon flavor states in this case is shown in Table 10.

Again, such an array of spin and flavor states will arise any time the three quark spatial wave function is of mixed symmetry. The set of these spin and flavor states is then often referred to by their total spin times flavor multiplicity, e.g. for three quarks one has the "70", composed of an S=3/2 SU(3) octet and an S=1/2 SU(3) singlet, octet, and decuplet.

TABLE 10: Multiplicity of the Baryon Orbital Excitations with Mixed Symmetry Spatial Wave Functions

N=No. of Quark Flavors	No. of Baryon Flavor States S=3/2				No. of Spin Times Flavor States
	S=3/2	S=1/2			
	Mixed	Antisym.	Mixed	Sym.	
1	0	0	0	1	2
2	2	0	2	4	20
3	8	1	8	10	70
4	20	4	20	20	168
5	40	10	40	35	330
6	70	20	70	56	572

Aside from the observed charmed baryons, which are candidates for being members of the L=0 ground state, only states composed of u, d and s quarks are known for baryons. Therefore, in discussing the observations of radially and orbitally excited baryonic levels,[52] we consider only states composed of three quarks. As indicated above, we refer to the multiplets of given L by their spin (S_z) times flavor multiplicity.

The first excited baryon level above the ground state is a 56, L=0 multiplet, i.e. a radial excitation of the 56, L=0 ground state. Its most familiar non-strange member is the Roper resonance, N*(1470). The radially excited counterpart of the 3-3 resonance is the Δ*(1690).

At slightly higher mass, on average, is a set of negative parity states which form a 70, L=1 orbital excitation. All seven of the non-strange resonances needed to fill this multiplet are known to exist with the right spins and isospins - no more and no less than the expected states.

Above the 70, L=1 there is another possible radial excitation of the ground state 56, L=0. However, most of the evidence for this is based on the N*(1780) with $J^P=1/2^+$ and confirmation of the whole multiplet awaits evidence for some of the other states.

In the same mass range there is a further established multiplet, a 56, L=2. Most, if not all of the six non-strange states sitting in this multiplet are found experimentally, including the long established N*(1688) with $J^P=5/2^+$ and the Δ*(1950) with $J^P=7/2^+$.

In the 2 GeV mass region there is fairly good evidence for a 70, L=3 set of states. In particular the established N*(2190) and N*(2140) with $J^P=7/2^-$ and $9/2^-$ respectively, rather uniquely fit into just such a multiplet.

At still higher mass there are the established $J^P=9/2^+$ N(2220) and the $11/2^+$ Δ*(2420). Even though essentially all the other states remain to be found, these two levels are very likely the first members of a 56, L=4 multiplet.

Thus we see a fairly extensive sequence of radial and orbital excitations in the baryon spectrum, just as in the case of the meson spectrum. A few more multiplets are quite possible in the mass range discussed up to now (e.g. a 56, L=2 radial excitation and a 70, L=1 radial excitation).

The established multiplets so far all have the property that L even corresponds to a flavor times spin multiplicity of 56 while those with L odd have a multiplicity of 70. While this is trivial for the ground state, or first orbital excitation, it is entirely non-trivial that we do not see, say, 70, L=0 and 70, L=2 multiplets below 2 GeV. (These are expected in a harmonic oscillator potential to be degenerate with the 56, L=2.) The full significance of this for the quark-quark force remains to be seen. In fact, there are recent suggestions that the empirical connection of 56's and 70's with L even and odd, respectively, may break down: this is based on a $5/2^-$ Δ^* near 1960 MeV which would seem to fit best in a 56, L=1 multiplet.[53]

In any case, there are further N* and Δ^* bumps (with unknown J^P) extending well into the 3 GeV region.[11] We have no reason to doubt that the baryon spectrum continues to much higher masses, albeit with broader, low elasticity, states making it almost impossible to isolate individual levels and their quantum numbers.

4. DECAY PROCESSES

Higher mass hadronic states are unstable with respect to the strong interactions. They generally decay to the lowest mass states with the same (net) flavor and other quantum numbers conserved in the strong interaction, typically by pion emission. Examples are $A_2 \xrightarrow{\pi} \rho \xrightarrow{\pi} \pi$, $K^*(1420) \xrightarrow{\pi} K^*(890) \xrightarrow{\pi} K$, and $N^*(1670) \xrightarrow{\pi} \Delta(1232) \xrightarrow{\pi} N$.

Occasionally a hadron will have a prominent electromagnetic decay into another hadron (or hadrons) with the same net flavor, when strong interaction quantum numbers and/or phase space inhibit a strong decay mode. $D^{*o} \xrightarrow{\gamma} D^0$, $F^* \xrightarrow{\gamma} F$, and $\omega \xrightarrow{\gamma} \pi^0$ are some outstanding examples of these electromagnetic decays.

A given hadron will then cascade down in mass by strong and/or electromagnetic decays until eventually it drops down to the lowest mass state(s) with baryon number and net flavor the same as the parent hadron. The state of lowest mass, characterized by a combination of quark flavors, then decays weakly except for the lowest mass (pseudoscalar) mesons composed of a quark and its antiquark, which decay electromagnetically, or if massive enough, by strong interactions. We shall now discuss these three types of decays - weak, electromagnetic, and strong - in more detail. We start with the weak decays.

4.1 Weak Decays

We view all weak decays of hadrons in terms of what is happening at the quark level. The various amplitudes and their strengths can be read off easily for the u, d, s and c quarks from the doublet structure discussed in Section 1 for the standard SU(2)×U(1) model[20] of the weak and electromagnetic interactions.

For example, the usual strangeness non-changing semi-leptonic hadron decays arise at the quark level as $d \to u + W^- \to u + e^- \bar{\nu}_e$. They have an amplitude proportional to $\cos\theta$ ("Cabibbo allowed") and are clearly characterized by $\Delta I=1$. On the other hand the semileptonic decay of the strange quark, $s \to u + W^- \to u + e^- \bar{\nu}_e$ or $u + \mu^- \bar{\nu}_\mu$, is the quark process responsible for all strange particle semi-leptonic decays. Its amplitude is proportional to $\sin\theta$ ("Cabibbo suppressed") and is characterized by the well known selection rules $\Delta S = \Delta Q = \pm 1$ and $\Delta I=1/2$. There are also non-leptonic strange particle decays. These presumably arise at the quark level as $s \to u + W^- \to u + d\bar{u}$. *A priori* this could be either $\Delta I=1/2$ or 3/2, but should always be "Cabibbo suppressed".

For charmed particles, the "Cabibbo allowed" decays at the quark level are $c \to s + W^+ \to s + e^+\nu_e$ or $s + \mu^+\nu_\mu$ for semileptonic decays and $c \to s + W^+ \to s + u\bar{d}$ for non-leptonic decays, respectively. The former are then characterized by the selection rules $\Delta C = \Delta S = \pm 1$, $\Delta I=0$, while the latter also have $\Delta C = \Delta S = \pm 1$, but $\Delta I=1$. The corresponding Cabibbo suppressed modes of charmed particles are generated at the quark level by $c \to d + e^+\nu_e$ or $d + \mu^+\nu_\mu$ and $c \to d + u\bar{d}$.

For strange particle decays the magnitude of observed semi-leptonic amplitudes agrees with that expected from the quark weak decay amplitudes. Just looking at the quark level processes, one might expect the nonleptonic and semileptonic decay rates to be comparable. In fact $s \to u + d\bar{u}$ should very naively occur at three times the rate (because of color) that $s \to u + e^- \bar{\nu}_e$ does. This is not true, as evidenced by the fact that the strange baryons decay about a thousand times more frequently in nonleptonic modes than in semileptonic ones.[11]

The amplitude for $\Delta S=\pm 1$ non-leptonic decays thus appears to be enhanced compared to the semi-leptonic amplitude.[54] Furthermore, it is the $\Delta I=1/2$ (octet in SU(3)) part of the overall non-leptonic interaction that is enhanced. While there are explanations of this enhancement, none gives a completely satisfactory quantitative description of the effect. It is particularly damaging to some explanations that no similar enhancement occurs for charm: the semi-leptonic (semi-muonic plus semi-electronic) branching ratio of the D's is ∿20 percent.[4] This is not so far from the 40 percent expected in the most naive quark level calculation where the decay rates for $c \to s + e^+\nu_e$, $c \to s + \mu^+\nu_\mu$ and $c \to s + u\bar{d}$ are in the ratio 1:1:3.

Up to this time all D decays which have been seen[4] are in accord with the standard model, with the non-leptonic Cabibbo allowed selection rules (particularly $\Delta C = \Delta S = \pm 1$) being spectacularly verified in decays like $D^+ \to K^-\pi^+\pi^+$ and $D^0 \to K^-\pi^+$, $K^-\pi^+\pi^+\pi^-$, and

$D^0 \rightarrow K^-\pi^+\pi^0$. Cabibbo suppressed modes like $\pi^+\pi^-$, $\pi^+\pi^+\pi^-$, etc. are not seen down to levels of ∿10 percent of corresponding Cabibbo allowed modes.[4] This is entirely consistent with the relative rate of $\tan^2\theta \approx 0.05$ expected in the standard model.

With the discovery[19] of the Υ it has become fairly clear that there is a fifth quark, and the question immediately arises of how it behaves with respect to the electromagnetic and weak interactions. The possibilities are essentially limitless if we do not restrict the discussion to particular weak gauge groups and particular representations of those groups. In the following we only consider the gauge group[20] SU(2)×U(1), with fermions in either doublet or singlet representations.[55]

With only four quarks, the most general classification of the left-handed u, d, s and c quarks into doublets is just

$$\begin{pmatrix} u \\ d_\theta \end{pmatrix}_L , \quad \begin{pmatrix} c \\ s_\theta \end{pmatrix}_L ,$$

the standard model with θ the Cabibbo angle, as discussed in Section 1. All angles except for one which are involved in going from the "bare" quark doublets to a representation in which the quark mass matrix is diagonal can be absorbed in the definition of the quark fields. The single remaining angle can be chosen to be that defining an orthogonal transformation among the d and s quarks: it is just what is called the Cabibbo angle.

For the right-handed quarks, we could also contemplate putting the quarks in right-handed doublets (with, in general, another angle, θ', characterizing the rotation between right-handed quarks). However, experiment tells us that the transitions $u_R \overset{W^\pm}{\leftrightarrow} d_R$, $u_R \overset{W^\pm}{\leftrightarrow} s_R$ and $c_R \overset{W^\pm}{\leftrightarrow} d_R$ can only have a few percent[21] of full strength (characterized by putting the corresponding quarks unmixed in right-handed doublets). The restrictions on the strength of the first two pairs comes from neutron beta decay, strange particle decays, and the y distributions at moderate energies in deep inelastic scattering. The third pair is restricted by the lack of observation of decays of the D mesons which involve no net strangeness in the final state (e.g. all pionic modes). There is even preliminary evidence that the only remaining pairing, $c_R \overset{W^\pm}{\leftrightarrow} s_R$, cannot have full strength. This follows from the y distribution of dimuon events in $\bar{\nu}$ deep inelastic scattering (presumably due to charm production off $\bar{s}$ quarks and subsequent semi-muonic decay) as observed in the CDHS experiment[56] at the SPS.

Thus, if we had only u, d, s and c quarks, they cannot be assigned to right-handed doublets, as no pairing of u or c to d or s, or combination, has the full charged current strength required for such a doublet. Therefore with four quarks we would assign them to be right-handed singlets under SU(2)×U(1). This is just the so-called standard model.[20]

Now let us assume that Υ involves a fifth quark and its corresponding antiquark. We do not want it to be a left-handed singlet, for this would generally mean[57] that there would be flavor changing neutral currents - something on which there are stringent limits in the case of strangeness and charm.[4] So our fifth quark needs a partner, in order that it can be put in a left-handed doublet. We call these two new quarks t and b, with the Υ being either a $\bar{b}b$ or $\bar{t}t$ vector meson. There are now basically two alternatives: the six quarks are all in left-handed doublets and right-handed singlets, or in left-handed doublets and right-handed doublets. Assuming there are exactly six (and no more) quarks, we consider these possibilities in turn.

First, if the right-handed quarks are all singlets, then only the left-handed quarks are non-trivial. They are to be in doublets which can be written

$$\begin{pmatrix} u \\ d' \end{pmatrix}_L , \quad \begin{pmatrix} c \\ s' \end{pmatrix}_L , \quad \begin{pmatrix} t \\ b' \end{pmatrix}_L ,$$

where d', s', and b' are orthogonal mixtures of d, s, and b. The mixing can be parametrized by three real angles in this case (neglecting a complex angle[58] which leads to CP violation), and their values will completely fix how the quarks decay weakly.

There are again some important restrictions coming from experiment. The combination of muon decay (strength of the weak interaction), neutron decay and strange particle decays (strengths of $u\overset{W^\pm}{\longleftrightarrow}d$ and $u\overset{W^\pm}{\longleftrightarrow}s$, respectively) tells us that d' must be rather close to the $d\cos\theta + s\sin\theta$ of Cabibbo. The square of the coefficient of b (in d') is thereby limited[21] to be $\lesssim 0.004$. Furthermore, charm decays into strange particles imply that s' contains a non-trivial s component. There is also a more theoretical argument which tightly restricts the amount of d and s which can be together in b' by demanding that the K_1^0-K_2^0 mass difference turn out to be of the right magnitude.

Altogether, these arguments indicate that d', s' and b' are dominantly d, s, and b respectively, as the names would indicate. In fact, at the moment nothing rules out d' and s' being very close to the Cabibbo mixtures and b' being almost entirely b. In the limit of b' = b, particles containing the lighter of t and b become stable with respect to their weak decay!

The more likely scenario[21,59], however, is that there is some b mixed into s', and only a tiny bit in d'. Then if $m_t > m_b$, the t quark decays weakly mostly to the b quark, which then decays to the c quark. On the other hand, if $m_t < m_b$, we would have the b quark decaying weakly mostly to the t quark, which then decays weakly mostly to s. Either way, hadrons containing a b quark undergo two successive weak decays before the resulting hadrons contain only "old" quarks (u, d, and s). This might well provide a very characteristic two lepton decay signature for such hadrons, which would help greatly in their discovery. When produced in neutrino induced reactions this leads to various multilepton (≥ 3) final states, but the detailed rates depend crucially on the various mixing angles.[21,59]

Second, if the six quarks are in both left- and right-handed doublets, they must have the following form:

$$\begin{pmatrix} u \\ d' \end{pmatrix}_L , \begin{pmatrix} c \\ s' \end{pmatrix}_L , \begin{pmatrix} t \\ b' \end{pmatrix}_L$$

and

$$\begin{pmatrix} u \\ b'' \end{pmatrix}_R , \begin{pmatrix} c \\ s'' \end{pmatrix}_R , \begin{pmatrix} t \\ d'' \end{pmatrix}_R$$

Here the Cabibbo-like mixing angles for both the left- and right-handed quarks must be small. The situation leading to this for the left-handed quarks was just discussed. For the right-handed ones it follows from the limits on the strengths of $u_R \leftrightarrow d_R$, $u_R \leftrightarrow s_R$ and, less restrictively, $c_R \leftrightarrow d_R$ discussed in the case of four quarks. This forces u to be paired dominantly with b, and thence, c with s, leaving t and d, which by process of elimination form the third doublet.

But, given that either b or t is to have a mass of ~ 5 GeV (from the ϒ mass), the assignment to right-handed doublets given above leads to spectacular predictions for inelastic neutrino or antineutrino scattering at presently available energies. For, depending on which quark is in the ϒ, there should be a threshold above which there is large production of t(b) quarks by $\nu(\bar{\nu})$ on valence d(u) quarks in the nucleon. This will give:[60]

1. A rise in $\sigma_T(\nu N)/E_\nu$ $(\sigma_T(\bar{\nu}N)/E_{\bar{\nu}})$, or correspondingly, a rise in $\sigma_T(\nu N)/\sigma_T(\bar{\nu}N)$ $(\sigma_T(\bar{\nu}N)/\sigma_T(\nu N))$;

2. An "anomalous y distribution", behaving like $(1-y)^2$ for $\nu d_R \rightarrow \mu^- t_R$ and flat for $\bar{\nu}u_R \rightarrow \mu^+ b_R$;

3. A rise in dilepton events in $\nu(\bar{\nu})$ induced reactions[61] if hadrons involving t(b) have semileptonic branching ratios anything like those containing the c quark.

All these phenomena must happen together. While there is some experimental controversy[62] on 1 and 2 in the case of antineutrinos, there is no disagreement on 3 , where there is no indication of anything besides the production and decay of charm.[62]

It would seem that the second possibility of the six quarks in right-handed doublets is ruled out by experiment. Of course, one can avoid 1 , 2 and 3 at present energies by simply pushing the b and t quarks to an unaccessible mass, but then neither has anything to do with the ϒ. The other, easier, way out is to allow more than six quarks. Then the right-handed u ↔ b and/or d ↔ t pairings are no longer forced. The u_R, d_R, b_R and t_R could then be coupled mostly to still heavier quarks, and hence there would be no large neutrino or antineutrino production of the t or b off valence quarks. Whether nature chooses this rather peculiar pairing, seems unlikely, but we will have to wait and see.

4.2 Electromagnetic Decays

As with the weak interactions, we view the electromagnetic interactions of hadrons at the quark level. The current due to a quark with flavor index i is proportional to $Q_i\bar{u}_i\gamma_\mu u_i$, simply the Dirac current in space-time which is also diagonal in flavor. At the hadronic level such a current is capable of generating the variety of both electric and magnetic multipole transitions that are observed.

Between hadronic states that have internal quark angular momentum L=0, the electromagnetic current, taken as a sum of quark currents, has matrix elements which correspond to magnetic dipole transitions between the hadron states. These magnetic transitions include:

1. The static magnetic moments of the "stable" baryons;

2. The transition moments for $\Sigma \rightarrow \Lambda\gamma$ and $\Delta \rightarrow N\gamma$;

3. The transition moments for the decays of vector mesons to a pseudoscalar plus photon, e.g. $\omega \rightarrow \pi\gamma$, $\phi \rightarrow \eta\gamma$, $D^* \rightarrow D\gamma$.

We have a reasonable quantitative understanding of 1 , 2 , and 3 based on the quark level description of electromagnetic properties.[63] Among the "new" particles, rather dramatic exceptions to this are the decays $\psi \to X\gamma$ and $\psi' \to \chi(3455)\gamma$ if X(2830) and $\chi(3455)$ are identified as η_c and η'_c respectively. For then the observed magnetic dipole transitions from the vectors to corresponding pseudoscalars of the charmonium system are an order of magnitude too small when compared to predicted rates.[64]

For transitions from hadrons with internal $L' \neq 0$ to those with $L=0$ (e.g. $N^*(1520) \to N\gamma$, $\Delta^*(1950) \to N\gamma$), one has both electric and magnetic multipole transitions generated by the quark current. The structure of these amplitudes is understood in explicit quark models and in the more general framework of the transformation from constituent to current quarks, or Melosh transformation.[6,7] Both the signs and magnitudes of many amplitudes for $N^* \to \gamma N$ or $\Delta^* \to \gamma N$ are predicted correctly. For mesons, the best known transitions of this type are $\psi' \to \chi_J\gamma$. With heavy quarks and a non-relativistic situation these should be (related) electric dipole transitions for which $\Gamma(\psi' \to \chi_J\gamma) \propto (2J+1)k_\gamma^3$. Experiment is consistent with this, as well as giving absolute rates which agree with theory within a factor of two or better.[64]

4.3 Strong Decays

As is now widely recognized, strong interactions decays are of two rather distinct types, depending on whether the corresponding quark diagram is topologically connected ("Zweig allowed") or disconnected ("Zweig forbidden").[14] For meson decays, the requirement of having a connected quark diagram is equivalent to demanding that each quark line flow between two different mesons.

Processes corresponding to connected quark diagrams occur with typical strong interaction couplings and widths. Most of the hadron decays that one usually associates with the strong interactions, such as $\Delta \to \pi N$, $K^* \to \pi K$, and $\Sigma^* \to \bar{K}N$, are of this type.

Decays corresponding to disconnected quark diagrams do still occur, but with widths which are greatly suppressed. Among the old mesons we have $\phi \to \pi\rho$, suppressed in rate by a factor of order 10^2 while in the charmonium spectrum ψ decays are down a factor of 10^4 or more.

Considerable effort has gone into trying to understand the actual rates of "Zweig forbidden" decays quantitatively in terms of quantum chromodynamics, where one views the sum of these decays of a given state as occurring via annihilation of a quark and antiquark into gluons.[15] The quark and antiquark in a meson state with even charge conjugation can annihilate into a minimum of two gluons, whereas odd charge conjugation states result in a minimum of three (one is forbidden by color conservation). These gluons then dress themselves as hadronic matter in all possible ways with unit probability. The gluon couplings to the quarks are evaluated at a value of q^2 corresponding to the mass squared of the quark-antiquark hadronic state.[15]

This picture leads to a very clear systematics in the properties of "Zweig forbidden" decays:[15]

1. Widths should decrease with increasing mass, everything else being the same, since the square of the gluon coupling decreases (as $1/\log q^2$);

2. Odd charge conjugation states should have smaller widths than even charge conjugation ones because they decay via annihilation into more gluons, and hence the width involves another power of $\alpha_s(q^2)$ (the square of the gluon coupling divided by 4π) which is less than unity (at least for $q^2 \gtrsim 1$ GeV^2);

3. The absolute value of the widths can be used to compute $\alpha_s(q^2)$, provided we know the remaining factors in the decay rate, and compared with values extracted from knowledge of the quark-quark forces due to single gluon exchange and from asymptotic freedom corrections to deep inelastic scattering.

This systematics has had some success in the case of charmonium. Item 3 may be turned around and used to compute $\psi \to$ hadrons or $\psi' \to$ hadrons via "Zweig violating" processes. (Since the ψ is below $D\bar{D}$ threshold, it decays into "old" hadrons - hence the c and $\bar{c}$ quarks in the ψ must annihilate, yielding a disconnected quark diagram.) The resulting widths are comparable, considering the uncertainties in the parameters, with the experimentally measured ones.[15,64] Again, the apparently much larger decay widths into hadrons of the $\chi_0(3414)$ and $\chi_2(3552)$, with even charge conjugation, compared to the $\psi(3095)$, is explained by item 2. The $\chi_1(3508)$, although it has even charge conjugation, cannot annihilate into two massless gluons, so that its hadronic width, which is likely smaller than that of its L=1 companions χ_0 and χ_2, is understood.[64,65]

Unfortunately, there are also some major problems,[65] which revolve around the X(2830) and $\chi(3455)$, if these are η_c and η_c', respectively. For then the branching ratio for $\eta_c \to \gamma\gamma$ is $\gtrsim 0.5 \times 10^{-2}$, and that for $\eta_c' \to \gamma\psi$ is $\gtrsim 1/4$. This suggests total widths which are less than a few hundred keV for both η_c and η_c', and perhaps much smaller than that for η_c'. On the other hand, annihilation through two gluons fairly unambiguously predicts widths of several MeV for such even charge conjugation states. It remains to be seen whether this disagreement of theory and experiment represents a major flaw in the whole idea of explaining "Zweig forbidden" decays quantitatively in terms of QCD. Time, and probably the Υ system, will tell.

REFERENCES

* Work supported by ERDA.

1) See the reviews of deep inelastic scattering by F. J. Gilman, Proceedings of the XVII International Conference on High Energy Physics, J. R. Smith, ed. (Rutherford Laboratory, Chilton, Didcot, 1974), p. IV-149 and by C. H. Llewellyn-Smith, Proceedings of the 1975 International Symposium on Lepton and Photon Interactions at High Energies, W. T. Kirk, ed. (Stanford Linear Accelerator Center, Stanford, 1975), p. 709.

2) M. J. Alguard et al., Phys. Rev. Letters 37, 1261 (1976).

3) R. F. Schwitters, Proceedings of the 1975 International Symposium on Lepton and Photon Interactions at High Energy, W. T. Kirk, ed. (Stanford Linear Accelerator Center, Stanford, 1975), p. 5 and V. Lüth, topical conference of the SLAC Summer Institute on Particle Physics, July 11-22, 1977.

4) G. J. Feldman, topical conference of the SLAC Summer Institute on Particle Physics, July 11-22, 1977 and lectures in these proceedings.

5) A recent review of some aspects of exotics is found in R. Jaffe, topical conference of the SLAC Summer Institute on Particle Physics, July 11-22, 1977. Also see reference 22.

6) H. J. Melosh, Phys. Rev. D9, 1095 (1974).

7) Applications are reviewed in F. J. Gilman, Proceedings of the Summer Institute on Particle Physics, 1974, M. Zipf, ed., SLAC Report No. 179, Vol. 1, p. 307.

8) See the discussion of J. D. Bjorken, Proceedings of the Summer Institute on Particle Physics, 1975, M. Zipf, ed., SLAC Report No. 191, p. 85.

9) R. Blankenbecker, S. Brodsky and D. Sivers, Phys. Reports 23C (1976); S. D. Ellis and R. Stroynowski, Rev. Mod. Phys. 49, 753 (1977).

10) S. L. Adler, Phys. Rev. 177, 2426 (1969); J. S. Bell and R. Jackiw, Nuovo Cimento 51, 47 (1969).

11) Particle Data Group, Rev. Mod. Phys. 48, 51 (1976).

12) G. 't Hooft (unpublished); D. J. Gross and F. Wilczek, Phys. Rev. Letters 30, 1343 (1973); H. D. Politzer, Phys. Rev. Letters 30, 1346 (1973).

13) A review of quark confinement in QCD is given by L. Susskind, invited talk at the 1977 International Symposium on Lepton and Photon Interactions at High Energies, Hamburg, Germany, August 25-31, 1977.

14) S. Okubo, Physics Letters 5, 165 (1963); G. Zweig (unpublished); J. Iizuka, Supplement to Progress Theor. Physics 37-38, 21 (1966).

15) T. Appelquist and H. D. Politzer, Phys. Rev. Letters 34, 43 (1975); A. De Rujula and S. L. Glashow, Phys. Rev. Letters 34, 46 (1975).

16) The effect of the "color magnetic" interaction in hadron spectroscopy was first calculated in the MIT bag model by T. DeGrand et al., Phys. Rev. D12, 2060 (1975). See also T. A. DeGrand and R. L. Jaffe, Ann. Phys. (N.Y.)100, 425 (1976), T. A. DeGrand, Ann. Phys. (N.Y.)101, 496 (1976), and A. De Rujula et al., Phys. Rev. D12, 147 (1975). A review, particularly of work on charmonium masses is found in J. D. Jackson, Proceedings of the Summer Institute on Particle Physics, 1976, M. Zipf, ed., SLAC Report No. 198, p. 147.

17) See, for example, J. Ellis et al., Nucl. Phys. B111, 253 (1976). Also T. DeGrand et al., SLAC-PUB-1950, 1977 and references therein.

18) E. Eichten et al., Phys. Rev. Letters 34, 369 (1975).

19) S. W. Herb et al., Phys. Rev. Letters 39, 252 (1977); W. R. Innes et al., Phys. Rev. Letters 39, 1240 (1977).

20) S. Weinberg, Phys. Rev. Letters 19, 1264 (1967); A. Salam, in Elementary Particle Theory, E. Svartholm, ed. (Almquist and Wiksell, Stockholm, 1968), p. 367.

21) An extensive review of quarks and leptons and their weak interactions is found in H. Harari, lectures at the SLAC Summer Institute on Particle Physics, July 11-22, 1977. See references to other experimental and theoretical work therein.

22) R. Jaffe, Phys. Rev. D15, 267 and 281 (1977); R. Jaffe, MIT preprint MIT-CTP-657, 1977; see also reference 5.

23) L. B. Okun and M. B. Voloshin, JETP Letters 23, 333 (1976); A.De Rujula et al., Phys. Rev. Letters 38, 317 (1977). See also M. Bander et al., Phys. Rev. Letters 36, 695 (1976); C. Rosenzweig, Phys. Rev. Letters 36, 697 (1976); Y. Iwasaki, Prog. Theor. Phys. 54, 492 (1975).

24) J. Rosner, Phys. Rev. Letters 21, 950 (1968) and the review and references in Reference 47.

25) D. Horn and J. Mandula, Caltech preprint CALT-68-575, 1977.

26) H. Fritzsch and M. Gell-Mann, Proceedings of the XVI International Conference on High Energy Physics, J. D. Jackson and A. Roberts, eds. (National Accelerator Laboratory, Batavia, 1972), Vol. 2, p. 135; J. Kogut et al., Nucl. Phys. B114, 199 (1976); D. Robson, University of Liverpool preprint, 1977.

27) A neutral fermion "soul" for baryons was more popular before the advent of color when the spin-statistics theorem for quarks was otherwise in question. Related effects on the spectroscopy occur from the motion of a "bag" relative to the quarks: see A. Chodos et al., Phys. Rev. D9, 3471 (1974) and W. Bardeen et al., Phys. Rev. D11, 1094 (1975). Also, T. A. DeGrand and R. L. Jaffe, reference 16.

28) See R. Giles and S. H. Tye, SLAC preprint SLAC-PUB-1907, 1977.

29) G. Goldhaber et al., Phys. Rev. Letters 37, 255 (1976); I. Peruzzi et al., ibid. 569 (1976); G. Feldman et al., Phys. Rev. Letters 38, 1313 (1977); G. Goldhaber et al., Phys. Letters 69B, 503 (1977).

30) S. Yamada, invited talk presented at the 1977 International Symposium on Lepton and Photon Interactions at High Energies, Hamburg, Germany, August 25-31, 1977.

31) R. Brandelik et al., DESY preprint DESY 77/44, 1977; S. Yamada, Reference 30.

32) F. J. Gilman, Experimental Meson Spectroscopy-1974, D. Garelick, ed. (American Institute of Physics, New York, 1974), AIP Conference Proceedings, No. 21, p. 369.

33) R. Cashmore et al., Nucl. Phys. B92, 37 (1975).

34) Throughout these lectures we use the notation of putting a particles mass in MeV in parentheses after its name.

35) P. A. Rapidis et al., Phys. Rev. Letters 39, 526 (1977).

36) G. Brandenburg et al., Phys. Rev. Letters 36, 1239 (1976). See also Reference 42.

37) P. Estabrooks et al., SLAC preprint, SLAC-PUB-1886, 1977.

38) G. Cosme et al., Phys. Letters 67B, 231 (1977). See also Reference 40.

39) B. Esposito et al., Phys. Letters 68B, 389 (1977); C. Bacci et. al., Phys. Letters 68B, 393 (1977); G. Barbiellini et al., Phys. Letters 68B, 397 (1977). See also Reference 35.

40) F. Laplanche and C. Bemporad, invited talks at the 1977 International Symposium on Lepton and Photon Interactions at High Energies, Hamburg, Germany, August 25-31, 1977.

41) G. Brandenburg et al., Phys. Rev. Letters 36, 703 and 706 (1976); H. Otter et al., Nucl. Phys. B106, 77 (1976). See also R. K. Carnegie et al., Phys. Letters 63B, 235 (1976).

42) D. W. G. S. Leith, invited talk at the V International Conference on Experimental Meson Spectroscopy, Boston, April 28-30, 1977 and SLAC-PUB-1908, 1977. T. Lasinski, topical conference of the SLAC Summer Institute of Particle Physics, July 11-12, 1977 (unpublished).

43) R. Hemingway, topical conference of the SLAC Summer Institute on Particle Physics, July 11-22, 1977 (unpublished).

44) J. L. Basdevant and E. Berger, Argonne Preprint ANL-HEP-PR77-02, 1977; R. Aaron and R. S. Longacre, Phys. Rev. Letters 38, 1509 (1977).

45) G. Knies and M. L. Perl, invited talks presented at the 1977 International Symposium on Lepton and Photon Interactions at High Energies, Hamburg, Germany, August 25-31, 1977. J. Jaros, private communication.

46) See F. J. Gilman, Lectures at the SLAC Summer Institute on Particle Physics, July 11-22, 1977; and D. W. G. S. Leith, Reference 42.

47) See J. Rosner, Phys. Reports 11C, 189 (1974).

48) A. A. Carter et al., Phys. Letters 67B, 117 (1977) and references therein to previous work.

49) We employ the notation (C_0, C_1, S, T, ...) of M. K. Gaillard et al., Rev. Mod. Phys. 47, 277 (1977) or B. W. Lee et al., Phys. Rev. D15, 157 (1977) for charmed baryon states; but use as well Λ_c, Σ_c and Σ_c^*, where these indicate the states with the quark content of a Λ, Σ, and Σ^*, respectively with the s quark replaced by a c quark.

50) B. Knapp et al., Phys. Rev. Letters 37, 882 (1976).

51) E. G. Cazzoli et al., Phys. Rev. Letters 34, 1125 (1975).

52) The present state of baryon spectroscopy is reviewed by K. Lanius, Proceedings of the XVII International Conference on High Energy Physics, (JINR, Dubna, 1976), Vol. I, p. C45.

53) R. E. Cutkosky and R. E. Hendrick, Carnegie-Mellon University preprints COO-3066-81 and COO-3066-99, 1977.

54) Various aspects of weak decays are reviewed in M. K. Gaillard et al., Reference 49; J. D. Jackson, Reference 16; and H. Harari, Reference 21.

55) For a review and further references to recent developments in unified gauge theories, see S. Weinberg, invited talk at the VII International Conference on High Energy Physics and Nuclear Structure, Zurich, August 29-September 2, 1977 and Harvard preprint HUTP-77/A061, 1977.

56) M. Holder et al., Phys. Letters 70B, 396 (1977).

57) S. L. Glashow and S. Weinberg, Phys. Rev. D15, 1958 (1977). Also E. A. Paschos, Phys. Rev. D15, 1966 (1977).

58) M. Kobayashi and K. Maskawa, Prog. Theor. Phys. 49, 652 (1973); S. Pakvasa and H. Sugawara, Phys. Rev. D14, 305 (1976); L. Maiani, Phys. Letters 62B, 183 (1976); J. Ellis et al., Nucl. Phys. B109, 293 (1976).

59) J. Ellis et al., CERN preprint TH.2346-CERN, 1977.

60) A review of predictions of gauge theories of the weak and electromagnetic interactions and their comparison with experiment is found in R. M. Barnett, invited talk at the European Conference on Particle Physics, Budapest, Hungary, July 4-9, 1977 and SLAC-PUB-1961, 1977.

61) This was particularly emphasized by R. M. Barnett and F. Martin, SLAC preprint SLAC-PUB-1892, 1977. Also R. N. Cahn and S. D. Ellis, University of Michigan preprint, 1977.

62) See the reports for the Caltech, CDHS, HPWFR and BEBC neutrino experiments at the 1977 International Symposium on Lepton and Photon Interactions at High Energies, Hamburg, Germany, August 25-31, 1977.

63) See the discussion and experimental comparison by F. J. Gilman, Reference 46.

64) Aspects of charmonium spectroscopy, with a comparison between theory and experiment, is reviewed recently by J. D. Jackson, invited talk at the European Conference on Particle Physics, Budapest, Hungary, July 4-9, 1977 and CERN preprint TH.2351-CERN, 1977; K. Gottfried, invited talk at the 1977 International Symposium on Lepton and Photon Interactions at High Energies, Hamburg, Germany, August 25-31, 1977 and Cornell preprint CLNS-376, 1977; T. Appelquist, these proceedings.

65) M. Chanowitz and F. J. Gilman, Phys. Letters 63B, 178 (1976).

EXTENDED OBJECTS IN GAUGE FIELD THEORIES

G. 't Hooft

University of Utrecht

The Netherlands

1. INTRODUCTION AND PROTOTYPES IN 1+1 DIMENSIONS

With the exception of the electron, the neutrino, and some other elementary particles, all objects in our physical world are known to be extended over some region in space. Thus, the subject "extended objects" is too large to be covered in just four lectures. However, most of these extended objects may be considered as bound states of more elementary constituents. If we exclude all those, then a very interesting small set of peculiar objects remains: objects that cannot be considered as just a bunch of particles, but as some smeared, but also more or less localized, configuration of fields. These fields must obey very special types of field equations to allow such smeared lumps of energy ("solitons") to be stable.

These may be two alternative reasons for a lump of field-energy to be stable against decay:

i) A conservation law might tell us that the total mass-energy of the decay products would be larger than our object itself. It is easy to prove that this cannot happen in a <u>linear</u> (i.e. non interacting) theory: If electric charge is given by

$$Q = ie \int [\phi^* \partial_0 \phi - (\partial_0 \phi^*)\phi] \, d^3x \ , \tag{1.1}$$

and the energy by

$$E = \int (\partial_0 \phi^* \partial_0 \phi + \vec{\partial}\phi^* \vec{\partial}\phi + m^2 \phi^* \phi) \, d^3x \ , \tag{1.2}$$

then from

$$(\partial_o\phi^* \pm im\phi^*)(\partial_o\phi \mp im\phi) \geq 0 \ , \tag{1.3}$$

it follows that

$$E \geq \int [\partial_o\phi^*\partial_o\phi + m^2\phi^*\phi)\, d^3x \ \geq \ \frac{m|Q|}{e} \ , \tag{1.4}$$

whereas a bunch of charged particles at rest, with a total charge equal to Q, could have a total mass-energy not exceeding

$$m|\frac{Q}{e}| \ .$$

Consequently, a field configuration with $\vec{\partial}\phi \neq 0$ would not be stable by charge conservation alone. In interacting field theories however, one may construct many interesting stable lumps of fields simply because the above arguments do not apply when there are interactions. These have been considered by T.D. Lee and coworkers[1] and will not be treated in this series of talks.

ii) Topological stability. In these lectures we will only consider topologically stable structures. The statement that our objects are really field configurations implies that we consider field equations without bothering about the fact that small oscillations about the solutions ought to be quantized.

To explain topological stability we first turn our attention to models in one space - one time dimension[2]. Consider a simple scalar field ϕ satisfying

$$\partial_t^2\phi - \phi_x^2\phi + \frac{\partial}{\partial\phi} V(\phi) = 0 \ , \tag{1.5}$$

to which corresponds a Lagrangian:

$$\mathcal{L} = \frac{1}{2}(\partial_t\phi)^2 - \frac{1}{2}(\partial_x\phi)^2 - V(\phi) \ . \tag{1.6}$$

Eq. (1.5) corresponds to

$$\int \mathcal{L}[\phi+\eta]dxdt = \int \mathcal{L}[\phi]dxdt + \mathcal{O}(\eta^2) \ . \tag{1.7}$$

The two cases we consider are

a) $\quad V(\phi) = \frac{\lambda}{4!}(\phi^2 - F^2)^2 \quad$ (see Fig. 1) . $\qquad$ (1.8)

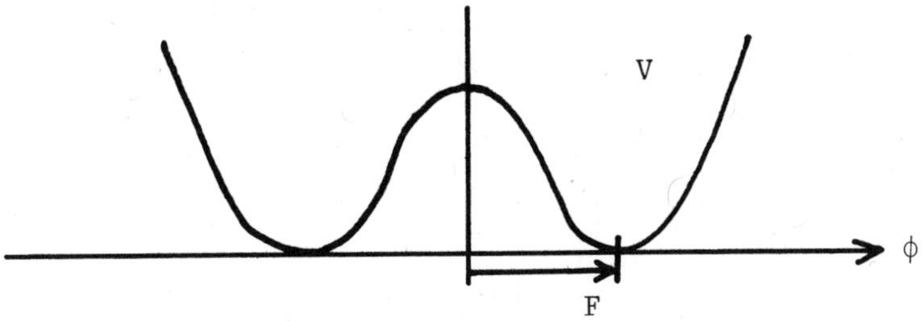

Fig. 1. Potential of the form $\lambda(\phi^2 - F^2)^2/4!$

It is customary then to write $\phi = F + \phi'$,

$$V = \frac{1}{2} m^2 \phi'^2 + \frac{g}{3!} \phi'^3 + \frac{\lambda}{4!} \phi'^4 , \qquad (1.9)$$

with

$$m^2 = \frac{1}{3} \lambda F^2 ; \qquad g = \lambda F . \qquad (1.10)$$

b) $V(\phi) = A(1 - \cos \frac{2\pi\phi}{F})$ ("Sine-Gordon" model- see Fig.2). (1.11)

Here we write

$$A = m^4/\lambda , \qquad F = \frac{2\pi m}{\sqrt{\lambda}} , \qquad (1.12)$$

so that

$$V = \frac{1}{2} m^2 \phi^2 - \frac{\lambda}{4!} \phi^4 + \dots \qquad (1.13)$$

In both cases m stands for the mass of the physical particle, and λ is the usual coupling constant-expansion parameter.

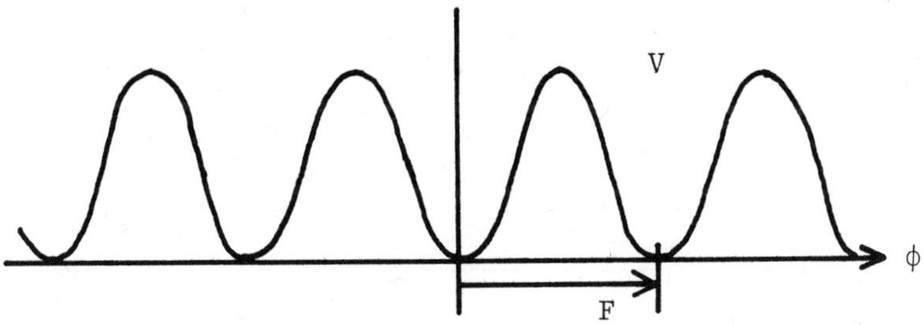

Fig. 2. Potential in "Sine-Gordon" model.

The energy of a field configuration is

$$H = \int \mathcal{H} d^3x \ ,$$
$$\mathcal{H} = \frac{1}{2}(\partial_t \phi)^2 + \frac{1}{2}(\partial_x \phi)^2 + V(\phi) \ . \tag{1.14}$$

The examples I gave here have the property that there is more than one way to make a zero-energy state[3] ("vacuum"):

$$\text{a)} \quad \phi = nF \ , \quad n = \pm 1 \ ,$$
$$\text{b)} \quad \phi = nF \ , \quad n = \ldots, -2, -1, 0, 1, 2, \ldots \tag{1.15}$$

The world in these models is assumed to be a one-dimensional line. We envisage the situation that at one side of the line ϕ has one value for n, at the other side a different value (see Fig. 3). We can have a transition region close to the origin which carries energy:

$$\phi \neq nF \ , \ n \text{ integer} \ ; \ \partial_x \phi \neq 0 \ . \tag{1.16}$$

Assume that the situation is stationary:

$$\partial_t \phi = 0 \ . \tag{1.17}$$

We wish a finite total energy; $\mathcal{H}$ must approach zero at $x \to \pm\infty$. So ϕ must approach one if its possible vacuum values at $x \to \pm\infty$. Thus we find a discrete set of allowable boundary conditions.

In a stationary solution of the equation (1.6) also the energy is an extremum. The lowest-energy configuration under the given boundary conditions also satisfies this equation. It is obviously stable. In our examples the solutions are easily given:

$$\partial_x^2 \phi = \partial V / \partial \phi \ , \tag{1.18}$$

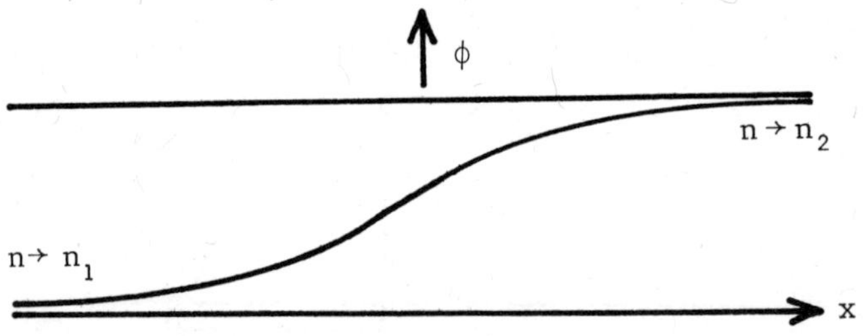

Fig. 3. A static form for ϕ with different boundary conditions at $x \to \pm\infty$.

$$\frac{d}{dx}\left[\frac{1}{2}(\partial_x\phi)^2 - V(\phi)\right] = 0 \ , \tag{1.19}$$

$$(\partial_x\phi)^2 = V(\phi) = 0 \quad \text{at} \quad x = \pm\infty, \tag{1.20}$$

therefore

$$\frac{dx}{d\phi} = [2V(\phi)]^{-\frac{1}{2}} \ . \tag{1.21}$$

Case a):

$$\phi(x) = F \tanh\frac{1}{2}mx \ , \tag{1.22}$$

Case b):

$$\phi(x) = \frac{2F}{\pi} \text{arctg } e^{mx} \ . \tag{1.23}$$

Note that, as $x \to \infty$,

$$F \tanh\frac{1}{2}mx \to F(1 - 2e^{-mx}) \ , \tag{1.24}$$

$$\frac{2F}{\pi} \text{arctg } e^{mx} \to F(1 - \frac{2}{\pi} e^{-mx}) \ , \tag{1.25}$$

thus the vacuum value is reached exponentially, with the mass of the physical particle in the exponent (see Fig. 3). The total energy of these objects ("solitons") is

$$E = \int\left[\frac{1}{2}(\partial_x\phi)^2 + V(\phi)\right]dx = 2\int V(\phi)dx \ , \tag{1.26}$$

Case a):

$$E = 2m^3/\lambda \ ,$$

Case b):

$$E = 8m^3/\lambda \ .$$

They behave as particles in every sense; for instance one can show that the relations between energy, momentum and velocity are as indicated by relativity theory.

We distinguish two types of stability requirements:

1) Topological stability. Since the boundary conditions in our case form a discrete set there can be no continuous transition

towards the vacuum boundary condition (which is the one for a bunch of ordinary elementary particles).

2) Stability against scaling. Let us return for a moment to n spacelike dimensions (instead of one). In the stationary case we always have

$$H = S + V\ , \quad S = \int \frac{1}{2}(\partial_x \phi)^2 d^n x\ , \quad V = \int V(\phi) d^n x\ . \tag{1.27}$$

The energy must be stationary under any infinitesimal variation. Let us try one special variation:

$$\phi(x) \rightarrow \phi(\Lambda x)\ , \tag{1.28}$$

then

$$S \rightarrow \Lambda^{n-2}\, S\ ; \quad V \rightarrow \Lambda^n V\ . \tag{1.29}$$

We must have

$$\frac{\Lambda\partial}{\partial\Lambda}\,(S + V) = (n-2)\,S + nV = 0\ . \tag{1.30}$$

But S and V are both positive. That is only compatible with (1.30) in the case $n = 1$. Since the decomposition (1.27) is possible in all scalar field theories, scalar theories have only stationary solitons in one space-dimension. In two or more dimensions it would always be energetically favorable for a system to shrink until it becomes pointlike. That is not an extended object by definition.

Finally, we emphasize that description of particles in terms of quantized fields in most cases is only possible if the coupling is not too strong:

$$\lambda/m^2 \ll 1 \tag{1.31}$$

(in 1+1 dimensions λ has the dimension of a mass-squared). Therefore, the mass of solitons, being proportional to m^3/λ, is always much greater than that of the original particles in the theory.

2. SOLITONS IN 2 SPACE DIMENSIONS AND STRINGS IN 3 SPACE DIMENSIONS

There is another reason why scalar theories have no topological solitons in 2 or more space dimensions. We can only have topological stability if the boundary condition differs from: $\phi \to$ constant at $|x| \to \infty$. Imagine that we have several field components $\vec{\phi}$ and assume that $V(\vec{\phi})$ has a whole continuum of minima. For instance we could take n field components (n= number of space-like dimensions) and have $V(\vec{\phi})$ = minimal if $|\vec{\phi}| = F$. We could think of mapping

$$\vec{\phi}(\vec{x}) \to F \frac{\vec{x}}{|\vec{x}|} \quad \text{at} \quad |\vec{x}| \to \infty , \tag{2.1}$$

as a boundary condition. This would imply topological stability. But then

$$|\vec{\partial}\vec{\phi}| \to F/|\vec{x}| \quad , \tag{2.2}$$

and

$$2S = \int |\vec{\partial}\vec{\phi}|^2 \, d^n x \to \int F^2 d^n x/x^2 \ , \tag{2.3}$$

diverges at large x unless n= 1. This infrared divergence is closely related to the existence of massless Goldstone bosons in such a theory.

The only way known at present to re-obtain a topologically stable soliton in more than one spacelike dimension is to add a gauge field:

$$\partial_i \vec{\phi} \to D_i \vec{\phi} = (\partial_i + gA_i^a T^a)\vec{\phi} \ . \tag{2.4}$$

The space-components of the vector field A can be arranged such that $(D\vec{\phi})^2 \to 0$ more rapidly than $1/x^2$, and this way we can approach a physical vacuum so rapidly that the total energy converges. (Note also that by adding a gauge field the massless Goldstone boson disappears, so the theory has better infrared convergence. There may still be a massless vector boson but that is usually less harmful, as we will see.)

The simplest theory with a soliton in 2 space dimensions is scalar quantum-electrodynamics in the Higgs mode[4,5]:

$$\mathcal{L} = -\frac{1}{4} F_{\mu\nu}F_{\mu\nu} - (\partial_\mu - iqA_\mu)\psi^*(\partial_\mu + iqA_\mu)\psi - \frac{\lambda}{2}(\psi^*\psi - F^2)^2 . \tag{2.5}$$

Here q is the unit of charge. The 3+1 dimensional analogue of this model is well known in physics: it describes the superconductor. ψ is the simplified field ("order parameter") that essentially describes two-electron bound states with total spin zero (hence it is a scalar boson field). A_μ is the ordinary vector potential. $q = 2e$.

Inside the superconductor the photon behaves as a massive particle and electromagnetic fields are of short range. Long-range magnetic fields are forbidden because in order to create them we need potential differences (according to Maxwell's laws) and those do not occur inside a superconductor. But when placed in a strong magnetic field a superconductor will find an excited state in which it can allow magnetic flux to penetrate. This flux will go through the superconductor in narrow tubes. The flux in such a tube turns out to be quantized. If we consider a plane orthogonal to these tubes then we get equations for extended particle-like objects in this plane. They are the 2 dimensional solitons to be considered now. The flux tubes ("vortices", "strings"), in the three dimensional world are derived from them by adding a third coordinate (parallel to the axis) on which the fields do not depend.

The vacuum is described by

$$|\psi| = F . \tag{2.6}$$

We construct our soliton just as in the previous section; we choose to approach a vacuum rapidly at $|\vec{x}| \to \infty$, but in such a way that continuity requires a transition region at $x \to 0$. Write

$$x_1 = r \cos \theta ,$$

$$x_2 = r \sin \theta ,$$

then we choose as $r \to \infty$, $\psi \to Fe^{i\theta}$.

Continuity requires that ψ produces a zero somewhere near the origin of x-space. Since the vacuum value of $|\psi|$ is F, it costs energy to produce such a zero: we will obtain a topologically stable lump of energy near the origin of x-space. Let us look at the lowest-energy configuration. Assume it is time-independent and $A_o = 0$. As a gauge condition we choose:

$$\partial_i A_i = 0 , \tag{2.7}$$

and, by symmetry arguments:

$$\begin{aligned} \psi &= \chi(r)e^{i\theta} \quad , \\ A_i &= \varepsilon_{ij}x_j A(r) \quad . \end{aligned} \tag{2.8}$$

One can easily check that the above is a self-consistent ansatz for a solution of the field equations.

To obtain the equation for χ and A it is easiest to reexpress the Lagrangian in terms of χ and A:

$$\int \mathcal{L}\, d^2\vec{x} = \int_0^\infty 2\pi r dr \left[-\frac{1}{2}r^2\left(\frac{dA}{dr}\right)^2 - \frac{1}{2}\left(\frac{d\chi}{dr}\right)^2 - \left(\frac{1}{r}+qAr\right)^2\chi^2 - \frac{\lambda}{2}(\chi^2-F^2)^2\right] - \pi r^2 A^2 \Big|_0^\infty \tag{2.9}$$

Assuming $|A_i(\infty)| \to 0$ we may ignore the boundary term. As for boundary conditions at $r \to 0$ we must require

$$\begin{cases} \chi \to 0 & , \\ A \text{ stays finite} & . \end{cases} \tag{2.10}$$

At $r \to \infty$ the third and fourth terms in (2.9) converge only if

$$\begin{cases} \chi \to F & , \\ A \to -1/qr^2 & . \end{cases} \tag{2.11}$$

Since our system is assumed to be stationary ($\partial/\partial t = 0$), and the time-components of A_μ vanish, the energy will be just minus the Lagrangian.

$$E = \int_0^\infty \mathcal{E}(r)dr \quad , \tag{2.12}$$

$$\mathcal{E}(r) = 2\pi r\left[\frac{1}{2}r^2\left(\frac{dA}{dr}\right)^2 + \frac{1}{2}\left(\frac{d\chi}{dr}\right)^2 + \left(\frac{1}{r} + qAr\right)^2\chi^2 + \frac{1}{2}\lambda(\chi^2 - F^2)^2\right] \quad . \tag{2.13}$$

We must find the minimum of this energy under the above boundary conditions. The coupled differential equations,

$$\frac{d}{dr}\left(\frac{\partial\mathcal{E}}{\partial(dA/dr)}\right)=\frac{\partial\mathcal{E}}{\partial A}\ ,$$
$$\frac{d}{dr}\left(\frac{\partial\mathcal{E}}{\partial(d\chi/dr)}\right)=\frac{\partial\mathcal{E}}{\partial \chi}\ , \tag{2.14}$$

are not exactly soluble. The solution is sketched in Fig. 4. As in the 1+1 dimensional case, we expect that the fields at large distance from the origin deviate only exponentially from the vacuum values, with the masses of the two massive particles (vector boson and Higgs particle) in the exponents. Thus we obtain a well behaved, topologically stable soliton, particle-like in 2+1 dimensions, string- or vortex-like in 3+1 dimensions.

What will the mass of this particle be? Without solving the equations explicitly, we can make some scaling arguments. Put

$$F^2 = m_H^2/2\lambda = m_W^2/2q^2\ , \tag{2.15}$$

where m_H and m_W are, respectively the Higgs and the vector boson masses, and

$$r=\bar{r}/m_W\ ,\ \chi=\bar{\chi}\,m_W/q\ ,\quad A=\bar{A}\,m_W^2/q\ ,\quad \lambda/q^2=\beta\ , \tag{2.16}$$

then

$$E=\frac{2\pi m_W^2}{q^2}\int \bar{r}d\bar{r}\left\{\frac{1}{2}\bar{r}^2\Big(\frac{d}{d\bar{r}}\bar{A}\Big)^2+\frac{1}{2}\Big(\frac{d}{d\bar{r}}\bar{\chi}\Big)^2+\Big(\frac{1}{\bar{r}}+\bar{A}\bar{r}\Big)^2\bar{\chi}^2 + \frac{1}{2}\beta\Big(\bar{\chi}^2-\frac{1}{2}\Big)^2\right\}$$
$$=\frac{2\pi m_W^2}{q^2}\,C_1(\beta)\ . \tag{2.17}$$

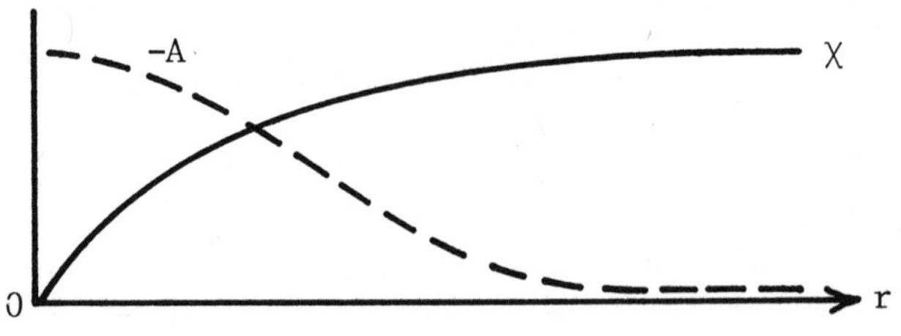

Fig. 4. Sketch of the solutions to Eq. (2.14).

Note the coupling constant downstairs. Of course, in 2+1 dimensions, q^2 has the dimension of a mass, hence m_W^2, not m_W. In 3+1 dimensions, the energy of a string is proportional to its length, and (2.17) gives the energy per unit of length, which is in units of mass-squared.

In the beginning of this section we anticipated that this vortex would be a magnetic flux tube. Let us now verify that by computing the magnetic field.

At spatial infinity the vector potential is found to be, from (2.11),

$$A_i \to -\varepsilon_{ij}x_j/qr^2 \quad . \tag{2.18}$$

Therefore

$$\oint A_i dx^i \to 2\pi/q \ . \tag{2.19}$$

So, indeed, we have a magnetic field going along the vortex with total flux $2\pi/q$, this in spite of the fact that the photon had become massive which would imply that electromagnetic fields are only short-range. There is no way of producing a broken fraction of that flux, therefore magnetic flux in a superconductor is quantized, by units

$$2\pi/q = \pi/e \ . \tag{2.20}$$

There is clearly also a physical reason for our vortex or string to be stable: the magnetic flux it contains cannot spread out in the superconductor and it is exactly conserved.

3. THE BOUNDARY CONDITION - HOMOTOPY CLASSES

The above was a pedestrian way to obtain the vortex in a superconductor. We will now reformulate the boundary condition more precisely in order to be able to produce more complicated objects later.

Remember we chose our boundary condition to be $\psi \to e^{i\theta}F$ at $r \to \infty$. We could have chosen

$$\psi \to e^{in\theta}F \ ;$$

every choice of n would yield a particular configuration with lowest energy, dependent on n. If there would have been any way to make a continuous transition from one n to another, without violating $|\psi| = F$ at large distance from the origin, then that would cause an instability for whatever state had the higher energy. But there is no way to make this continuous transition. We say that the different choices of n are homotopically different[2]. The boundary has the topological shape of a circle. (In N dimensions it is the S_{N-1} sphere.) The vectors

$$\begin{pmatrix}\psi_1\\ \psi_2\end{pmatrix} \quad \text{with} \quad |\psi| = F$$

also form a circle. Two continuous mappings of a circle onto a circle are called "homotopic" if one can continuously be transformed into the other. All classes of mappings that are homotopic to each other form a "homotopy class". For the mappings of a circle onto a circle, or any S_{N-1} sphere onto an S_{N-1} sphere, these homotopy classes are labeled by an integer n running from $-\infty$ to ∞.

In order to formulate the boundary condition in the general case we introduce two types of vacuum: a <u>supervacuum</u> is a region of space or space-time where all vector fields vanish and all scalar fields have their vacuum value: $|\vec{\phi}| = F$ and are pointing in a preassigned direction in isospace, for instance the z-direction:

$$\begin{pmatrix}\psi_1\\ \psi_2\end{pmatrix} = \begin{pmatrix}F\\ 0\end{pmatrix} \quad \text{or} \quad \begin{pmatrix}\phi_1\\ \phi_2\\ \phi_3\end{pmatrix} = \begin{pmatrix}0\\ 0\\ F\end{pmatrix}, \text{ etc.}$$

Now in a gauge theory, like electromagnetism, we are allowed to make gauge rotations; for instance in the model of the previous section:

$$\begin{cases} A_\mu \to A_\mu - \frac{1}{q}\partial_\mu \Lambda(x) , \\ \psi \to e^{i\Lambda(x)}\psi . \end{cases} \tag{3.1}$$

After this, in general space-time-dependent, gauge rotation we obtain non-vanishing vector-potentials and the scalar fields will point in arbitrary direction in isospace. Physically, we still have a vacuum. This we will call a normal vacuum, but no longer supervacuum.

Any vacuum configuration is defined by specifying the gauge rotation $\Omega(x,t)$ that produces it from the supervacuum. Sometimes, if there is no complete spontaneous symmetry breaking, Ω is only defined up to a space-time independent constant rotation. We will ignore that for a moment.

Returning to the model of the previous section, based on the gauge group U(1), the gauge rotations Ω are determined by a point on the unit circle, and so the general vacuum is determined by an angle that may vary as a function of space-time. Now, infinity in two dimensions is characterized by an angle θ (the direction in which we go to infinity), so the boundary condition is defined by a mapping

$$\Omega(\theta)$$

of the unit circle onto an angle θ. As we saw before, this mapping must be in one of a discrete set of homotopy classes, labeled by $-\infty < n < +\infty$:

$$\Omega(\theta) = e^{in\theta} \quad . \tag{3.2}$$

Because n must stay integer there cannot be continuous transitions from a state containing n solitons of the same type into a state containing fewer solitons. Physically this is again the statement that magnetic flux going through a two-dimensional surface is conserved.

Now we are in a position that we can easily generalize our model into a similar model but with a non-Abelian gauge field instead of electromagnetism: the non-Abelian superconductor. We assume that the gauge rotations (3.1) and (3.2) are replaced by real, orthogonal, 3×3 matrix rotation $\Omega(x)$, with determinant one. They form the non-Abelian group SO(3). We assume, that, as in the U(1) superconductor, a Higgs field or set of Higgs fields with integer isospin break the symmetry spontaneously and completely. Does this model allow for stable flux tubes as its Abelian counterpart does? Do we have solitons in two-dimensional space?

To analyse this question we merely have to investigate the homotopy classes of the mappings of SO(3) matrices in the periodic space of angles θ. It is well known that the group SO(3) is locally isomorphic with SU(2), the group of 2×2 unitary matrices with determinant one. The correspondence is made by identifying 3-vectors with traceless, 2×2 tensors. An SU(2) matrix acting on the tensor corresponds to an SO(3) rotation of the 3-vector.

The SU(2) matrices are easier to handle. They can be decomposed as

$$\begin{aligned} \Omega &= a_o + i \sum_{\ell=1}^{3} a_\ell \sigma_\ell \quad , \\ \Omega^\dagger &= a_o^* - i \sum_{\ell=1}^{3} a_\ell^* \sigma_\ell \quad . \end{aligned} \tag{3.3}$$

The requirements $\Omega\Omega^\dagger = 1$ and $\det \Omega = 1$ correspond to:

a_0, a_ℓ are real and

$$\sum_{\ell=0}^{3} a_\ell^2 = 1 \; . \qquad (3.4)$$

Clearly, the SU(2) matrices form an S_3 sphere (the surface of a sphere embedded in 4 dimensions).

The SO(3) matrices do not form an S_3 sphere, however. The reason is that an SU(2) rotation

$$\Omega = \begin{pmatrix} -1 & 0 \\ 0 & -1 \end{pmatrix}$$

on spinors leaves 2×2 tensors invariant. Consequently, two opposite points on the S_3-sphere correspond to the same SO(3) matrix (see Fig. 5). It is this feature that will allow a non-trivial homotopy class to emerge.

The homotopy classes here are entirely different from those in the U(1) model: any closed contour drawn on an S_3 sphere can be deformed continuously to a dot (see Fig. 6a), so the mappings of SU(2) onto the unit circle are characterized by only one homotopy class. But if we are only concerned with SO(3) matrices one may construct one different homotopy class: all contours that start on one point of the S_3 sphere and close by ending up at the point opposite to A (see Fig. 6b). These are the only two homotopy classes in this case, this in contrast with the situation in the ordinary superconductor where we had an infinity of homotopy classes.

What is the consequence for our SO(3)-superconductor? The existence of one non-trivial homotopy class implies that we can

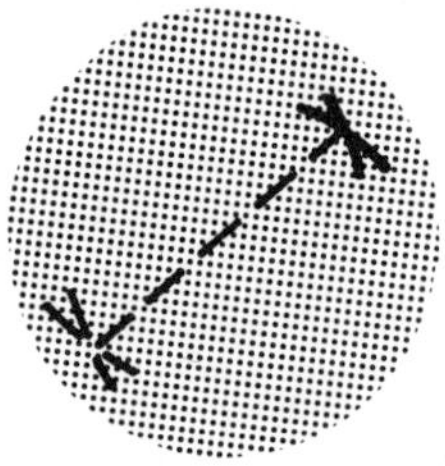

Fig. 5. The opposite points on S_3-sphere are identified.

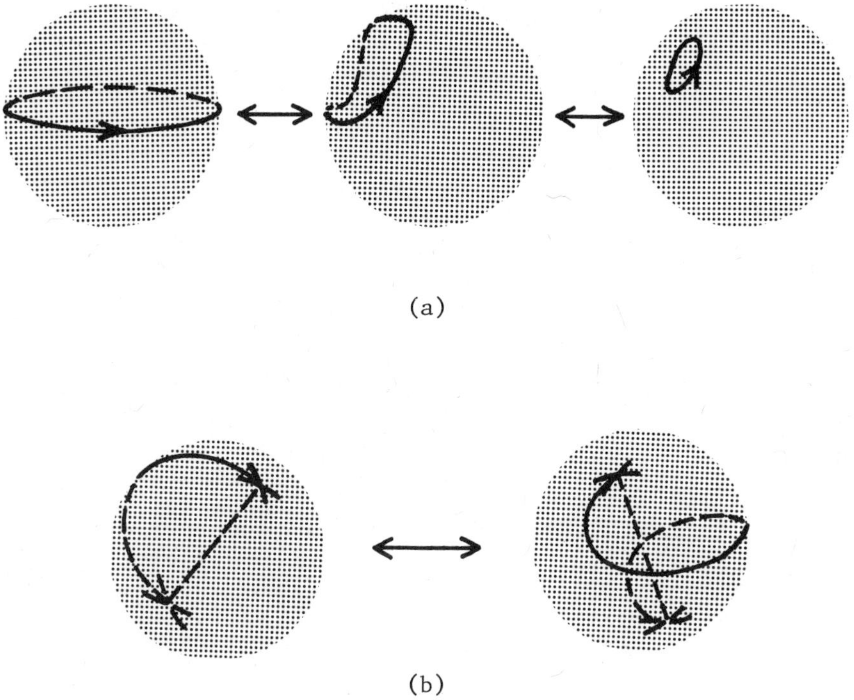

Fig 6. Two (and the only two) homotopy classes of closed paths for SO(3). Paths in (b) cannot be deformed into the paths in (a).

construct a non-trivial boundary condition for a stable soliton in 2 dimensions. So our model does admit a stable flux tube. However, now there is no longer an additive flux conservation law: if two of these solitons come together then the boundary of the total system is given by a contour in SO(3) space that goes through the S_3 sphere twice. That contour is in the trivial homotopy class, together with the boundary of a supervacuum. So the pair of solitons is not topologically stable which means they may annihilate each other. Thus, only single flux tubes are stable. Pairs of vortices, regardless their orientation, may annihilate each other, forming a shower of ordinary particles.

4. STRINGS AND MONOPOLES

Next, let us discuss a model very similar to the previous one. Again our gauge group is SO(3), but now we assume only one Higgs triplet, breaking it down to U(1). This is the so-called Georgi-Glashow model[6] introduced five years ago as a possible candidate for the weak interactions. It contains a massless U(1) photon, a massive charged "intermediate vector boson", and a neutral massive Higgs particle. Designed to avoid the necessity of neutral currents and charm, at the price of having to introduce heavy leptons, this model was doomed to become obsolete as a realistic description of observed weak interactions. However, being one of the very few truly unifying models it has some peculiar features which we now wish to focus on.

To create a situation exactly as in the previous model, we consider an ordinary superconductor[7] in a world whose high-energy behavior is described by the Georgi-Glashow model. The order parameter then provides for breaking of the residual U(1) symmetry.

According to the previous section this superconductor differs from the ordinary superconductor by the fact that two flux tubes pointing in the same direction may annihilate each other, because the gauge group is SO(3), not U(1). But physically this looks very odd, since one would expect magnetic flux to be conserved. The only possible explanation of the breaking up of (double) flux tubes is that a pair of magnetic monopoles is formed. They carry magnetic charge $\pm 4\pi/e$.

This way one is naturally led to accept the idea that magnetic monopoles occur in the Georgi-Glashow model. They are not confined to the superconductor, which we had only introduced for pedagogical reasons.

Let us now focus on the near vacuum surrounding this magnetic monopole. At one side, say the positive z-direction, we assume a double magnetic flux tube to emerge. Around this flux tube we do not have a supervacuum but a gauge rotation of that, which is in SU(2) notation:

$$\Omega(z \to +\infty, \phi) = \begin{pmatrix} e^{i\phi} & 0 \\ 0 & e^{-i\phi} \end{pmatrix} \tag{4.1}$$

where ϕ is the angle around the z-axis. At the other side is no magnetic flux coming in, so

$$\Omega(z \to -\infty, \phi) = \text{indep. } (\phi) \ . \tag{4.2}$$

According to homotopy theory no discontinuity in Ω is needed elsewhere (except at the location of the monopole, where no Ω is needed because it is not a vacuum configuration). Indeed we may construct Ω everywhere as follows:

$$\Omega(\theta,\phi) = \cos\frac{1}{2}\theta\begin{pmatrix} e^{i\phi} & 0 \\ 0 & e^{-i\phi}\end{pmatrix} + \sin\frac{1}{2}\theta\begin{pmatrix} 0 & i \\ i & 0\end{pmatrix} . \tag{4.3}$$

The Higgs field in this model is an isovector, which in a super-vacuum takes the form

$$\begin{pmatrix} 0 \\ 0 \\ F \end{pmatrix} . \tag{4.4}$$

In the surroundings of a single monopole it is

$$\Omega(\theta,\phi)\begin{pmatrix} 0 \\ 0 \\ F \end{pmatrix} = \frac{F}{|\vec{x}|}\begin{pmatrix} x \\ y \\ z \end{pmatrix} \tag{4.5}$$

where it is understood that the Ω of Eq. (4.3) has first been written in SO(3) notation.

It is clear that the boundary condition (4.5) for the Higgs field leaves a stable soliton at the center. It is also clear from the previous discussion that this soliton will carry a magnetic charge equal to $4\pi/e$.

We skip further discussion of the monopole in these notes because we have little to add to the existing literature[6,8,9]. Just note that, as was the case for other solitons, its mass is inversely proportional to the coupling parameter:

$$M = \frac{m_W C_2(\lambda/e^2)}{\alpha} , \qquad \alpha = e^2/4\pi \tag{4.6}$$

5. INSTANTONS

5.1. Construction

We have seen extended objects in one, two and three dimensions. None of the models of the previous sections, in which these objects occur are of direct interest in high energy physics. The models of interest are two different types of gauge theories that are most extensively studied these days: the SU(2) × U(1) gauge theory for the weak and electromagnetic interactions, "spontaneously broken" to U(1) by an isodoublet Higgs field[10,11], and a pure gauge theory based on SU(3) for the strong interactions[12].

The weak interaction gauge group has essentially a four component Higgs field because the Higgs doublet is complex. Therefore we expect a topologically stable object not in three but in four dimensions. Why? The boundary of a four dimensional world is topologically an S_3 sphere. Also the complex Higgs doublet

$$\begin{pmatrix}\phi_1\\ \phi_2\end{pmatrix} , \quad \text{with} \quad |\phi|^2 = F^2 ,$$

form an S_3 sphere. The mappings of S_3 spheres on S_3 spheres are again characterized by non-trivial homotopy classes, labeled by an integer n that can run from $-\infty$ to ∞. Thus we can construct boundary conditions for a four dimensional space under which configurations with n soliton-like objects are stable.

In contrast to the cases in lower dimensions, also pure gauge theories allow for topologically stable soliton-like objects in four dimensions. This is most easily understood when we consider the simplest non-Abelian group SU(2). Remember that one way to define a vacuum at the boundary of a region is to specify the gauge rotations $\Omega(\vec{x})$ that produce this vacuum out of the supervacuum. The gauge rotations Ω themselves form an S_3 sphere (see section 3), so here too we have the non-trivial homotopy classes at the boundary. In fact, the S_3 sphere of the gauge rotations Ω themselves, is the same as the S_3 sphere of the complex doublets

$$\begin{pmatrix}\phi_1\\ \phi_2\end{pmatrix} = \Omega\begin{pmatrix}F\\ 0\end{pmatrix} ,$$

so one can say that the 4 dimensional solitons in the pure gauge theory are essentially the same as those in the theories broken by Higgs isodoublets.

In order to distinguish our 4-dimensional objects from the 3-dimensional solitons and the strings and surfaces, we give them a new name: "instantons". The name emphasizes that if one of the four dimensions is time, then these objects occupy only a short time-interval (instant).

We will now argue that there is an important quantity that is sensitive to the presence of instantons:

$$G^a_{\mu\nu}\tilde{G}^a_{\mu\nu} \quad ,$$

with

$$\tilde{G}^a_{\mu\nu} = \frac{1}{2}\,\varepsilon_{\mu\nu\alpha\beta}G^a_{\alpha\beta} \quad , \tag{5.1}$$

where $G^a_{\mu\nu}$ is the usual covariant curl of the gauge vector field A^a_μ:

$$G^a_{\mu\nu} = \partial_\mu A^a_\nu - \partial_\nu A^a_\mu + g\varepsilon_{abc}A^b_\mu A^c_\nu \quad . \tag{5.2}$$

$G\tilde{G}$ is gauge invariant, and it is a pure divergence:

$$G\tilde{G} = \partial_\mu K_\mu \quad ,$$

$$K_\mu = 2\varepsilon_{\mu\nu\alpha\beta}A^a_\nu(\partial_\alpha A^a_\beta + \frac{g}{3}\,\varepsilon_{abc}A^b_\alpha A^c_\beta) \quad . \tag{5.3}$$

However, as one can verify easily, K_μ is not gauge invariant. Let us see what the consequence is of that.

Consider a region of space-time enclosed by a large S_3 sphere. It is irrelevant as yet whether the space-time is Euclidean or Minkowskian; we are only concerned about topological stability and not yet interested in any equation of motion for the fields, not even the space-time metric. Suppose we have a supervacuum at the surrounding sphere:

$$A^a_\mu = 0 \quad ,$$

therefore

$$K_\mu = 0 \quad . \tag{5.4}$$

Because of (5.4) at the boundary, and because of (5.3), we find by means of Gauss' law:

$$\int G^a_{\mu\nu}\tilde{G}^a_{\mu\nu}d^4x = 0 \tag{5.5}$$

regardless of what values the fields take inside the sphere.

Now we take a non-trivial homotopy class of gauge rotations to get a new vacuum boundary condition:

$$\Omega(x_1,\ldots,x_4) = \frac{x_4 - ix_i\sigma_i}{\sqrt{x_4^2 + (x_i)^2}} \quad . \tag{5.6}$$

If we write the vector field in an SU(2) notation,

$$A_\mu = -\frac{i}{2}\sigma_a A_\mu^a \ , \tag{5.7}$$

then the transformation law is

$$A'_\mu = \Omega A_\mu \Omega^{-1} - \frac{1}{g}\partial_\mu \Omega \cdot \Omega^{-1} \tag{5.8}$$

We find that the vacuum at the boundary has

$$A'^a_\mu = \frac{2}{g}\eta^a_{\mu\nu}\frac{x^\nu}{x^2} \ , \tag{5.9}$$

with

$$\begin{aligned} \eta^a_{\mu\nu} &= \varepsilon_{a\mu\nu} \quad \text{for} \quad \mu,\nu = 1,2,3 \ , \\ \eta^a_{\mu 4} &= -\delta_{a\mu} \quad \text{for} \quad \mu = 1,2,3 \ , \\ \eta^a_{4\nu} &= \delta_{a\nu} \quad \text{for} \quad \nu = 1,2,3 \ . \end{aligned} \tag{5.10}$$

Now although the field (5.9) is physically equivalent to the vacuum (for instance, $G^a_{\mu\nu}$ at the boundary vanishes), we nevertheless get a non-vanishing value for K_μ. This was possible because K_μ is not gauge-invariant:

$$K'_\mu = 16x_\mu/g^2|x|^4, \tag{5.11}$$

where

$$|x|^2 = x_4^2 + x_i^2 \quad .$$

If we now apply Gauss' law we find that inside the S_3 sphere $G_{\mu\nu}$ cannot vanish everywhere because

$$\int G\tilde{G}\, d^4x = \int \partial_\mu K_\mu d^4x = \int_{\text{boundary}} K_\perp \, d^3x = \frac{32\pi^2}{g^2} \quad . \tag{5.12}$$

Merely because the vacuum at our boundary is topologically distinct from the supervacuum we get a non-trivial physical field configuration inside.

Note that the result (5.12) holds true also if the SU(2) group considered would be embedded in a larger gauge group. This implies that instantons remain topologically stable even if the gauge group is enlarged. Remember the Abelian flux tubes that could become unstable if the U(1) group were part of a larger group; such a thing does not happen for instantons.

Now let us insert the field equations. It is easiest to consider first Euclidean space, where the boundary choice (5.6) has nice rotational properties. Is there a solution to the Euclidean field equations that approaches (5.9) at $x^2 \to \infty$ and stays regular at the origin? We try

$$A^a_\mu(x) = \frac{2}{g}\eta^a_{\mu\nu}x^\nu f(r) . \tag{5.13}$$

The field equation, $D_\mu G_{\mu\nu} = 0$, now reads

$$\frac{d^2}{(d \log r)^2}(r^2 f) = 4r^2 f - 12(r^2 f)^2 + 8(r^2 f)^3 . \tag{5.14}$$

This equation is scale-invariant, which makes it as easy to solve as the one-dimensional soliton, whose equation is translation-invariant. Besides the trivial solution, $f = 1/r^2$, we can have[13]

$$f(r) = \frac{1}{r^2 + \lambda} , \quad \lambda \text{ arbitrary} , \tag{5.15}$$

which has the required properties.

The total action for the solution is

$$S = -\frac{1}{4}\int G^a_{\mu\nu}G^a_{\mu\nu}d^4x . \tag{5.16}$$

Without actually computing S it is easy to derive an inequality for it. In Euclidean space $G^a_{\mu\nu}$ and $\tilde{G}^a_{\mu\nu}$ all have real components only. Therefore[13]

$$0 \leq (G^a_{\mu\nu} \pm \tilde{G}^a_{\mu\nu})^2 = 2G^a_{\mu\nu}G^a_{\mu\nu} \pm 2G^a_{\mu\nu}\tilde{G}^a_{\mu\nu} ; \tag{5.17}$$

$$S \leq \pm\frac{1}{4}\int G^a_{\mu\nu}\tilde{G}^a_{\mu\nu}d^4x \leq -\frac{8\pi^2}{g^2} , \tag{5.18}$$

because of (5.12), and we can choose either sign. For our solution,

$$G^a_{\mu\nu} = -\frac{4\lambda}{g}\frac{\eta^a_{\mu\nu}}{(x^2 + \lambda)^2} , \tag{5.19}$$

and we find

$$S = -8\pi^2/g^2 \ . \tag{5.20}$$

The inequality is saturated, because $G^a_{\mu\nu} = \tilde{G}^a_{\mu\nu}$ for our solution. Indeed, any solution of the (first order) equation $G_{\mu\nu} = \tilde{G}_{\mu\nu}$ automatically satisfies the second order equation $D_\mu G_{\mu\nu} = 0$ because $D_\mu \tilde{G}_{\mu\nu} = 0$ for any gauge field derived from a vector potential (5.2).

5.2. Euclidean Field Theory and Tunneling

A solution of the classical field equation in Euclidean space may seem to be of little physical relevance. What we really would like to consider are instanton events in Minkowski space. Consider again the S_3 boundary of a compact region in Minkowski space. The mapping $\Omega(x)$ at the boundary may be deformed a bit, as long as we stay in the same homotopy class. Thus, the instanton can be seen to correspond to a transition from a supervacuum at $t = -\infty$ (where $\Omega(\vec{x}) = 1$) to a gauge rotated vacuum at $t = +\infty$. There we can have

$$\Omega(\vec{x}) = \frac{\vec{x}^2 - 1 + ix^a\sigma^a}{\sqrt{(\vec{x}^2 - 1)^2 + \vec{x}^2}} \ . \tag{5.21}$$

At intermediate times there must be a region where $G_{\mu\nu} \neq 0$, because of Eq. (5.12). Because we have a vacuum at the beginning and at the end this field configuration cannot be a solution of the classical equations. Rather, it should be considered as a quantum-mechanical tunneling transition. How do we compute the amplitude of such a tunneling transition? It is instructive to compare this problem with the motion of a single particle through a one-dimensional potential well:

$$H = \frac{p^2}{2m} + V(x) \ , \tag{5.22}$$

in a region where $V(x) >> E$, assuming that Planck's constant is small. To first approximation

$$\frac{1}{|\psi|}\left|\frac{\partial\psi}{\partial x}\right| \approx \frac{1}{\hbar}\sqrt{2m(V-E)} \ . \tag{5.23}$$

The amplitude for a tunneling process is proportional to $e^{\tilde{S}}$ with

$$\tilde{S} = -\frac{1}{\hbar}\int_{x_a}^{x_b} \sqrt{2m(V-E)}\, dx \quad . \tag{5.24}$$

On the other hand, if a trajectory is in an energetically allowed region, $E > V$, then the wave function oscillates. The number of oscillations is given by

$$\frac{1}{2\pi\hbar}\int_{x_a}^{x_b} p dx = \frac{1}{2\pi\hbar}\int \sqrt{2m(E-V)}\, dx = \frac{1}{2\pi\hbar} R \quad . \tag{5.25}$$

R is related to the action integral:

$$R = \int p dx = \int p\dot{x} dt = \int (H+L)\, dt = \int (E+L)\, dt \quad .$$

If we normalize the energy to zero then

$$R = \int L dt = S \ . \tag{5.26}$$

This is the total action of a motion from x_a to x_b with given energy. Note that the tunneling process is given by (5.24) which is the same expression except that the sign of $V-E$ is flipped. The sign of the potential in:

$$m\frac{d^2x}{dt^2} = \frac{\partial V}{\partial x} \quad , \tag{5.27}$$

is also flipped if we replace t by it. If there is no classically allowed movement at zero energy from x_a to x_b at <u>real</u> times then there is an allowed transition for <u>imaginary</u> times. We find that the total action $\tilde{S}$ for this imaginary transition determines the tunneling amplitude[14]. In quantum field theory that corresponds to replacing Minkowski space by Euclidean space.

The physical tunneling amplitude from the supervacuum to a gauge rotated vacuum in a different homotopy class is governed to first approximation by the total action $\tilde{S}$ of a classical solution in Euclidean space. The instanton has, as we saw, $\tilde{S} = -8\pi^2/g^2$. So, we will find amplitudes proportion to

$$e^{-8\pi^2/g^2} \quad . \tag{5.28}$$

The superposition of the supervacuum and the gauge rotated vacua gives rise to different possibilities for the true ground state of Hilbert space, labeled by an arbitrary angle Θ. Discussion of this phenomenon is to be found in Refs. 15 and 16.

5.3. Symmetry Breaking through Instantons

The above holds for a gauge theory without fermions. Instantons there rearrange the ground state of the theory. They do not show up as special events[17].

A new phenomenon occurs if fermions are coupled to the theory. Consider some gauge-invariant fermion current J_μ. If there is a strong background gauge field $G^a_{\mu\nu}$, then there will be a vacuum polarization:

$$J_\mu \propto (G^a_{\mu\nu})^2 \quad . \tag{5.29}$$

That effect is computed in a triangular Feynman diagram (see Fig.7). One of the external legs denotes the space-time point x where $J_\mu(x)$ is measured. The other two legs are the two gauge photon external lines.

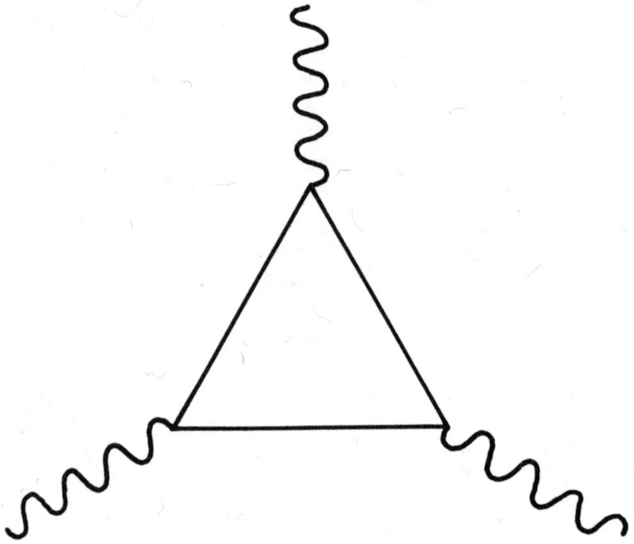

Fig. 7. Triangular anomaly diagram.

Let us assume, without yet specifying J, that it is conserved according to the Noether theorem. Then our diagram, which we could call

$$\Gamma_{\mu\alpha\beta}(q,k,k-q) \qquad (5.30)$$

should satisfy

$$q_\mu \Gamma_{\mu\alpha\beta} = 0 \quad . \qquad (5.31)$$

Gauge invariance further dictates that only transverse photons are coupled:

$$k_\alpha \Gamma_{\mu\alpha\beta} = 0 \quad ; \qquad (5.32)$$

$$(k-q)_\beta \Gamma_{\mu\alpha\beta} = 0 \ . \qquad (5.33)$$

Now the triangle diagram, when computed, shows an infinity that must be subtracted. It can happen that there are only two subtraction constants to be chosen. Eqs. (5.31)-(5.33) form three conditions for these two constants. They can be incompatible. In that case we have to keep gauge invariance in order to keep renormalizability. Consequently (5.31) is altered[18,19]. The result can be written as

$$\partial_\mu J_\mu = \frac{-ig^2 N}{16\pi^2} \cdot G^a_{\mu\nu} \tilde{G}^a_{\mu\nu} \quad . \qquad (5.34)$$

Here N is an integer determined by the details which were left out in my discussion. This is called the Adler-Bell-Jackiw-anomaly.

Now consider Eq. (5.12)[20]. We find that an integer number of charge units Q belonging to the current J_μ are consumed by the instanton:

$$\Delta Q = \int d^4x \ \partial_\mu J_\mu = 2N \ . \qquad (5.35)$$

In the strong-interaction gauge theory, a current J with this property is the chiral charge (total number of quarks with helicity + minus quarks with helicity -). Since the right hand side of (5.34) is a singlet under flavor SU(3) it is the ninth axial vector current which is nonconserved through the instanton. This probably explains why there is no SU(3) singlet (or SU(2) singlet) pseudoscalar particle as light as the pion[21,22].

Since the vector currents remain conserved, the instanton appears to flip the helicity of one of each type of fermion involved (see Fig. 8). This symmetry breaking effect of the instanton can be written in terms of an effective interaction Lagrangian. The original chiral symmetry was

$$U(N)_{\text{left}} \times U(N)_{\text{right}} \ . \tag{5.36}$$

It is broken into

$$SU(N)_{\text{left}} \times SU(N)_{\text{right}} \times U(1)_{\text{vector}} \ . \tag{5.37}$$

An effective Lagrangian that does this breaking is

$$\Delta\mathcal{L}^{\text{eff}} = C \exp\left(\frac{-8\pi^2}{g^2} + i\Theta\right) \det_{st} \left[\bar{\psi}_s (1+\gamma_5)\psi_t\right] + \text{h.c.} \tag{5.38}$$

Actually the effective interaction is more complicated when color indices are included[22], and C still contains powers of g. In the exponent, Θ is an arbitrary angle associated with the symmetry breaking.

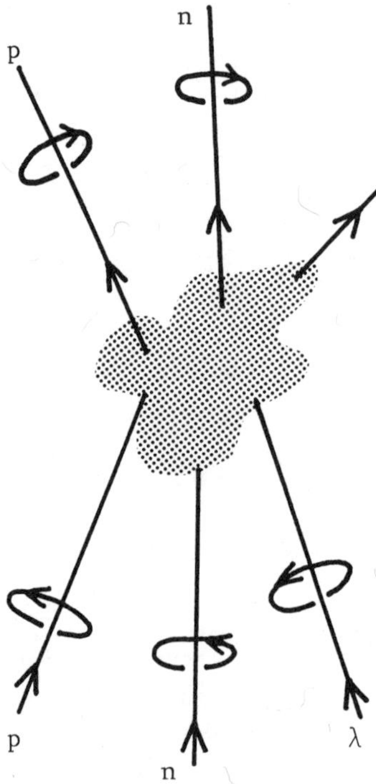

Fig. 8. The instanton appears to flip the helicity of one of each type of fermion involved.

Now let us turn to the instantons in the weak and electromagnetic guage theory SU(2) × U(1), with leptons, and non-strange, strange and charmed quarks which have color indices. This theory is anomaly-free, according to its commercials, as far as such is needed for renormalizability. Indeed, in triangle diagrams where all three external lines are gauge photons of whatever type, the anomalies (must) cancel. But if one of the external lines is one of the familiar conserved currents like baryon or lepton current (to which no known photons are coupled) then there are anomalies. We find, by inserting the anomaly equation (5.34) with the correct values for N, that one instanton gives

$$\Delta Q_E = 1 \quad \text{(electron-number nonconservation)} ,$$

$$\Delta Q_M = 1 \quad \text{(muon-number nonconservation)} ,$$

$$\Delta Q_N = 3 \quad \text{(non-strange quark nonconservation)} ,$$

$$\Delta Q_C = 3 \quad \text{(strange/charm quark nonconservation)} .$$

In total, two baryon and two lepton units are consumed by the instanton (strange and non-strange may mix through the Cabibbo angle). Thus we can get the decays

$$PN \to e^{\pm}\overset{\mp}{\nu}_\mu \quad \text{or} \quad \mu^{\pm}\overset{\mp}{\nu}_e ,$$

$$NN \to \bar{\nu}_e\bar{\nu}_\mu ,$$

$$PP \to e^+\mu^+ , \text{ etc.}$$

The order of magnitude of these decays is given by

$$[e^{-8\pi^2/g^2}]^2 = e^{-16\pi^2/e^2\sin^{-2}\Theta_W} = e^{-4\pi\cdot 137\cdot\sin^2\Theta_W} . \tag{5.39}$$

With a Weinberg angle $\sin^2\Theta_W = .35$, and assuming that the weak intermediate vector boson determines the scale, then we get lifetimes of the order of (give or take many orders of magnitude)

$$\tau \simeq 10^{225} \text{ sec} , \tag{5.40}$$

corresponding to one deuteron decay in 10^{137} universes.

6. SOME REFLECTIONS ON CLASSICAL SOLUTIONS AND THE QUARK CONFINEMENT PROBLEM

Clearly, the study of non-trivial classical field configurations gives us a welcome extension of our knowledge and understanding of field theory beyond the usual perturbation expansion. One non-perturbative problem is particularly intriguing: quark confinement in QCD. There has been wide-spread speculation that the new classical solutions are somehow responsible for this phenomenon. Some authors believe that some plasma of instantons or instanton-like objects does the trick. We will not go that far. We will however exhibit some simple models with interesting properties. They will clearly show that a "phase transition" towards permanent confinement is not at all an absurd idea (it is actually much harder to make models with "nearly confinement" of quarks).

Our first model is based on an SU(3) gauge theory in 3+1 dimensions. The gauge group here is neither "color" nor "flavor", so let us call it "horror" (or: "terror"), for reasons that will become clear later (the model does not describe the observations on quarks very well).

Let us assume that the SU(3) symmetry is spontaneously completely broken by a conventional Higgs field. The Higgs field must be a "non-exotic" representation of horror (mathematically: a representation of SU(3)/Z(3)), e.g. an octet or decuplet representation.

In the following we will show that this model admits, like the U(1), and the SU(2) analogues, string-vortices, but these are again different. After that we will introduce quarks and show that they bind to the strings in the combination that we actually see in hadrons[23].

To see the topological properties of vortices in this model we consider, as we did in section 3, a two dimensional space, enclosed by a boundary which has the topology of a circle. The vacuum at this boundary is specified by a gauge rotation $\Omega(\theta)$, so the number of topologically stable structures is given by the number of homotopy classes of the mapping $\theta \rightarrow \Omega(\theta)$. The group SU(3) has an invariant subgroup Z(3) given by the three elements I, $e^{2\pi i/3}I$ and $e^{-2\pi i/3}I$. Our physical fields have been required to be invariant under Z(3). So, the mappings with $\Omega(2\pi) = e^{\pm 2\pi i/3}\Omega(0)$ are acceptable. They form two homotopy classes besides the trivial one: $\Omega(2\pi) = \Omega(0)$. See Fig. 9. We conclude that there are stable vortices. Oppositely oriented vortices are different: their boundaries are in different homotopy classes. But two vortices oriented in the same direction are in the same class as one vortex in the other direction (see Fig. 10), because

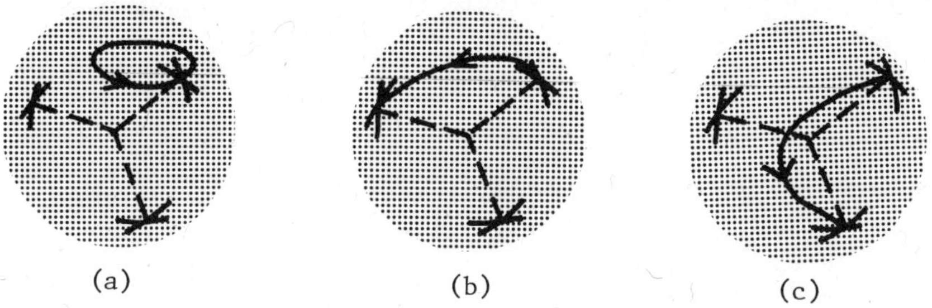

Fig. 9. Three homotopy classes of the mapping $\theta \to \Omega(\theta)$.

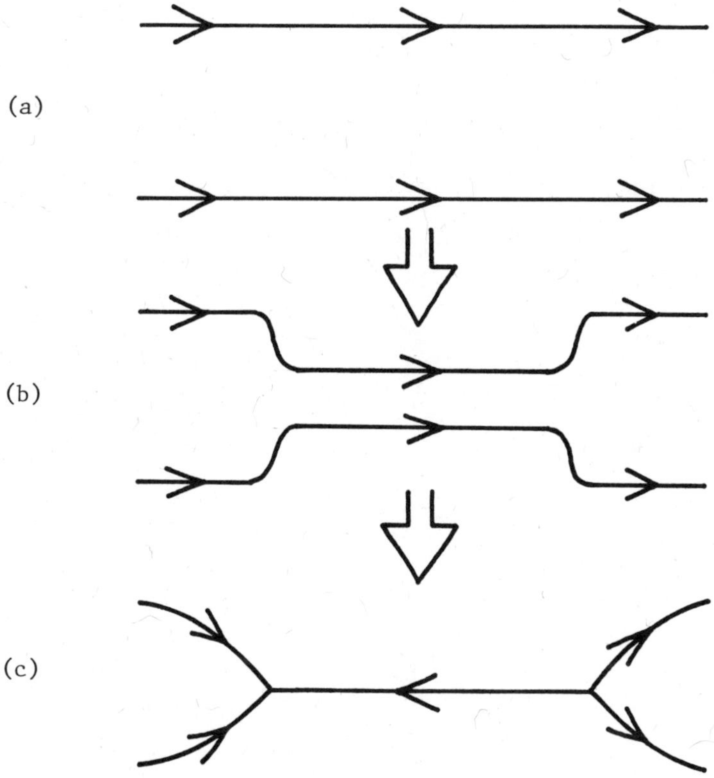

Fig. 10. Decay of two parallel vortices into a single vortex in opposite direction.

$e^{4\pi i/3} = e^{-2\pi i/3}$, so two parallel vortices will decay into an energetically more favorable single vortex in the other direction. Note that three-string connections are possible.

Quarks are now introduced as the end-points of strings. They could be compared with magnetic monopoles inside a superconductor. Note that quarks differ from antiquarks and because of the topological properties of these strings they only occur in the form of "hadronic" bound states (see Fig. 11).

We could go into lengthy details about actual construction of quark monopoles in the model, for instance by embedding SU(3)/Z(3) in a large gauge group that does not contain Z(3) itself. Then these monopoles themselves are allowed as classical solutions.

But the model is not very viable from a physical point of view because:

(i) The "horror" degrees of freedom are not known to exist; where are the "horror"-vector bosons etc.?

(ii) The Dirac monopoles will have monopole charges of the order of $2\pi/g_{horror}$. So, if quarks should behave as approximately free partons inside hadrons, g_{horror} should be so large that no perturbation expansion is likely to make sense, and

(iii) unlike monopoles (see section 4), quarks are light, relativistic, particles.

(iv) It isn't Q.C.D.

One thing on the other hand is very clear in this model: because the strings are topologically stable quarks are absolutely and permanently confined.

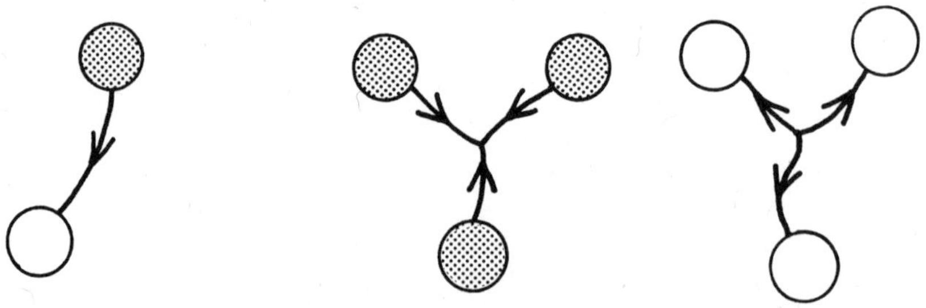

Fig. 11. Allowed quark and antiquark bound states.

The next model we consider is entirely different from the previous one. It is pure SU(N) color gauge theory (quantum-chromodynamics) but in 2 space - one time dimension. We claim that there is an elegant way to formulate the quark confinement mechanism here, although the dynamics is very complicated. For pedagogical reasons let us first add a set of Higgs fields to the system, which must be a representation of SU(N)/Z(N) as in the previous model. There is a complete spontaneous symmetry breakdown and obviously no quark confinement. Let us also leave out the quarks now. Since we have only 2 space dimensions the model has no flux tubes but instead soliton particles. The boundary of this space can be in one of N homotopy classes, so if we have more than N/2 solitons they may decay: the number of solitons minus anti-solitons is only conserved modulo N. We now make a very important step[24]: define an operator field $\phi(\vec{x},t)$ that destroys one soliton (or creates one anti-soliton) at $(\vec{x},t)$. There is not enough time now to formulate a precise definition of $\phi(\vec{x},t)$. It suffices to state that in particular the Green's functions

$$<\phi(0)\phi^{\dagger}(x)>_0 \tag{6.1}$$

and

$$<\phi(x_1)\phi(x_2) \ldots \phi(x_N)>_0 \tag{6.2}$$

can be defined accurately in terms of ordinary perturbation theory.[25] We can require that $\phi(\vec{x},t)$ produces or destroys only "bare" solitons: the vacuum (not supervacuum) remains at all $\vec{x}'$ with $|\vec{x}' - \vec{x}| \geq \varepsilon$ but a large field configuration is produced where $|\vec{x}' - \vec{x}| < \varepsilon$; then $\varepsilon \to 0$. Then (6.1) and (6.2) satisfy the Wightman axioms. We could assume that they may be more or less reproduced by an effective Lagrangian:

$$\mathcal{L}(\phi,\phi^*) = -\partial_\mu\phi^*\partial_\mu\phi - M^2\phi^*\phi - \frac{\lambda}{N!}(\phi^N + (\phi^*)^N) . \tag{6.3}$$

Here M is the calculable mass of the soliton. There will be many interaction terms; the one we wrote down is necessary to make (6.2) different from zero. Observe a Z(N) symmetry that leaves (6.1), (6.2) and (6.3) invariant:

$$\begin{aligned} \phi &\to e^{2\pi i/N}\phi , \\ \phi^* &\to e^{-2\pi i/N}\phi^* . \end{aligned} \tag{6.4}$$

Now what would happen if we vary the Higgs potential such that its second derivative at the origin turns from negative to positive? The field ϕ remains well defined. The soliton mass M^2 goes to zero if the Higgs mass goes to zero. It is therefore natural to assume that after the transition M^2 in (6.3) has changed sign. Our effective Lagrangian now could be

$$\mathcal{L}(\phi,\phi^*) = -\partial_\mu\phi^*\partial_\mu\phi - \frac{g}{2}(\phi^*\phi - F^2)^2 - \frac{\lambda}{N!}(\phi^N + (\phi^*)^N) \ . \qquad (6.5)$$

We imagine that then our Z(N) symmetry (6.4) will be spontaneously broken! We stress that we do not yet understand the details of the dynamics, but <u>if</u> this assumption is correct then the original model confines its quarks! Why? There are now N different vacuum states for the field ϕ:

$$\langle\phi\rangle_1 = e^{2\pi i/N}\langle\phi\rangle_2 = \dots \qquad (6.6)$$

If there are two different vacuum configurations in one plane then we obtain "Bloch walls" separating them. These Bloch walls can be considered to be closed strings. It is not hard to see that if we now add the quarks to the original model, then these will form end-points of the strings ("Dirac monopoles"). Since the strings carry energy per unit of length these quarks will be permanently confined by a linear potential.

We like to look at the transition from the SU(N) gauge theory to the Z(N) scalar theory as a dual transformation. The solitons in one theory are the elementary particles of the other. We note a peculiar antagonism: if the SU(N) symmetry is spontaneously broken, the Z(N) symmetry is intact and vice versa. After the dual transformation from the unbroken SU(N) theory to the broken Z(N) theory, quark confinement in the latter is obvious (once we accept the symmetry breaking antagonism).[26] What about 3+1 dimensional quantumchromodynamics? Our final speculation is that the "horror" model discussed in the beginning of this section is in a similar way the dual transform of quantumchromodynamics. The dual transformation in 3+1 dimensions is much more complicated than in 2+1 dimensions, but again, one can show that certain dual Green's functions corresponding to the dual gauge field may be constructed. The subject is obscured by the fact that we are not able to construct Lagrangians from the Green's functions, and that the horror coupling constant is inversely proportional to the color coupling constant so that it is impossible to compare the two perturbation series. However, if these dual transformations can be given a more sound footing then the quark confinement phenomenon will no longer look as mysterious as it does now.

REFERENCES

1) T.D. Lee and G.C. Wick, Phys. Rev. D9, 2291 (1974); T.D. Lee and M. Margulies, Phys. Rev. D11, 1591 (1975).

2) S. Coleman, in New Phenomena in Subnuclear Physics, International School of Subnuclear Physics "Ettore Majorana", Erice 1975, Part A, ed. A. Zichichi (Plenum Press, New York, 1977).

3) "State" here does not mean "state in Hilbert space" but rather a stationary classical field configuration.

4) H.B. Nielsen and P. Olesen, Nucl. Phys. B61, 45 (1973); H.B. Nielsen in Proceedings of the Adratic Summer Meeting on Particle Physics, Rovinj, Yugoslavia 1973, ed. M. Martinis et al. (North Holland, Amsterdam, 1974).

5) B. Zumino, in Renormalization and Invariance in Quantum Field Theory, NATO Adv. Summer Inst., Capri 1973, ed. E.R.Caianiello (Plenum Press, New York, 1974).

6) H. Georgi and S.L. Glashow, Phys. Rev. Lett. 32, 438 (1974).

7) For simplicity our order parameter is chosen to carry a single charge q=e, so that single flux tubes are quantized by units $2\pi/e$.

8) G. 't Hooft, Nucl. Phys. B79, 276 (1974); Nucl. Phys. B105, 538 (1976); A. Polyakov, JETP Lett. 20, 194 (1974).

9) B. Julia and A. Zee, Phys. Rev. D11, 2227 (1975).

10) S. Weinberg, Phys. Rev. Lett. 19, 1264 (1967).

11) G. 't Hooft, Nucl. Phys. B35, 167 (1971).

12) J. Kogut and L. Susskind, Phys. Rev. D9, 3501 (1974); K.G. Wilson, Phys. Rev. D10, 2445 (1974).

13) A.A. Belavin et al., Phys. Lett. 59B, 85 (1975).

14) S. Coleman, in The Why's of Subnuclear Physics, Erice lecture notes, Erice 1977.

15) R. Jackiw and C. Rebbi, Phys. Rev. Lett. 37, 172 (1976).

16) C. Callen, R. Dashen and D. Gross, Phys. Lett. 63B, 334 (1976).

17) It is to be remarked however that the instanton effects violate parity conservation if Θ is not a multiple of π. Parity violating events could then be searched for.

18) S.L. Adler, Phys. Rev. 177, 2426 (1969).

19) J.S. Bell and R. Jackiw, Il Nuovo Cimento A60, 47 (1969).

20) The factor i should be added in (5.12) since we are now dealing with a Minkowski metric where time components of vectors are taken to be imaginary.

21) H. Fritzsch, M. Gell-Mann and H. Lentwyler, Phys. Lett. 47B, 365 (1973).

22) G. 't Hooft, Phys. Rev. Lett. B37, 8 (1976), and G. 't Hooft, Phys. Rev. D14, 3432, (1976).

23) F. Englert, Lectures given at the Cargèse Summer School, July 1977.

24) S. Mandelstam, Phys. Rev. D11, 3026 (1975).

25) G. 't Hooft, to be published.

26) Compare the dual transformation in the 2 dimensional Ising model: the ordered phase is transformed into the disordered phase and vice versa.

CLASSICAL AND SEMI-CLASSICAL SOLUTIONS OF THE YANG-MILLS THEORY*†

R. Jackiw and C. Nohl

Massachusetts Institute of Technology, Cambridge, U.S.A.

C. Rebbi

Brookhaven National Laboratory, Upton, U.S.A.

This Review summarizes what is known at present about classical solutions to Yang-Mills theory both in Euclidean and Minkowski space. The quantal meaning of these solutions is also discussed. Solutions in Euclidean space expose multiple vacua and tunnelling of the quantum theory. Those in Minkowski space provide a semi-classical spectrum for a conformal generator.

1. INTRODUCTION

The strategy recently evolved for extracting new information from a quantum field theory consists first of ignoring the quantal character of the operator fields, and solving field equations as non-linear partial differential equations for classical functions. These solutions are then studied by various semi-classical approximation methods to yield information about the quantized theory[1].

Classical solutions can be conveniently characterized by their space-time dependence. Most familiar are time- and space-independent solutions; these are constants which satisfy the field equations. The quantum significance of such fields has been known for some time: when they describe stable configurations of finite [zero] energy, these constants are approximations to the vacuum expectation value of the quantum field and frequently signal spontaneous symmetry breaking.

Attention has been drawn also to time-independent but space-dependent solutions, such as the kink in one spatial dimension and the monopole in three. These - the solitons - are stable finite-energy field configurations and their quantal significance is known: classical soliton solutions signal the existence, in the quantum theory, of coherent bound states which describe heavy particles, the quantum solitons. The classical soliton energy is a weak coupling approximation to the quantum soliton's mass, while the Fourier transform of the classical soliton field approximates the matrix element of the quantum field between one-soliton momentum eigenstates.

Time- and space-dependent solutions are much harder to come by, simply because the non-linear partial differential equations are sufficiently complicated to prevent a complete analysis. A prominent exception is the sine-Gordon theory where time-dependent periodic and scattering multi-soliton solutions have been obtained. A semi-classical quantization provides information about the bound states and the S matrix for quantum solitons. Moreover, the model turns out to be sufficiently transparent so that a complete quantum-mechanical solution is possible. The exact results provide an important check on the approximate ones: the two agree where expected - for weak coupling. [In fact the WKB bound state spectrum turns out to be exact!] No such success has been achieved for realistic models in three dimensions. However most recently some very interesting space- and time-dependent solutions of the Yang-Mills equations have been found. Their quantum meaning is now being explored, and some of our [tentative] ideas regarding them will be presented in Section 4.

The study of the physics of solitons has exposed several fascinating effects, which should be briefly recalled. For weak coupling strength g, there are three scales of interaction strength. The interactions of the ordinary particles of the theory are weak, $O(g)$; the solitons however interact strongly $O(g^{-1})$; the interaction between solitons and ordinary particles is of intermediate strength, $O(g^{\circ})$. New types of conserved quantum numbers have been discovered. These insure the stability of the quantum soliton, but do not arise from local conservation laws of the Noether variety; rather they reflect topological properties of the field configurations. Furthermore, a startling phenomenon has been found: conversion of bosons to fermions, and correspondingly conversion of internal symmetry degrees of freedom into spin degrees of freedom. Finally, the coupling of Dirac fermions to the solitons has produced peculiar zero-energy bound states with profound effects on the theory.

While we have clearly learned that a quantum field theory gives rise to a much richer variety of phenomena than previously seen in perturbative Feynman-Dyson expansions, the fact remains that in theories which are presently intensely studied as possible candidates for a fundamental theory of natural processes - the Yang-Mills gauge models of strong interactions or of unified weak and electromagnetic interactions - no soliton solutions have been found. Indeed there are non-existence theorems which indicate that something different must be done, if one wants to apply semi-classical ideas to these models[2].

There is yet a further type of classical solution that can be considered; a solution not of the original equations, but rather of modified equations which are obtained by replacing time by imaginary time $t = x^0 \to -ix_4$ [and similarly changing the time components of all tensors]. What then is the quantal significance of these Euclidean fields? For an immediate answer recall that practical calculations in quantum field theory are most frequently performed in Euclidean space which is reached by a Wick rotation. Thus one may expect classical Euclidean fields to contain some information about the quantum theory. More precisely, one may formulate the quantum theory in Euclidean space by a functional integral or by an operator method, compute various semi-classical amplitudes, and continue back to Minkowski space. Moreover, there is a physical reason for finding Euclidean [imaginary-time] solutions. It is well-known that such classical solutions signal, in the corresponding quantum theory, the occurrence of tunnelling - i.e. there is motion which, though classically forbidden, is allowed quantum mechanically.

Imaginary-time solutions to pure Yang-Mills theory have indeed been found; in these lectures we review both the original example of Belavin, Polyakov, Schwartz and Tuypkin[3], and the later generalizations. The physical import of the original solution has been now established; it gives evidence of a rich non-perturbative structure to the quantum theory. The following description emerges.

For the classical SU(2) Yang-Mills theory in the gauge $A_a^0 = 0$, $a = 1, 2, 3$, the classical vacua - that is, the classical zero-energy configurations - are gauge potentials, which themselves are pure gauges. The gauge functions $g(\vec{r})$ are time-independent 2×2 unitary matrices; they need be characterized further to obtain a complete description of the classical vacuum. An important gauge function is the constant one, leading to a vanishing gauge potential. One need not consider those $g(\vec{r})$ which are singular functions of $\vec{r}$, nor those that do not tend to a constant for large $\vec{r}$, since the corresponding gauge potentials are separated by an infinite energy barrier from the vanishing one and are presumably irrelevant to the physical sector which includes the vanishing potential. The remaining gauge functions - those that do tend to a constant at spatial infinity - are mappings from 3-dimensional space [augmented by the point at infinity] to the SU(2) group, and as such can be arranged into homotopically inequivalent classes characterized by the integers $n = 0, \pm 1, \ldots$. Gauge functions belonging to distinct homotopy classes cannot be continuously deformed into each other. Pure gauge potentials constructed from the gauge functions can thus be classified by the integers, n. Gauge potentials within a class are gauge-deformable one into another by gauge transformation built from gauge functions of the trivial class, $n = 0$; these gauge transformations are called "small" gauge transformations. On the other hand two gauge potentials which belong to different classes can be gauge-deformed into each other only by gauge transformations built from gauge functions of some non-trivial class, $n \neq 0$; these are called the "large" gauge transformations.

In the quantum theory, there are distinct states $|n\rangle$ describing the gauge potential in each characteristic class n. No physical significance is given to the degrees of freedom associated with small gauge transformations. Indeed conventional gauge fixing procedures in the $A_a^0 = 0$ gauge are tantamount to imposing Gauss's law on the physical states and as a consequence the states $|n\rangle$ are invariant under small gauge transformations. Under the large gauge transformation $\mathcal{G}$, $|n\rangle$ transforms to $|n+1\rangle$, and a linear superposition must be taken to obtain a gauge invariant description for physical processes. However there is no requirement that the physical states be invariant under a large gauge transformation; gauge invariance is still achieved if the action of $\mathcal{G}$ on a physical state produces a phase. Therefore the linear superposition must be of the form $|\theta\rangle = \sum_n e^{in\theta}|n\rangle$ and $\mathcal{G}|\theta\rangle = e^{-i\theta}|\theta\rangle$.

To complete the description of the quantal low-lying states, we must ascertain whether the states $|\theta\rangle$ are degenerate in energy, or whether tunnelling splits them. It is here that the pseudo-particle solution becomes relevant. One notes that this solution interpolates, as its imaginary time parameter passes from $-\infty$ to ∞, between a vanishing gauge potential - one that evidently belongs to the $n = 0$ class - and a gauge potential belonging to the $n = 1$ class. We conclude that in the quantum theory there is tunnelling; the energy levels acquire a θ dependence and exhibit a band spectrum.

This Bloch wave picture is dramatically altered when massless fermions are included in the theory. The anomaly of the axial vector current renders the conserved chiral charge $\tilde{Q}_5$ gauge non-invariant under large gauge transformations. Specifically one finds $\mathcal{G}\tilde{Q}_5\mathcal{G}^{-1} = \tilde{Q}_5 + 2$. It is impossible to diagonalize simultaneous H, $\mathcal{G}$ and $\tilde{Q}_5$. Physical considerations require that energy eigenstates diagonalize $\mathcal{G}$; hence they are chirally non-invariant. Indeed chiral transformations shift θ; since they also commute with H, the θ dependence of energy eigenvalue disappears and tunnelling is suppressed.

Detailed calculations based on the above physical picture have been performed and were described by 't Hooft[4]. The original pseudoparticle solution is used in an approximate evaluation of the Euclidean functional integral. The physical significance of more general Euclidean solutions has not as yet been established; they appear to give insignificant corrections to the amplitudes described by 't Hooft. Nevertheless, we feel that it is important to undertake a detailed study of all solutions to Yang-Mills theory, for several reasons. Firstly, it is self-evident that any information about the theory will be helpful in establishing its physical content. Let us recall, especially, that computations of the dominant effects are not completely satisfactory since they suffer from uncontrollable infra-red divergences which reflect the infra-red instability of the theory. Secondly, our analysis of this system has put us in contact with parallel developments in pure mathematics. That there should be a conjunction of interests between modern mathematics and physics is truly a gratifying circumstance, and we are happy to be participating in it. The collaborative physical-mathematical efforts yield a new understanding of the axial-vector-current anomaly through the Atiyah-Singer index theorem. Also they give some hints that the Yang-Mills equations are considerably simpler than appear at first, and may even in some sense be linear. Further understanding of these possibilities shall certainly be substantial progress towards a solution of the theory.

2. PSEUDOPARTICLE CONFIGURATIONS

We begin these lectures by a study of solutions to the Yang-Mills field equations in Euclidean four-space. We shall consider an SU(2) gauge group and represent the potentials and field strengths as anti-hermitian matrices in the space of infinitesimal generators, with the gauge coupling constant e scaled out.

$$\frac{1}{e} A^{\mu} = A^{\mu}_{a} \frac{\sigma^a}{2i} \tag{2.1a}$$

$$\frac{1}{e} F^{\mu\nu} = F^{\mu\nu}_{a} \frac{\sigma^a}{2i} \tag{2.1b}$$

$$F^{\mu\nu} = \partial^{\mu} A^{\nu} - \partial^{\nu} A^{\mu} + [A^{\mu}, A^{\nu}] \tag{2.1c}$$

σ^a [a = 1, 2, 3] are the Pauli matrices, and summation over repeated indices is implied.

The Yang-Mills field equations

$$D^{A}_{\mu} F^{\mu\nu} = \partial_{\mu} F^{\mu\nu} + [A_{\mu}, F^{\mu\nu}] = 0 \tag{2.2}$$

follow from the requirement that the action S be stationary.

$$S = -\frac{1}{2}\int d^4x\,\mathrm{tr}F^{\mu\nu}F_{\mu\nu} \quad . \tag{2.3}$$

2.1. Topological Considerations

As in Ref. 3, we shall proceed by establishing a lower bound on S. The bound is saturated if the fields satisfy a set of first order non-linear differential equations, which of course imply the second order Eqs. (2.2). Throughout these Lectures we shall consider only solutions to the Yang-Mills equations which minimize the action by saturating its lower bound. No other solutions with finite action have been found in the Euclidean domain, and one may conjecture that the class we consider is exhaustive of all finite action solutions.

We define the dual $*F^{\mu\nu}$ of the field strength by the totally antisymmetric tensor $\varepsilon_{\mu\nu\rho\sigma}(\varepsilon_{1234} = 1)$.

$$*F^{\mu\nu} = \frac{1}{2}\varepsilon^{\mu\nu\rho\sigma}F_{\rho\sigma} \quad . \tag{2.4}$$

The inequality

$$-\frac{1}{4}\int d^4x\,\mathrm{tr}(F^{\mu\nu} \pm *F^{\mu\nu})(F_{\mu\nu} \pm *F_{\mu\nu}) \geqslant 0 \tag{2.5}$$

together with the algebraic identity $F^{\mu\nu}F_{\mu\nu} = *F^{\mu\nu}*F_{\mu\nu}$, establishes a lower bound on S.

$$S \geqslant \left|\frac{1}{2}\int d^4x\,\mathrm{tr}*F^{\mu\nu}F_{\mu\nu}\right| \quad . \tag{2.6}$$

We shall soon show that if S is finite the right hand side of this inequality does not depend on the detailed features of the field configuration, but only on general topological properties of the boundary values of the potentials A^{μ}. More precisely, it will emerge that the requirement of finite action separates all possible field configurations into equivalence classes of potentials which can be continuously distorted into each other. Within each class the quantity

$$q = -\frac{1}{16\pi^2}\int d^4x\,\mathrm{tr}*F^{\mu\nu}F_{\mu\nu} \tag{2.7}$$

takes a definite integer value, called the "Pontryagin index" of the field configuration. Postponing the proof of these statements for a moment, we see that the bound on the action

$$S \geqslant 8\pi^2 |q| \tag{2.8}$$

is saturated if

$$F^{\mu\nu} = \pm *F^{\mu\nu} \quad , \tag{2.9}$$

i.e., if the field is self-dual or anti-self-dual. Eq. (2.9) implies the Yang-Mills equations (2.2). This follows from $*F^{\mu\nu} = \frac{1}{2}\varepsilon^{\mu\nu\rho\sigma}F_{\rho\sigma}$ and the Bianchi identity satisfied by $*F^{\mu\nu}$,

$$D^A_\mu *F^{\mu\nu} = \partial_\mu *F^{\mu\nu} + [A_\mu, *F^{\mu\nu}] = 0 \quad . \tag{2.10}$$

What we learn here is that the self-duality or anti-self-duality condition is the equation for the absolute minimum of the action within a definite Pontryagin class. In the literature the self-dual field configurations with Pontryagin index $q = 1$ are often referred to as pseudoparticles [or instantons]; those with $|q|>1$ are called multi-pseudoparticle configurations.

The proof that q is a topological invariant, to which we now return, proceeds as follows. Notice first that $\mathrm{tr}*F^{\mu\nu}F_{\mu\nu}$ is a divergence

$$\mathrm{tr}*F^{\mu\nu}F_{\mu\nu} = \partial_\mu X^\mu \tag{2.11}$$

with

$$X^\mu = 4\varepsilon^{\mu\alpha\beta\gamma}\mathrm{tr}(\frac{1}{2} A_\alpha\partial_\beta A_\gamma + \frac{1}{3} A_\alpha A_\beta A_\gamma) \quad . \tag{2.12}$$

Assuming for the moment that A^μ has no singularities, the integral in Eq. (2.7) can be replaced by a boundary surface integral,

$$-\frac{1}{16\pi^2}\int d^4x\,\mathrm{tr}*F^{\mu\nu}F_{\mu\nu} = \lim_{R\to\infty} -\frac{1}{16\pi^2}\int_\sigma d^3\sigma n_\mu X^\mu \tag{2.13}$$

where σ is a surface enclosing a sphere of radius R and n^μ is its outward normal. The requirement that the action be finite is satisfied when $F^{\mu\nu}$ approach zero faster than $|x|^{-2}$ as $|x|\to\infty$. The field configuration then becomes integrable and one can find a unitary matrix g(x) such that, for $|x|\to\infty$.

$$A^\mu = g^{-1}\partial^\mu g + O(|x|^{-2}) \quad . \tag{2.14}$$

$g(x)$ defines a mapping from the points of any boundary surface σ into the manifold G of the SU(2) group. This mapping determines q; substituting Eq. (2.14) into Eqs. (2.12) and (2.13), we find

$$q = \lim_{R\to\infty} \frac{1}{24\pi^2} \int_\sigma d^3\sigma \varepsilon^{\mu\alpha\beta\gamma} n_\mu \mathrm{tr}\{(g^{-1}\partial_\alpha g)(g^{-1}\partial_\beta g)(g^{-1}\partial_\gamma g)\}$$

$$= \lim_{R\to\infty} \frac{1}{24\pi^2} \int_\sigma d^3\sigma \frac{\partial(g)}{\partial(\sigma)} = \lim_{R\to\infty} \frac{1}{24\pi^2} \int_G d^3 g \tag{2.15}$$

where d^3g is the invariant volume element of the SU(2) group. Thus we see that q counts how many times the volume of the group manifold G is covered by $g(x)$ as we let the argument x^μ span the whole surface at infinity. The fact that A^μ must approach a pure gauge at infinity is motivated by the requirement of finite action, and then it is obvious that no continuous deformation of A^μ preserving the finiteness of the action can modify the value of q.

The equivalence classes of potentials A^μ that can be continuously distorted into each other are in one-to-one correspondence with the equivalence classes of mappings $\sigma\to G$ which can be continuously related. Because σ is topologically equivalent to a three-dimensional sphere S^3, these classes are also in one-to-one correspondence with the elements of the third homotopy group of G.

A realization of the boundary conditions leading to $q = 1$ that will be very convenient for our analysis is obtained by demanding that as $|x|\to\infty$ $A^\mu \to g^{-1}\partial^\mu g$, with

$$g(x) = \frac{\alpha\cdot x}{\sqrt{x^2}} \qquad g^{-1}(x) = \frac{\bar{\alpha}\cdot x}{\sqrt{x^2}} \tag{2.16}$$

$$\alpha^\mu = (-i\vec{\sigma}, I) \qquad \bar{\alpha}^\mu = (\alpha^\mu)^\dagger = (i\vec{\sigma}, I) \quad . \tag{2.17}$$

Let us define

$$A_o^\mu = g^{-1}\partial^\mu g = -2i\sigma^{\mu\nu} x_\nu / x^2 \tag{2.18}$$

$$\sigma^{\mu\nu} = *\sigma^{\mu\nu} = \frac{1}{4i}(\bar{\alpha}^\mu \alpha^\nu - \bar{\alpha}^\nu \alpha^\mu) \ ; \ \sigma^{i4} = \frac{\sigma^i}{2} \ , \ \sigma^{ij} = \varepsilon^{ijk}\frac{\sigma^k}{2} \quad . \tag{2.19}$$

Inserting A^{μ} into Eq. (2.12) we find

$$X^{\mu} = - 8x^{\mu}/x^{4} \tag{2.20}$$

and

$$\lim_{R\to\infty} - \frac{1}{16\pi^{2}} \int_{\Omega} R^{3} d\Omega \, \frac{X^{\mu}x_{\mu}}{R} = 1 \tag{2.21}$$

which shows that a regular field A^{μ} approaching A_{o}^{μ} at infinity has $q = 1$.

This computation also shows that we must recognize an important point. The field strength associated with the pure gauge potential A_{o}^{μ} of course vanishes, so that q, for this case, must be zero. Written as a surface integral, q receives a contribution + 1 from the boundary term at infinity. This must be cancelled by some other surface contribution at finite x^{μ}. Indeed, A_{o}^{μ} is singular also at the origin, as is apparent from Eq. (2.18), and the contribution to q coming from a small surface enclosing the origin is - 1. We retain from this example the fact that the Pontryagin index, expressed as a surface integral over the group volume [Eq. (2.15)], may receive contributions from all the singularities of the gauge potential A^{μ}. This will be very relevant in the following.

2.2. One-Pseudoparticle Solution

From the discussion presented above we also see that a field configuration with $q = 1$ is obtained if A_{o}^{μ} is multiplied by a function $f(x^{2})$ such that $f(0) = 0$, $f(\infty) = 1$. Let us consider then

$$A^{\mu} = - 2if(x^{2}) \, \frac{\sigma^{\mu\nu}x_{\nu}}{x^{2}} \quad . \tag{2.22}$$

Symmetry considerations, which will be expanded later, suggest that the functional form of this Ansatz is compatible with the self-duality equation, leading to an equation for $f(x^{2})$. From Eq. (2.22) we evaluate

$$F^{\mu\nu} = 4i(f - f^{2}) \, \frac{\sigma^{\mu\nu}}{x^{2}}$$

$$+ \frac{4i}{x^{2}}(f' - \frac{f}{x^{2}} + \frac{f^{2}}{x^{2}})(\sigma^{\mu\rho}x_{\rho}x^{\nu} - \sigma^{\nu\rho}x_{\rho}x^{\mu}) \quad . \tag{2.23}$$

The matrix valued tensor $\sigma^{\mu\nu}$ is self-dual and the condition $F^{\mu\nu} = {}^*F^{\mu\nu}$ is satisfied if and only if

$$x^2 f' - f + f^2 = 0 \quad . \tag{2.24a}$$

This equation is solved by

$$f(x^2) = \frac{x^2}{\lambda^2 + x^2} \tag{2.24b}$$

with λ^2 being an arbitrary scale. The resulting field strength is

$$F^{\mu\nu} = \frac{4i\lambda^2}{(\lambda^2 + x^2)^2} \sigma^{\mu\nu} \quad . \tag{2.25}$$

If the function $f(x^2)$ is subject instead to the boundary conditions $f(0) = 1$, $f(\infty) = 0$, the ensuing field configuration will have $q = -1$. The anti-self-duality condition $F^{\mu\nu} = -{}^*F^{\mu\nu}$ can then be imposed and it is solved if

$$f(x^2) = \frac{\lambda^2}{\lambda^2 + x^2} \tag{2.26}$$

leading to

$$F^{\mu\nu} = \frac{4i\lambda^2}{(\lambda^2 + x^2)^2 x^2} [\sigma^{\mu\nu} x^2 - 2\sigma^{\mu\rho} x^\nu x_\rho + 2\sigma^{\nu\rho} x^\mu x_\rho] \quad . \tag{2.27}$$

Notice that the right hand sides of (2.24b) and (2.26) add to 1 and that the sum of the corresponding gauge potentials is the pure gauge A_0^μ.

Field configurations with $q = -1$, but without a singularity at the origin, or with $q = +1$, and with a singularity at the origin, can be obtained by replacing the matrices $\sigma^{\mu\nu}$ in Eqs. (2.18) to (2.27) by matrices $\bar{\sigma}^{\mu\nu}$ defined as

$$\bar{\sigma}^{\mu\nu} = -{}^*\bar{\sigma}^{\mu\nu} = \frac{1}{4i} (\alpha^\mu \bar{\alpha}^\nu - \alpha^\nu \bar{\alpha}^\mu) \quad ,$$

$$\bar{\sigma}^{i4} = -\sigma^{i4} \quad , \quad \bar{\sigma}^{ij} = \sigma^{ij} \quad . \tag{2.28}$$

The matrices $\bar{\sigma}^{\mu\nu}$ are obtained from $\sigma^{\mu\nu}$ by a parity inversion of the 4th axis and are anti-self-dual.

2.3. Symmetries of the One-Pseudoparticle Solution

The action density, $\mathcal{A}$, of the pseudoparticle

$$-\frac{1}{2}\,\mathrm{tr}F^{\mu\nu}F_{\mu\nu} = \frac{48\lambda^4}{(\lambda^2 + x^2)^4} \tag{2.29}$$

is spherically symmetric. This suggests that the field configuration itself may be symmetric under O(4) rotations. The expressions for A^μ and $F^{\mu\nu}$ are not explicitly symmetric; for instance, the right hand side of Eq. (2.25) remains invariant under an O(4) rotation, whereas $F^{\mu\nu}$ must transform as a second rank tensor. However, the apparent non-symmetry may be compensated by an appropriate gauge transformation. Let us consider a combined rotation in the Euclidean and SU(2) spaces, generated by

$$\Theta = \frac{1}{2}(M^{\mu\nu} + \sigma^{\mu\nu})\theta_{\mu\nu} \quad , \tag{2.30a}$$

where $M^{\mu\nu}$ denotes the operators that effect space rotations; i.e., $M^{\mu\nu}$ contains both an orbital component, which acts upon the position dependence of a field, and a spin component, which acts upon its tensor indices. $\sigma^{\mu\nu}$ are the matrices of Eq. (2.19); one shows that they [as well as the $\bar{\sigma}^{\mu\nu}$] obey commutation relations identical to those of $M^{\mu\nu}$. It is easy to verify that

$$[\Theta\, , A^\mu] = [\Theta\, , F^{\mu\nu}] = 0 \tag{2.30b}$$

for the fields of the pseudoparticle, which proves the rotational symmetry of the configuration. [The commutator notation is symbolic; it does not represent quantum mechanical commutation, but rather the infinitesimal action of the transformation.]

Notice that the group of combined O(4) and SU(2) transformations is a group of covariance of the self-duality equations; Eq. (2.22) gives the most general Ansatz invariant under these rotations. This explains why the Ansatz is compatible with the self-duality constraint.

Because the theory we are considering is covariant under the full O(5,1) group of conformal transformations, any conformal transformation of the pseudoparticle will still solve the self-duality equations. Of course, as we have just verified for the O(4) subgroup of rotations, some conformal transformations may not give a different solution, but just a gauge transform of the original one. We investigate now whether the group of symmetries of the pseudoparticle is actually larger than O(4).

The conformal group in Euclidean four-space has, as infinitesimal generators, the operators $M^{\mu\nu}$, P^{μ}, $K^{\mu} = \mathcal{I} P^{\mu} \mathcal{I}$ and D, which generate, respectively, rotations, translations, special conformal transformations and dilatations, with $\mathcal{I}$ the inversion operator. Of these, a dilatation changes the scale λ of the pseudoparticle and is thus not a candidate for an invariance. The effect on $\mathcal{S}$ of the remaining generators is readily determined

$$[P^{\mu}, \mathcal{S}] = \frac{8x^{\mu}}{\lambda^2 + x^2} \mathcal{S} \tag{2.31a}$$

$$[K^{\mu}, \mathcal{S}] = \frac{-8\lambda^2 x^{\mu}}{\lambda^2 + x^2} \mathcal{S} \tag{2.31b}$$

While it is clear that translations and special conformal transformations are not symmetries of the solution, it is apparent that $\mathcal{S}$ is invariant under the transformations generated by

$$R^{\mu} = \frac{1}{2} \left(\frac{1}{\lambda} K^{\mu} + \lambda P^{\mu}\right) \quad . \tag{2.32}$$

The commutators of R^{μ} and $M^{\mu\nu}$ close into the algebra of an O(5) subgroup of the conformal group.

The proof that the field configuration itself is also O(5) symmetric is slightly less trivial than for the O(4) subgroup; one must perform a space-dependent gauge transformation together with an O(5) conformal transformation to achieve form invariance of A^{μ} and $F^{\mu\nu}$. One verifies[5] that A^{μ} and $F^{\mu\nu}$ are symmetric under the combined conformal and gauge transformation generated by

$$\mathcal{R}^{\mu} = R^{\mu} + \frac{\sigma^{\mu\nu} x_{\nu}}{\lambda} \tag{2.33}$$

2.4. O(5) Formalism

The symmetry of the pseudoparticle under the O(5) subgroup of conformal transformations can be made manifest by an extension of the formalism,[5] which is very convenient for computational purposes and which we shall now illustrate.

The O(5) subgroup of conformal transformations can be realized as an ordinary group of rotations, if we introduce coordinates

r_a, $a = 1, \ldots, 5$, $r_a r^a = 1$, related to the Euclidean coordinates x^μ by a projective transformation.

$$r^\mu = \frac{2\lambda x^\mu}{\lambda^2 + x^2}$$

$$r^5 = \frac{\lambda^2 - x^2}{\lambda^2 + x^2} \quad . \tag{2.34}$$

Eqs. (2.34) effect a projection of Euclidean four-space onto the surface of an hypersphere, S^4, imbedded in a five-dimensional Euclidean space. The rotations of this hypersphere induce, via Eqs. (2.34), conformal transformations belonging to the O(5) group on the coordinates x^μ. Gauge fields A^μ [which can be considered as the components of a matrix valued form $A = A_\mu dx^\mu$] transform in a mapping of manifolds like the derivatives of a function [which are the components of the differential $df = \partial f/\partial x^\mu \, dx^\mu$]. Differentials over the surface of S^4 are conveniently expressed by tangential derivatives

$$df = \hat{\partial}_a f dr^a$$

$$\hat{\partial}_a = \frac{\partial}{\partial r^a} - r_a \left(r^b \frac{\partial}{\partial r^b}\right) \quad . \tag{2.35}$$

Correspondingly, we introduce a five component gauge field[6] $\hat{A}_a$, which is related to the usual four component gauge field A_μ by

$$A_\mu(x) = \hat{A}_a(r) \frac{\partial r^a}{\partial x^\mu} \tag{2.36a}$$

Eq. (2.36a) gives four relations; to specify the five-component object $\hat{A}_a$ completely, we need one more. This is clearly

$$r^a \hat{A}_a = 0 \tag{2.36b}$$

which insures that the gauge potentials are, like the derivatives, transverse. [The metric on the hypersphere is Euclidean; hence, there is no distinction between upper and lower indices.]

With $\hat{A}_a$ we construct covariant derivatives over the hypersphere

$$\hat{D}^{\hat{A}}_a = \hat{\partial}_a + \hat{A}_a \tag{2.37}$$

and the field strength is related as usual to the non-commutativity of these. We must, however, take into account the non-commutativity of the tangential derivatives themselves,

$$[\hat{\partial}_a, \hat{\partial}_b] - r_a\hat{\partial}_b + r_b\hat{\partial}_a = 0 \quad . \tag{2.38}$$

We therefore define $\hat{F}_{ab}$ as the value the left hand side of Eq. (2.38) takes when $\hat{\partial}_a$ is replaced by the covariant derivative $\hat{D}^{\hat{A}}_a$.

$$\hat{F}_{ab} = \hat{\partial}_a\hat{A}_b - \hat{\partial}_b\hat{A}_a - r_a\hat{A}_b + r_b\hat{A}_a + [\hat{A}_a, \hat{A}_b] \quad . \tag{2.39}$$

Notice that $\hat{F}_{ab}$ is anti-symmetric and tangential; it has six independent components.

The configuration of Eq. (2.22) is invariant under combined space and SU(2) rotations because the algebra of the $\sigma^{\mu\nu}$ matrices is isomorphic to the algebra of the rotation generators $M^{\mu\nu}$. The addition of $\sigma^{\mu\nu}$ to $M^{\mu\nu}$ in Eq. (2.30) produces, upon commutation with $\sigma^{\mu\nu}x_\nu$, terms which compensate the variation of the expression due to space rotations. This suggests that the O(5) symmetry might be made manifest if the gauge and tensorial degrees of freedom of the fields could be combined into an expression of the form $\Sigma_{ab}r^b$, where now the Σ_{ab} matrices obey the algebra of the generators of O(5) rotations. This algebra cannot be realized by 2×2 matrices, but we can still achieve a manifestly O(5) invariant expression for the pseudoparticle by the device of putting together a pseudoparticle and an anti-pseudoparticle in an extended gauge system.

Let us define 4×4 matrices

$$\Sigma^{\mu\nu} = \begin{pmatrix} \sigma^{\mu\nu} & 0 \\ 0 & \bar{\sigma}^{\mu\nu} \end{pmatrix} \tag{2.40}$$

and extend the SU(2) gauge theory to a theory with SU(2)×SU(2)$\approx$O(4) gauge group. The potentials and field strengths are represented by 4×4 matrices.

$$A^\mu = \frac{1}{i} A^\mu_{\alpha\beta}\, \Sigma^{\alpha\beta} \tag{2.41a}$$

$$F^{\mu\nu} = \partial^\mu A^\nu - \partial^\nu A^\mu + [A^\mu, A^\nu] \quad . \tag{2.41b}$$

The expression

$$A^{\mu} = \frac{-2i\Sigma^{\mu\nu}x_{\nu}}{\lambda^2 + x^2} \tag{2.42}$$

is block diagonal and contains, in the upper diagonal block, the SU(2) gauge field of the pseudoparticle, and in the lower diagonal block the field of the anti-pseudoparticle [c.f. Eqs. (2.22), (2.24) and the remarks preceding Eq. (2.28)]. The right-hand-side of Eq. (2.42) is neither self-dual nor anti-self-dual [although the individual blocks are] and has q = 0. It does, however, satisfy the Yang-Mills equations of motion. The advantage of having combined the pseudoparticle and anti-pseudoparticle into a single expression is that, within the space of 4×4 matrices, we can find a representation of the algebra of O(5) generators, and as we now proceed to show, the field configuration of Eq. (2.42) can be cast in a manifestly O(5) symmetric form.

We obtain a matrix representation of the O(5) algebra by enlarging the set of matrices $\Sigma^{\mu\nu}$ to a set Σ^{ab}, with the four new matrices $\Sigma^{\mu 5}$ given by

$$\Sigma^{\mu 5} = \frac{1}{2}\begin{pmatrix} 0 & -i\bar{\alpha}^{\mu} \\ i\alpha^{\mu} & 0 \end{pmatrix} . \tag{2.43}$$

The ten independent matrices Σ^{ab} have the commutation relations of the O(5) infinitesimal generators. In the hyperspherical formalism it is now possible to write down a field configuration, with O(5) gauge group, which is invariant under combined space and global gauge rotations; it is given by the <u>Ansatz</u>

$$\hat{A}_a = i\alpha\Sigma_{ab}r^b \tag{2.44}$$

where the only freedom is in the constant α. The corresponding field strength is

$$\hat{F}_{ab} = -i(\alpha^2 + 2\alpha)(\Sigma_{ab} - r_a r_c \Sigma_{cb} - \Sigma_{ac} r_c r_b) \tag{2.45}$$

and the Yang-Mills equations of motion $\hat{D}_a^{\hat{A}}\hat{F}^{ab} = 0$ are satisfied if $\alpha(\alpha + 1)(\alpha + 2) = 0$. The two solutions $\alpha = 0$ and $\alpha = -2$ correspond to pure gauge fields; $\alpha = -1$ gives a non trivial field configuration.

$$\hat{A}_a = -i\Sigma_{ab}\mathbf{r}^b \tag{2.46a}$$

$$\hat{F}_{ab} = i(\Sigma_{ab} + r_a \Sigma_{bc} r^c - r_b \Sigma_{ac} r^c) \quad . \tag{2.46b}$$

This expression has manifest O(5) symmetry. It remains for us to verify whether it describes the field configuration of Eq. (2.42). We notice first that the potentials A^μ of Eq. (2.42) belong to the subalgebra spanned by the O(4) generators $\Sigma^{\mu\nu}$, whereas $\hat{A}_a$ in Eq. (2.46a) involves all the Σ^{ab} matrices. If they describe the same system, it must be possible to gauge transform $\hat{A}_a$ to the O(4) gauge space, that is to gauge $\hat{A}^a_{\alpha 5}$ to zero. This can be done by the gauge transformation

$$\hat{A}_a \to \hat{A}'_a = U^{-1}\hat{A}_a U + U^{-1}\hat{\partial}_a U \tag{2.47}$$

where

$$U = \exp[i \frac{\cos^{-1} r_5}{\sqrt{1 - r_5^2}} \Sigma_{\alpha 5} r^\alpha] \quad . \tag{2.48}$$

If we now evaluate the explicit form of $\hat{A}'_a$ and insert it into Eq. (2.36a), we obtain precisely the A^μ field of Eq. (2.42). This proves that $\hat{A}_a = - i\Sigma_{ab} r^b$ does, indeed, describe the fields of a pseudoparticle and an anti-pseudoparticle.

As we have mentioned before, the hyperspherical formalism is very convenient for computations involving the field of the pseudoparticle. We illustrate how to evaluate the eigenvalues and eigenfunctions of the operator that describes the propagation of a massless scalar, iso-vector field coupled to the pseudoparticle[7]. For further application of the formalism to various systems of spinor and vector fields see Refs. 5, 7, 8 and 9.

We represent a scalar iso-vector field by anti-hermitian matrices in the space of group generators.

$$\phi = \phi_a \frac{\sigma^a}{2i} \tag{2.49}$$

and define a corresponding field $\hat{\phi}$ over the surface of the hypersphere by

$$\hat{\phi}(r) = \frac{\lambda^2 + x^2}{2\lambda} \phi(x) \quad . \tag{2.50}$$

The weight factor on the right-hand-side of this equation guarantees that the conformal transformations of the O(5) subgroup [in which

ϕ transforms as a field of dimension 1] are represented by scalar rotations of $\hat{\phi}$.

We have seen that to exploit the O(5) formalism it is necessary to imbed the SU(2) gauge group in a larger gauge group O(5). We therefore define a more general field

$$\hat{\Phi} = \frac{1}{i} \Phi_{ab} \Sigma^{ab} \quad , \tag{2.51}$$

Σ^{ab} being the matrices of Eqs. (2.40) and (2.43). We shall study the propagation of $\hat{\Phi}$ over the hypersphere, in the background provided by the pseudoparticle field $\hat{A}_a = - i\Sigma_{ab}r^b$. Notice that the gauge transformation U of Eq. (2.48) reduces $\hat{A}_a$ to a block diagonal form, indicating that the system consists of independent pseudoparticle and anti-pseudoparticle. We shall require that the field $\hat{\Phi}$ also reduce to block diagonal form after the gauge transformation

$$\hat{\Phi}' = U^{-1}\hat{\Phi}U$$

$$\hat{\Phi}'_{\alpha 5} = 0 \quad . \tag{2.52}$$

This condition insures that the upper and lower diagonal blocks of $\hat{\Phi}'$ describe scalar, iso-vector fields coupled to the pseudoparticle and anti-pseudoparticle, respectively.

The requirement that $\hat{\Phi}'$ be block diagonal can be expressed by the equation

$$[\hat{\Phi}', \Gamma^5] = 0 \quad , \tag{2.53}$$

where

$$\Gamma^5 = \begin{pmatrix} -I & 0 \\ 0 & I \end{pmatrix} \quad . \tag{2.54}$$

The matrices $\Gamma^\mu = 2i\Sigma^{\mu 5}\Gamma^5$ and Γ^5 transform as the five components of a vector Γ^a in rotations generated by the matrices Σ^{ab}. This implies that

$$U\Gamma^5 U^{-1} = \Gamma^a r_a \quad , \tag{2.55}$$

and Eq. (2.53) is therefore equivalent to the condition

$$[\hat{\Phi}, \Gamma^a r_a] = 0 \quad . \tag{2.56}$$

With these premises in mind, we now determine eigenfunctions and eigenvalues of the operator

$$\{(\hat{D}^{\hat{A}})^2 - 2\}\hat{\Phi} = \hat{\partial}_a\hat{\partial}^a\hat{\Phi} + \hat{\partial}_a[\hat{A}^a, \hat{\Phi}] + [\hat{A}_a, \hat{\partial}^a\hat{\Phi}]$$

$$+ [\hat{A}_a, [\hat{A}^a, \hat{\Phi}]] - 2\hat{\Phi} \tag{2.57}$$

with $\hat{A}_a = - i\Sigma_{ab}r^b$. After gauge transforming by U and projecting back to Euclidean space, the equation

$$\{(\hat{D}^{\hat{A}})^2 - 2 + \mu\}\hat{\Phi} = 0 \tag{2.58a}$$

takes the form

$$\{(D^A)^2 + \frac{4\lambda^2\mu}{(\lambda^2 + x^2)^2}\}\Phi = 0 \quad . \tag{2.58b}$$

As discussed in Refs. 8 and 10, Eq. (2.58b), although it contains space dependent coefficients, is more advantageous than the standard eigenvalue equation $\{(D^A)^2 + \mu\}\Phi = 0$ for the computations relevant to the pseudoparticle system. The equation (2.58a) has the useful property that, if $\hat{\Phi}$ is a solution with a definite eigenvalue μ, so, also, are

$$\hat{\Phi}_1 = \Gamma_a r^a\hat{\Phi} \quad , \tag{2.59a}$$

$$\hat{\Phi}_2 = \hat{\Phi}\Gamma_a r^a \quad . \tag{2.59b}$$

This can be verified directly or, more simply, by noticing that after the gauge transformation U all matrix valued fields appearing in $(\hat{D}\hat{A})^2$ take block diagonal form, and Eqs. (2.59) reduce to

$$\hat{\Phi}_1' = \Gamma^5\hat{\Phi}' = \begin{pmatrix}-I & 0\\ 0 & I\end{pmatrix}\hat{\Phi}' \tag{2.60a}$$

$$\hat{\Phi}_2' = \hat{\Phi}'\Gamma^5 = \hat{\Phi}'\begin{pmatrix}-I & 0\\ 0 & I\end{pmatrix} \quad . \tag{2.60b}$$

This shows in particular that it is consistent to constrain $\hat{\Phi}$ by (2.56). If $\hat{\Phi}$ satisfies this equation, the two projections

$$\hat{\Phi}_\pm = \frac{1 \mp \Gamma_a r^a}{2}\hat{\Phi} \tag{2.61}$$

describe the fields coupled to the pseudoparticle and anti-pseudo-particle, respectively.

Eq. (2.57) has a very simple group theoretical meaning, as we now proceed to show. Substituting into it the explicit form of $\hat{A}_a$, we obtain

$$\hat{\partial}_a \hat{\partial}^a \hat{\Phi} - 2i[\Sigma_{ab}, r^b \hat{\partial}^a \hat{\Phi}] - [\Sigma_{ab} r^b, [\Sigma^{ac} r_c, \hat{\Phi}]]$$

$$+ (\mu - 2)\hat{\Phi} = 0 \quad . \tag{2.62}$$

If we define

$$L_{ab} = - i r_a \hat{\partial}_b + i r_b \hat{\partial}_a \quad , \tag{2.63}$$

then

$$\hat{\partial}_a \hat{\partial}^a = - \frac{1}{2} L_{ab} L^{ab} \quad . \tag{2.64}$$

Moreover L_{ab}, Σ_{ab} and the operator J_{ab} where

$$J_{ab}\hat{\Phi} = L_{ab}\hat{\Phi} + [\Sigma_{ab}, \hat{\Phi}] \tag{2.65}$$

all satisfy the algebra of the infinitesimal generators of the O(5) group. Using the fact that $\hat{\Phi}$ is a linear combination of Σ_{ab} matrices it is straightforward to show

$$[\Sigma_{ab}, [\Sigma^{ab}, \hat{\Phi}]] = 12\hat{\Phi} \tag{2.66}$$

so that, from Eqs. (2.64), (2.65) and (2.66), we obtain

$$- \frac{1}{2} J_{ab} J^{ab} \hat{\Phi} = \hat{\partial}_a \hat{\partial}^a \hat{\Phi} - 2i[\Sigma_{ab}, r^b \hat{\partial}^a \hat{\Phi}] - 6\hat{\Phi} \quad . \tag{2.67a}$$

If $\hat{\Phi}$ satisfies Eq. (2.56), it must be of the form $\hat{\Phi} = 1/i \, \hat{\Phi}_{ab} \Sigma^{ab}$ with $r_a \hat{\Phi}^{ab} = \hat{\Phi}_{ab} r^b = 0$, which implies

$$[\Sigma_{ab} r^b, [\Sigma_{ac} r^c, \hat{\Phi}]] = 2\hat{\Phi} \quad . \tag{2.67b}$$

Putting all these results together, we see that Eq. (2.58) is equivalent to

$$\frac{1}{2} J_{ab} J^{ab} \hat{\Phi} = (\mu + 2)\hat{\Phi} \tag{2.68}$$

i.e., $\hat{\Phi}$ must be an eigenfunction of the operator

$$C^{(1)} = \frac{1}{2} J_{ab} J^{ab} \qquad (2.69)$$

which is the first Casimir operator of the O(5) group generated by J_{ab}. If λ is the eigenvalue of $C^{(1)}$, then $\mu = \lambda - 2$.

The eigenfunctions of $C^{(1)}$ are easily found as follows. Let us consider any tensor harmonic $Y^{(\lambda,\lambda',m)}_{a_1\ldots a_N}(\omega)$ of the O(5) rotation group [λ and λ' are the eigenvalues of the two Casimir operators of O(5), m stands for the "magnetic" quantum numbers, ω denotes the hyperspherical angles and $a_1\ldots a_N$ are tensor indices], and let us saturate all the indices $a_1\ldots a_N$ with Γ^a matrices to form the covariant

$$\mathcal{Y}^{(\lambda,\, \lambda',\, m)} = Y^{(\lambda,\, \lambda',\, m)}_{a_1\ldots a_N} \Gamma^{a_1} \ldots \Gamma^{a_N} \quad . \qquad (2.70)$$

Then it is straightforward to show that

$$J_{ab}\, \mathcal{Y}^{(\lambda,\, \lambda',\, m)} = (L_{ab} + S_{ab})\, \mathcal{Y}^{(\lambda,\, \lambda',\, m)} \quad , \qquad (2.71)$$

where the spin operators S_{ab} act on the tensor indices of $Y^{(\lambda,\lambda',m)}_{a_1\ldots a_N}$. If follows that all the covariants $\mathcal{Y}^{(\lambda,\lambda',m)}$ will be eigenfunctions of the operator $\frac{1}{2}J_{ab}J^{ab}$ with eigenvalue λ. A complete set of eigenfunctions for the expansion of an arbitrary field $\hat{\Phi}$ is given by the covariants $\mathcal{Y}^{n,m}_i$, which are linear combinations of the $\mathcal{Y}^{(\lambda,\lambda',m)}$.[7]

$$\mathcal{Y}^{n,m}_1 = i\Sigma^{ab} L_{ab} Y^{n,m}$$

$$\mathcal{Y}^{n,m}_2 = \Sigma^{ab} r_a \mathcal{Y}^{n,m}_b$$

$$\mathcal{Y}^{n,m}_3 = \Sigma^{ab} \hat{\partial}_a \mathcal{Y}^{n,m}_b$$

$$\mathcal{Y}^{n,m}_4 = i\varepsilon^{abcde}\, \Sigma_{ab} L_{cd} \mathcal{Y}^{n,m}_e \quad , \qquad (2.72)$$

where $Y^{n,m}$ is a scalar hyperspherical harmonic, with $\lambda = n(n + 3)$, and $\mathcal{Y}^{n,m}_a$, $n > 1$, are vector hyperspherical harmonics, as given by Adler,[6] with $\lambda = (n + 1)(n + 2)$. It is easy to check that the covariants $\mathcal{Y}^{n,m}_1$ do not satisfy Eq. (2.56), that the covariants $\mathcal{Y}^{n,m}_4$ do satisfy the constraint, and that only a definite linear combination of the covariants $\mathcal{Y}_2^{n,m}$ and $\mathcal{Y}_3^{n,m}$ does. This linear combination is most easily found by multiplying $\mathcal{Y}_4^{n,m}$ by $\Gamma_a r^a$. From

$$\Gamma_f \Sigma_{ab} = \frac{1}{2i} (\delta_{af}\Gamma_b - \delta_{bf}\Gamma_a) - \frac{1}{2} \varepsilon_{abfcd}\Sigma^{cd} \tag{2.73}$$

we obtain $\Gamma_a r^a \mathcal{Y}_4^{n,m} = 2(\mathcal{Y}_2^{n,m} - \mathcal{Y}_3^{n,m})$.

We conclude that the matrix valued fields $\hat{\Phi} = \mathcal{Y}_2^{n,m} - \mathcal{Y}_3^{n,m}$ and $\hat{\Phi} = \mathcal{Y}_4^{n,m}$ are the sought eigenfunctions of the operator $\{(\hat{D}A)^2 - 2\}$ with eigenvalues $\mu = n^2 + 3n$.

2.5. Multi-Pseudoparticle Solutions

A system of n pseudoparticles is, by definition, a field configuration which has Pontryagin index n and satisfies the self-duality constraint. Whereas it is straightforward to write down field configurations with arbitrary values of the Pontryagin index q [$A^\mu = f(x^2)g^{-n}\partial^\mu g^n$, with g and f as given in Eqs. (2.16) and (2.24), for instance, has q = n], it is not obvious that there exist self-dual configurations with q > 1. In this sub-section we exhibit explicit self-dual field configurations with arbitrary [integer] values of the Pontryagin index, and study their properties.

The search for multi-pseudoparticle solutions starts with an analysis of the single pseudoparticle potential

$$A^\mu = -2i\,\frac{\sigma^{\mu\nu}x_\nu}{\lambda^2 + x^2} \tag{2.74}$$

or

$$A^\mu = -2i\,\frac{\bar{\sigma}^{\mu\nu}x_\nu\lambda^2}{(\lambda^2 + x^2)x^2}. \tag{2.75}$$

These expressions are of the form

$$A^\mu = i\sigma^{\mu\nu}a_\nu \tag{2.76}$$

and

$$A^\mu = i\bar{\sigma}^{\mu\nu}a_\nu \tag{2.77}$$

with an appropriate four-vector field a^μ. Notice that A^μ is a matrix-valued vector field, with twelve independent components. Eq. (2.76) couples the space index of the gauge potential with its isospin indices [implicit in the matrix structure] so as to re-express the twelve components of A^μ in terms of the four components of the vector a^μ. We are dealing with representations of the $O(4) \approx SU(2) \times SU(2)$ group of rotations together with the SU(2) gauge group; thus Eqs. (2.76) and (2.77) represent definite couplings of tensorial components. This point will be elaborated in Section 3.

We try to generalize the pseudoparticle solutions by assuming an Ansatz[11] as in (2.76) or (2.77), and checking whether the self-duality constraint is solved with an appropriate choice of a^ν. From Eqs. (2.76) and (2.77) we find

$$F^{\mu\nu} = i[(\partial^\mu a_\rho - a^\mu a_\rho)\sigma^{\nu\rho} - (\partial^\nu a_\rho - a^\nu a_\rho)\sigma^{\mu\rho} - a_\rho a^\rho \sigma^{\mu\nu}] \quad (2.78)$$

with $\sigma^{\mu\nu}$ replaced by $\bar{\sigma}^{\mu\nu}$ if we start from $A^\mu = i\bar{\sigma}^{\mu\nu} a_\nu$. The expression for $*F^{\mu\nu}$ can be simplified using the self-duality (anti-self-duality) property of $\sigma^{\mu\nu}$ ($\bar{\sigma}^{\mu\nu}$). Using the identities

$$\varepsilon^{\mu\nu\alpha\beta}\sigma_{\nu\rho} = \frac{1}{2}\,\varepsilon^{\mu\nu\alpha\beta}\varepsilon_{\nu\rho\gamma\delta}\sigma^{\gamma\delta}$$

$$= -\,g^\mu_\rho\,\sigma^{\alpha\beta} - g^\alpha_\rho\,\sigma^{\beta\mu} - g^\beta_\rho\,\sigma^{\mu\alpha} \quad (2.79a)$$

$$\varepsilon^{\mu\nu\alpha\beta}\bar{\sigma}_{\nu\rho} = -\frac{1}{2}\,\varepsilon^{\mu\nu\alpha\beta}\varepsilon_{\nu\rho\gamma\delta}\bar{\sigma}^{\gamma\delta}$$

$$= g^\mu_\rho\,\bar{\sigma}^{\alpha\beta} + g^\alpha_\rho\,\bar{\sigma}^{\beta\mu} + g^\beta_\rho\,\bar{\sigma}^{\mu\alpha} \quad (2.79b)$$

we obtain from (2.76)

$$*F^{\mu\nu} = -\,i[(\partial_\rho a^\mu - a_\rho a^\mu)\sigma^{\nu\rho} - (\partial_\rho a^\nu - a_\rho a^\nu)\sigma^{\mu\rho}$$

$$+ \partial_\rho a^\rho \sigma^{\mu\nu}] \quad (2.80)$$

or from (2.77)

$$*F^{\mu\nu} = i[(\partial_\rho a^\mu - a_\rho a^\mu)\bar{\sigma}^{\nu\rho} - (\partial_\rho a^\nu - a_\rho a^\nu)\bar{\sigma}^{\mu\rho}$$

$$+ \partial_\rho a^\rho \bar{\sigma}^{\mu\nu}] \ . \quad (2.81)$$

The self-duality constraint $F^{\mu\nu} = *F^{\mu\nu}$ can now be converted into equations for the vector field a^μ. In deriving these we must pay attention to the fact that the matrices $\sigma^{\mu\nu}$ ($\bar{\sigma}^{\mu\nu}$) are not independent. The proper procedure is to multiply the equation $F^{\mu\nu} - *F^{\mu\nu} = 0$ by $\sigma^{\alpha\beta}$ ($\bar{\sigma}^{\alpha\beta}$) and use the identities

$$\mathrm{tr}\sigma^{\alpha\beta}\sigma^{\mu\nu} = \frac{1}{2}(g^{\alpha\mu}g^{\beta\nu} - g^{\alpha\nu}g^{\beta\mu} + \varepsilon^{\alpha\beta\mu\nu}) \tag{2.82a}$$

$$\mathrm{tr}\bar{\sigma}^{\alpha\beta}\bar{\sigma}^{\mu\nu} = \frac{1}{2}(g^{\alpha\mu}g^{\beta\nu} - g^{\alpha\nu}g^{\beta\mu} - \varepsilon^{\alpha\beta\mu\nu}) \quad . \tag{2.82b}$$

Thus starting from the <u>Ansatz</u> $A^{\mu} = i\sigma^{\mu\nu}a_{\nu}$ we obtain

$$\partial^{\mu}a^{\nu} + \partial^{\nu}a^{\mu} - 2a^{\mu}a^{\nu} = \frac{1}{2}g^{\mu\nu}(\partial_{\alpha}a^{\alpha} - a_{\alpha}a^{\alpha}) \tag{2.83}$$

as the self-duality condition. If $A^{\mu} = i\bar{\sigma}^{\mu\nu}a_{\nu}$, the self-duality condition leads to two equations.

$$\partial^{\mu}a^{\nu} - \partial^{\nu}a^{\mu} = \varepsilon^{\mu\nu\rho\sigma}\partial_{\rho}a_{\sigma} \tag{2.84a}$$

and

$$\partial_{\mu}a^{\mu} + a_{\mu}a^{\mu} = 0 \quad . \tag{2.84b}$$

From now on, we shall consider the two cases separately. Eq. (2.83) constitutes a set of nine non-linear first order differential equations for the four components of a^{μ}. The integrability conditions give rise to a set of constraints involving the six independent components of $f^{\mu\nu} \equiv \partial^{\mu}a^{\nu} - \partial^{\nu}a^{\mu}$, the algebraic complexity of which makes it difficult to analyze them completely; however, since the number of independent equations is large, it is plausible that the only solution is $f^{\mu\nu} = 0$. We assume this to be true. If $f^{\mu\nu} = 0$, then a^{μ} may be written in the form $a^{\mu} = -\partial^{\mu}a$. Inserting this into Eq. (2.83), we get

$$\partial_{\mu}\partial_{\nu}a = \frac{g_{\mu\nu}}{4}\Box a \tag{2.85a}$$

or

$$\partial_{\mu}\Box a = 0 \quad . \tag{2.85b}$$

Eq. (2.85b) requires $\Box a$ to be constant, and (2.85a) shows that a is at most a quadratic polynomial in x, $a = \alpha[\lambda^2 + (x-y)^2]$, so that

$$a^{\mu} = \frac{-2(x-y)^{\mu}}{\lambda^2 + (x-y)^2} \quad . \tag{2.86}$$

Thus for $A^\mu = i\sigma^{\mu\nu}a_\nu$ we recover only the single pseudoparticle solution.

We turn now to the set of coupled differential equations (2.84). First we decompose a^μ into transverse and longitudinal parts,

$$a^\mu = \partial^\mu \log\rho + b^\mu \quad , \quad \partial_\mu b^\mu = 0 \quad . \tag{2.87}$$

Eq. (2.84a) states that the Abelian "field strength" $f^{\mu\nu} = \partial^\mu b^\nu - \partial^\nu b^\mu$ derived from the potential b^μ is self-dual,

$$f^{\mu\nu} = *f^{\mu\nu} \tag{2.88a}$$

while Eq. (2.84b) becomes

$$\frac{1}{\rho}(\partial_\mu + b_\mu)(\partial^\mu + b^\mu)\rho = 0 \quad . \tag{2.88b}$$

A general solution of (2.88a) is obtained as follows. Owing to its transversality, b^μ may always be written as a divergence of an anti-symmetric tensor $h^{\mu\nu}$. But $h^{\mu\nu}$ has six independent components, while b^μ has only three, so three conditions may be imposed; it is convenient to demand that $h^{\mu\nu}$ be anti-self-dual. Consequently b^μ is represented as follows.

$$b^\mu = \partial_\nu h^{\mu\nu} \tag{2.89a}$$

$$*h^{\mu\nu} = -\, h^{\mu\nu} \quad . \tag{2.89b}$$

It is then trivial to verify that Eq. (2.88a) reduces to the requirement that $h^{\mu\nu}$ be a harmonic function. Thus the non-linear self-duality equation has been linearized with the help of the Ansatz $A^\mu = i\bar{\sigma}^{\mu\nu}a_\nu$: we are to choose any harmonic, anti-self-dual tensor $h^{\mu\nu}$, form b^μ and solve the linear equation (2.88b). However a non-trivial global problem still remains. The functions $h^{\mu\nu}$ and ρ necessarily have singularities which may induce singularities in the potential A^μ. These singularities must be arranged so that the gauge invariant quantity tr $F^{\mu\nu}F_{\mu\nu}$ is non-singular. Later, in considering small deformations of a given potential, we shall encounter singularities of exactly the same type, which appear as pure gauge artifacts. However, we do not know how to arrange for this to happen in the general case; indeed it is not clear that this is possible for non-vanishing $h^{\mu\nu}$.

So we proceed with the assumption that a^μ is a gradient. Upon setting b^μ to zero Eq. (2.88a) is of course satisfied, and

Eq. (2.88b) reduces to

$$\frac{1}{\rho}\Box\rho = 0 \quad . \tag{2.90}$$

It is still true that the harmonic function ρ will possess singularities, but now it is easy to find a form for them so that gauge invariant quantities are non-singular. We take

$$\rho = \sum_{i=1}^{m} \frac{\lambda_i^2}{(x - y_i)^2} \quad . \tag{2.91}$$

Note that with this superposition of poles Eq. (2.90) is satisfied everywhere, even at the poles, due to the prefactor ρ^{-1}.

Summarizing, we have found that the formula

$$A^{\mu} = i\bar{\sigma}^{\mu\nu}\partial_{\nu}\log\rho \tag{2.92}$$

with ρ as in Eq. (2.91) gives origin to a self-dual field configuration.[12,13,14]

We must still verify that the singularities introduced in A^{μ} by the poles of ρ are pure gauges and evaluate the Pontryagin index of the field configuration. Near a singularity, which we take for convenience at the origin, A^{μ} behaves as

$$A^{\mu} \approx i\bar{\sigma}^{\mu\nu}\partial_{\nu}\log\frac{1}{x^2} = -\,2i\bar{\sigma}^{\mu\nu}x_{\nu}/x^2 \quad . \tag{2.93}$$

Comparing with Eq. (2.18) [with $\sigma^{\mu\nu}$ replaced by $\bar{\sigma}^{\mu\nu}$] we see that the singularity at the origin is indeed of a pure gauge form. Also, if we evaluate the Pontryagin index q as a surface integral, this singularity will contribute one unit to it [cf. also the discussion after Eqs. (2.21), (2.27)]. The behavior of the field at infinity, where $\rho \approx 1/x^2 \sum_{i=1}^{m} \lambda_i^2$, is still of the form given by Eq. (2.93). We conclude that the Pontryagin index is m-1, because the m singularities at $x = y_i$, $i = 1, \ldots m$, contribute +m to q, whereas a contribution -1 comes from the surface at infinity.

The value of q may also be found using the elegant formula

$$\mathrm{tr}\,{}^*F^{\mu\nu}F_{\mu\nu} = \Box\,\Box\log\rho \tag{2.94}$$

which can be derived when A^{μ} is given by Eq. (2.92). Eq. (2.94) shows in particular that $\mathrm{tr}{}^*F^{\mu\nu}F_{\mu\nu}$ is not changed when ρ is multiplied by a factor $(x-y)^2$. Thus q may be evaluated as

$$q = -\frac{1}{16\pi^2}\int d^4x \,\Box\Box \log P \tag{2.95a}$$

where P is a polynomial of degree 2m - 2, and this immediately gives

$$q = -\lim_{R\to\infty}\frac{1}{16\pi^2}\int d\Omega R^2 R_\mu \partial^\mu \Box \log(R^{2m-2} + \ldots) = m - 1 \quad . \tag{2.95b}$$

The number of parameters appearing in the expression of the self-dual field A^μ with Pontryagin index n is 5n + 4. This is apparent from Eqs. (2.91) and (2.92): the parameters are the 5m = 5n + 5 scales λ_i and positions y_i^μ, minus one overall scale, which can be modified by an additive change of $\log\rho$. The number 5n + 4 is surprising if one thinks that the field configuration is obtained putting together n pseudoparticles, each characterized by a position and a scale. One may indeed consider a limit where the (n+1)th scale λ_{n+1} and coordinate y_{n+1}^μ go to infinity simultaneously with $\lim \lambda_{n+1}^2/y_{n+1}^2 = 1$, in which case ρ takes the form[13]

$$\rho = 1 + \sum_{i=1}^{n} \frac{\lambda_i^2}{(x - y_i)^2} \quad . \tag{2.96}$$

For small λ_i, i = 1, ..., n, one can then identify the y_i^μ's as approximate positions of peaks in the action density, with width λ_i^2. A conformal transformation of the field configuration [which of course preserves the self-duality of $F^{\mu\nu}$] would re-introduce the more general form of Eq. (2.91).

It may be verified indeed that the class of field configurations represented by Eqs. (2.91) and (2.92) is closed under conformal transformations in the following sense[12]. In a finite special conformal transformation where

$$x^\mu \to \tilde{x}^\mu = \frac{x^\mu - c^\mu x^2}{1-2c\cdot x + c^2x^2} \tag{2.97a}$$

we let $\rho(x)$ transform as a scalar density of dimension +1, i.e.

$$\rho(x) \to \tilde{\rho}(x) = \frac{1}{1-2c\cdot x + c^2x^2}\,\rho(\tilde{x}) \quad . \tag{2.97b}$$

For infinitesimal $c^\mu = \varepsilon^\mu$,

$$\delta\rho = (2\varepsilon\cdot x\, x^\alpha - x^2\varepsilon^\alpha)\partial_\alpha\rho + 2\varepsilon\cdot x\rho \tag{2.97c}$$

and, with simple algebra, one verifies that the induced transformation of a^μ takes the form

$$\delta a^\mu = \delta\partial^\mu \log\rho = \delta_c a^\mu + 2\varepsilon^\mu \tag{2.97d}$$

where $\delta_c a^\mu$ is the conformal variation of a vector field of dimension +1. If we replace a^ν with $a^\nu + \delta_c a^\nu + 2\varepsilon^\nu$ in the Ansatz $A^\mu = i\bar{\sigma}^{\mu\nu}a_\nu$ and perform then an infinitesimal gauge transformation

$$\delta_g A^\mu = \partial^\mu\Theta + [A^\mu,\Theta] \tag{2.98a}$$

with parameter

$$\Theta = 2i\varepsilon_\alpha x_\beta \bar{\sigma}^{\alpha\beta} \tag{2.98b}$$

again it is a matter of straightforward algebra to check that the total variation of A^μ is precisely the conformal change $\delta_c A^\mu$ of a vector field of dimension 1.

Summarizing, a conformal transformation of the gauge potential A^μ of Eq. (2.92) can be obtained by changing ρ first according to Eq. (2.97) and then performing a suitable gauge transformation. But, starting from Eqs. (2.91) and (2.97) with an explicit computation we find

$$\tilde{\rho}(x) = \sum_{i=1}^{m} \frac{\tilde{\lambda}_i^2}{(x - \tilde{y}_i)^2} \tag{2.99a}$$

$$\tilde{\lambda}_i^2 = \frac{\lambda_i^2}{1+2c\cdot y_i + c^2 y_i^2} \tag{2.99b}$$

$$\tilde{y}_i^\mu = \frac{y_i^\mu + c^\mu y_i^2}{1+2c\cdot y_i + c^2 y_i^2} \; . \tag{2.99c}$$

We see therefore that a conformal transformation changes any field configuration of the class defined by Eqs. (2.91) and (2.92) into another field configuration of the same class, modulo a gauge transformation. The superpotential $\tilde{\rho}(x)$ of the new field configuration is obtained from the old one by a conformal transformation of the scales λ_i^2 and the positions y_i^μ. In particular, the function ρ of Eq. (2.96) can be considered as the limiting form of the more general ρ of Eq. (2.91) obtained when one of the singularities has moved to infinity; it is transformed into a ρ of the more general class by a conformal transformation.

2.6. Small Deformations of Self-Duality Condition

The realization that all the 5n + 4 parameters present in the expression A^μ are necessary to have an explicit representation of the conformal group still leaves open the possibility that some of the parameters are unphysical, i.e., that the values of the λ_i^2 and y_i^μ may be modified by a gauge transformation. We know after all that the single pseudoparticle configuration depends on five physical parameters, whereas the present analysis gives a number of parameters equal to nine for n = 1. We shall see later that there are indeed situations where some of the constants y_i^μ and λ_i^2 may be modified by performing a gauge transformation on the fields, but it is convenient to postpone the study of this residual gauge freedom. Instead, we consider now the problem of finding the most general infinitesimal deformation of the fields which preserves the self-duality of $F^{\mu\nu}$.[15] This analysis will also provide an answer to the question of the residual gauge freedom; as will become apparent, in general all the 5n + 4 parameters are physical.

A small variation of the potential A^μ

$$A^\mu \to A^\mu + \delta A^\mu \tag{2.100a}$$

generates a variation of $F^{\mu\nu}$

$$F^{\mu\nu} \to F^{\mu\nu} + \delta F^{\mu\nu} \tag{2.100b}$$

$$\delta F_{\mu\nu} = D_\mu^A \delta A_\nu - D_\nu^A \delta A_\mu$$

$$D_\mu^A \delta A_\nu = \partial_\mu \delta A_\nu + [A_\mu, \delta A_\nu] \ . \tag{2.100c}$$

We take A^μ to be given by Eqs. (2.91) and (2.92) and require that $\delta F^{\mu\nu}$ be self dual. The most general δA^μ can be represented by

$$\delta A^\mu = i\bar{\sigma}^{\alpha\beta} X^\mu_{\alpha\beta} \tag{2.101}$$

where $X^{\mu}_{\alpha\beta}$ is anti-symmetric and anti-self-dual in the indices $\alpha\beta$. But it is not convenient to consider an expression as general as (2.101), because all infinitesimal gauge transformations of A^{μ}

$$\delta_{\text{gauge}} A_{\mu} = D^{A}_{\mu} i\bar{\sigma}^{\alpha\beta}\omega_{\alpha\beta} \tag{2.102}$$

would appear as uninteresting solutions of

$$\delta F^{\mu\nu} = *\delta F^{\mu\nu} . \tag{2.103}$$

We fix the gauge by requiring that δA^{μ} be of the form

$$\delta A^{\mu} = i\bar{\sigma}^{\mu\nu}\partial_{\nu}\frac{\delta\rho}{\rho} + i\bar{\sigma}_{\alpha\beta}\rho\partial_{\nu}Y^{\mu\nu\ \alpha\beta} \tag{2.104a}$$

where $Y^{\mu\nu\ \alpha\beta}$ is a tensor field, anti-symmetric and anti-self-dual in both pairs of indices $\mu\nu$, $\alpha\beta$ and constrained by

$$Y_{\mu\nu}{}^{\mu\nu} = 0 . \tag{2.104b}$$

This apparently arbitrary choice of a guage is motivated by the fact that it simplifies the algebra. The first term in the right-hand-side of Eq. (2.104a) is included because we want to find among the infinitesimal deformations those induced by a variation of the parameters of ρ. The condition $Y_{\mu\nu}{}^{\mu\nu} = 0$ removes one of the nine independent components of $Y^{\mu\nu\ \alpha\beta}$ so as to leave the correct number of variable functions - nine - in the Ansatz of Eq. (2.104).

The tensor $Y^{\mu\nu\ \alpha\beta}$ can be decomposed into a symmetric, traceless and an anti-symmetric part.

$$Y^{\mu\nu\ \alpha\beta} = S^{\mu\nu\ \alpha\beta} + A^{\mu\nu\ \alpha\beta} \tag{2.105a}$$

$$S^{\mu\nu\ \alpha\beta} = S^{\alpha\beta\ \mu\nu} \tag{2.105b}$$

$$A^{\mu\nu\ \alpha\beta} = - A^{\alpha\beta\ \mu\nu} = \frac{1}{4}(g^{\mu\alpha}V^{\nu\beta} - g^{\nu\alpha}V^{\mu\beta} + g^{\nu\beta}V^{\mu\alpha} - g^{\mu\beta}V^{\nu\alpha}) . \tag{2.105c}$$

$V^{\alpha\beta}$ is anti-symmetric and anti-self-dual. After non-trivial algebraic manipulations that make use of many identities satisfied by anti-self-dual quantities, one finds that Eqs. (2.100), (2.103) and (2.104) imply

$$\Box\, S^{\mu\nu\,\alpha\beta} = 0 \quad . \tag{2.106}$$

All non-trivial solutions of this equation introduce in δA^μ singularities which cannot be removed by a gauge transformation, and therefore we set $S^{\mu\nu\,\alpha\beta} = 0$. When $S^{\mu\nu\,\alpha\beta}$ vanishes, one finds that the anti-self-duality of $F^{\mu\nu}$ implies for $\delta\rho$ and $V^{\mu\nu}$ the equations

$$\Box\, \rho V^{\mu\nu} = 0 \tag{2.107a}$$

$$\Box\, \delta\rho + 2\rho\partial_\mu V^{\mu\nu}\partial_\nu\rho = 0 \quad . \tag{2.107b}$$

These are solved by

$$V^{\mu\nu} = \frac{1}{\rho}\sum_{i=1}^{n+1} \frac{k_i^{\mu\nu}}{(x - y_i)^2} \tag{2.108a}$$

$$\delta\rho = \sum_{i=1}^{n+1}\sum_{\substack{j=1\\ j\neq i}}^{n+1} \frac{2}{(y_i - y_j)^2}\,\frac{k_i^{\mu\nu}\lambda_j^2(x-y_i)_\mu(x-y_j)_\nu}{(x-y_i)^2(x-y_j)^2} + \delta\tilde{\rho} \tag{2.108b}$$

where the $k_i^{\mu\nu}$ are constant anti-self-dual tensors and $\delta\tilde{\rho}$ is the variation induced by a change of the parameters of ρ.

Inserting $V^{\mu\nu}$ and $\delta\rho$ as given by Eqs. (2.108) into Eqs. (2.104) and (2.105), one finds an expression for δA^μ which is singular near the poles y_i^μ of ρ. It is possible to show, however, that the singularity can be removed by a suitable gauge transformation,[15] so that δA^μ represents an acceptable infinitesimal deformation of the self-dual field configuration.

It is very interesting to observe that upon performing an infinitesimal gauge transformation [according to Eq. (2.102)] with parameter

$$\omega^{\alpha\beta} = \frac{1}{4}\rho V^{\alpha\beta} \tag{2.109a}$$

the infinitesimal variation of the potential becomes

$$\begin{aligned}\delta A'^\mu &= \delta A^\mu + \delta_{\text{gauge}} A^\mu \\ &= i\bar{\sigma}^{\mu\nu}[\partial_\nu(\frac{\delta\rho}{\rho}) + \partial^\lambda(\rho V_{\lambda\nu})]\end{aligned} \tag{2.109b}$$

which is of the form

$$\delta A'^{\mu} = i\bar{\sigma}^{\mu\nu}\,\delta a_{\nu} \quad . \tag{2.109c}$$

In this gauge, the infinitesimal deformation appears as a first order variation of the original Ansatz, where δa^{ν} consists of both a gradient term and a divergenceless term. The gauge transformation leading to this form of the potential is singular and $\delta A'^{\mu}$ behaves as $|x-y_i|^{-3}$ near the poles of ρ. Because of these singularities, the representation of the infinitesimal deformations provided by $\delta A'^{\mu}$ is not very useful to study the finite physical deformations, but it is extremely convenient for an analysis of the residual gauge freedom.

To expose possible gauge artifacts among the infinitesimal deformations $\delta A'^{\mu}$, we perform an additional gauge transformation with parameter $\omega'^{\alpha\beta}$ and inquire whether

$$\delta A''^{\mu} = \delta A'^{\mu} + \delta'_{\text{gauge}} A^{\mu} \tag{2.110}$$

can still be of the form

$$\delta A''^{\mu} = i\bar{\sigma}^{\mu\nu}(\delta a_{\nu} + \delta'_{\text{gauge}} a_{\nu}) \quad . \tag{2.111}$$

With some algebra, one finds that the form of the Ansatz is preserved only if

$$\omega'^{\alpha\beta} = \rho\tilde{\omega}^{\alpha\beta} \tag{2.112a}$$

with

$$\begin{aligned}\tilde{\omega}^{\alpha\beta} &= 2x^{\alpha}A^{\beta\gamma}x_{\gamma} - 2x^{\beta}A^{\alpha\gamma}x_{\gamma} + x^{2}A^{\alpha\beta} \\ &\quad + B^{\alpha}x^{\beta} - B^{\beta}x^{\alpha} + \varepsilon^{\alpha\beta\gamma\delta}B_{\gamma}x_{\delta} + C^{\alpha\beta}\end{aligned} \tag{2.112b}$$

where B^{α} is a constant vector, $A^{\alpha\beta}$ and $C^{\alpha\beta}$ are constant self-dual and anti-self-dual tensors, respectively. The variation of $\delta'_{\text{gauge}}a^{\nu}$ is then given by

$$\delta'_{\text{gauge}}a^{\nu} = -\,4\tilde{\omega}^{\nu\alpha}\partial_{\alpha}\rho - \frac{4}{3}\,\rho\partial_{\alpha}\tilde{\omega}^{\nu\alpha} \tag{2.113a}$$

which in terms of $V^{\alpha\beta}$ and ρ, reads

$$\delta'_{\text{gauge}}\rho V^{\alpha\beta} = 4 \sum_{i=1}^{n+1} \frac{\lambda_i^2}{(x - y_i)^2} \tilde{\omega}^{\alpha\beta}(y_i) \tag{2.113b}$$

$$\frac{1}{\rho}\,\delta'_{\text{gauge}}\rho = \frac{4}{3} \sum_{i=1}^{n+1} \frac{\lambda_i^2}{(x - y_i)^2} (x^\alpha - y_i^\alpha)\partial^\beta\, \tilde{\omega}_{\alpha\beta}(y_i) \quad . \tag{2.113c}$$

Since $\tilde{\omega}^{\alpha\beta}$ contains ten independent constants, we conclude that ten of the independent components of the tensors $k_i^{\alpha\beta}$ in Eq. (2.108a) can be modified by a gauge transformation. Therefore the dimensionality of the space of physical small deformations of a given solution is 8(n+1)-10-1 [-1 because of the arbitrariness of an overall rescaling of the λ_i's], = 8n-3, which, by coninuity, must also be the dimensionality of [a connected component of] the full manifold of solutions. The number 8n-3 has a nice interpretation: the n-pseudoparticle solution appears parametrized by the positions, scales and relative group orientations of the pseudoparticles.

Notice that if we start from any of the 5n+4 solutions described by the Ansatz $A^\mu = i\bar{\sigma}^{\mu\nu}\partial_\nu \log\rho$ and perform an infinitesimal gauge transformation, Eq. (2.113b) tells us that we shall not preserve the functional form of the Ansatz unless the positions of the poles y_i^μ satisfy

$$\tilde{\omega}^{\alpha\beta}(y_i) = 0 \quad . \tag{2.114}$$

But this is the equation of a definite circle [or of a straight line as a limiting case] in 4-space, and therefore if the poles y_i^μ are more than three in number, and in general positions, then all the 5n+4 parameters represent physical degrees of freedom. On the other hand, when the poles y_i^μ lie on a circle, one can perform a gauge transformation which moves the singularities around the circle [see Eq. (2.113c)]. Through three points one can always draw a circle, so that if n=2, one of the 5n+4 = 14 parameters is always a gauge artifact, and the 2-pseudoparticle solution depends on 13 gauge invariant parameters. Through two points one can draw a three-dimensional variety of circles, and therefore four of the 5+4 = 9 parameters describing the single pseudoparticle within this Ansatz are gauge artifacts, in agreement with the fact that a single pseudoparticle is characterized by only five parameters, position and scale. These considerations indicate that for $n > 3$ there certainly exist solutions to the self-duality equations beyond the ones given by Eqs. (2.91) and (2.92); as yet they have not been found.

3. FURTHER MATHEMATICAL DEVELOPMENTS

3.1. Spinorial Formalism

As mentioned earlier, the solutions to the various equations that have been discussed take the form they do as a consequence of the coupling of internal degrees of freedom, [SU(2)], to kinematical degrees of freedom, [O(4)]. For Yang-Mills theory, this coupling can be made explicit in the context of a spinorial formalism, which we describe in this Section. The formalism is also important since it exposes features of the self-duality condition which are used as a point of departure for an analysis by methods of algebraic geometry. Furthermore, with the help of this formalism, we shall be able to simplify considerably the Dirac equation for zero-eigenvalue modes of a fermion, with arbitrary iso-spin, and to solve it completely for iso-spin ½ and 1.[16]

The spinorial formulation begins with the observation that the O(4) invariants of interest in Euclidean four-space may be designated by SU(2)×SU(2) representation labels. Also, the internal SU(2) gauge group gives rise to such labels. Hence all objects with which we are concerned are SU(2) multi-spinors, and equations are simplified when the various SU(2) groups are cunningly coupled to each other.

In this formalism all objects carry spinor labels A, B, C ..., which take on two values and describe the spin and iso-spin degrees of freedom. An anti-symmetric metric tensor with two upper indices is defined by

$$\varepsilon^{AB} = \begin{pmatrix} 0 & 1 \\ -1 & 0 \end{pmatrix} = i\sigma^2 \quad . \tag{3.1a}$$

The negative inverse of this matrix is a metric tensor with lower indices.

$$\varepsilon_{AB} = \begin{pmatrix} 0 & 1 \\ -1 & 0 \end{pmatrix} = \varepsilon^{AB} \quad . \tag{3.1b}$$

A spinor may have lower or upper indices, which can be raised or lowered with the metric tensors according to the following rules [repeated indices are summed].

$$\xi^A = \varepsilon^{AB}\, \xi_B \tag{3.2a}$$

$$\xi_B = \xi^A\, \varepsilon_{AB} \quad . \tag{3.2b}$$

Covariant summations always involve one upper and one lower index. Note that $\xi^A{}_A = -\xi_A{}^A$. For every pair of indices we may define a symmetric and anti-symmetric part

$$\xi_{AB} = \frac{1}{2}\,\varepsilon_{AB}\xi_C{}^C + \frac{1}{2}\,\xi_{\underset{\smile}{AB}} \tag{3.3}$$

where the symbol $\underset{\smile}{AB}$ denotes the symmetric sum $\xi_{AB} + \xi_{BA}$. More generally for a multi-indexed object, $\xi_{A_1A_2\ldots A_n}$, symmetric in $A_2\ldots A_n$,

$$\xi_{\underset{\smile}{A_1A_2}\ldots A_n} \equiv \xi_{A_1A_2\ldots A_n} + \xi_{A_2A_1\ldots A_n} + \ldots$$

$$+ \xi_{A_nA_2\ldots A_1} \quad \text{(n terms)} . \tag{3.4}$$

An O(4) two-component spinor is described by a spinor with one index. To every O(4) tensor with indices μ, ν, ... , there corresponds a spinor with index pairs AA', BB', ... The rule of association is given through the α matrices defined in (2.17).

$$\xi_\mu(\alpha^\mu)_{AA'} = \xi_{AA'} \tag{3.5a}$$

$$\xi_\mu(\bar{\alpha}^\mu)_{A'A} = \xi^{AA'} \quad . \tag{3.5b}$$

[That the two definitions are consistent with (3.2) is easily established from the properties of the Pauli matrices.] The O(4) covariants may be regained from the spinors by projecting with the appropriate α matrix. The above holds also for derivatives $\partial_\mu \leftrightarrow \partial_{AA'}$.

Iso-spinor indices are represented as follows. Iso-spin ½ objects are described by one-index spinors. For iso-spin 1, a two-index spinor, symmetric in the indices, is used. In general an iso-spin T object is described by a totally symmetric spinor with 2T indices, so that there are 2T + 1 independent components. The correspondence between the conventional description and the spinorial one is immediate for iso-spin ½ - the two coincide. For unit iso-spin the correspondence is

$$\xi^a\left(\frac{\sigma_a}{2i}\right)_{UV} = -\,\xi^V{}_U \quad . \tag{3.6}$$

A consequence is that ξ_{VU} and ξ^{VU} are symmetric in $U \leftrightarrow V$ and that $\varepsilon_{abc}\xi^b\bar{\xi}^c$ corresponds $\xi_{\underset{\smile}{UW}}\bar{\xi}^W{}_{\underset{\smile}{V}}$. The relations for higher iso-spin are more complicated, and will not be given here.

3.2. Gauge Field Equations in Spinorial Formalism

The gauge potential A_a^μ is described by $A_{AA';UV}$; the gauge field $F_a^{\mu\nu}$, by $F_{AA',BB';UV}$ which is anti-symmetric in the interchange $A\leftrightarrow B$, $A'\leftrightarrow B'$. Both expressions are symmetric in $U\leftrightarrow V$, and the formula relating the two is

$$F_{AA',BB';UV} = \partial_{AA'}A_{BB';UV} - \partial_{BB'}A_{AA';UV} + A_{AA';U\underbrace{W}A_{BB';}{}^{W}{}_{V}} \quad . \tag{3.7}$$

Due to its anti-symmetry properties, F may be split into two parts

$$F_{AA',BB';UV} = \frac{1}{2}\,\varepsilon_{AB}\overset{+}{F}_{A'B';UV} + \frac{1}{2}\,\varepsilon_{A'B'}\overset{-}{F}_{AB;UV} \tag{3.8}$$

where $\overset{+}{F}_{A'B';UV}$ is symmetric in $A'\leftrightarrow B'$, and $\overset{-}{F}_{AB;UV}$ is symmetric in $A\leftrightarrow B$; these are just the self-dual and anti-self-dual parts of the gauge field, as is seen by noting that the definition $*F^{\mu\nu} = \frac{1}{2}\,\varepsilon^{\mu\nu\,\alpha\beta}F_{\alpha\beta}$ becomes in the spinorial formalism

$$*F_{AA',BB';UV} = F_{AB',BA';UV} \tag{3.9a}$$

or

$$*\overset{+}{F}_{A'B';UV} = \overset{+}{F}_{A'B';UV} \tag{3.9b}$$

$$*\overset{-}{F}_{AB;UV} = -\,\overset{-}{F}_{AB;UV} \quad . \tag{3.9c}$$

It follows from (3.7) and (3.8) that

$$\overset{+}{F}_{A'B';UV} = \partial_{\underbrace{AA'}}A^{\underbrace{A}}{}_{B';UV} + A_{AA';U\underbrace{W}}A^{\underbrace{A}}{}_{B';}{}^{W}{}_{\underbrace{V}} \tag{3.10a}$$

$$\overset{-}{F}_{AB;UV} = \partial_{A\underbrace{A'}}A_{B}{}^{\underbrace{A'}}{}_{;UV} + A_{AA';U\underbrace{W}}A_{B}{}^{A'}{}_{;}{}^{W}{}_{\underbrace{V}} \quad . \tag{3.10b}$$

The self-duality condition demands that F^- vanishes. Hence a self-dual gauge potential satisfies

$$\partial_{A\underbrace{A'}}A_{B}{}^{\underbrace{A'}}{}_{;UV} + A_{AA';U\underbrace{W}}A_{B}{}^{A'}{}_{;}{}^{W}{}_{\underbrace{V}} = 0 \quad . \tag{3.11}$$

The conformal solution to this equation is

$$A_{AA';UV} = \frac{1}{2} \varepsilon_{AU} \partial_{VA'} \log\rho$$

$$\frac{1}{\rho} \Box \rho = 0 \quad . \tag{3.12}$$

Thus far we have merely transcribed into new formalism results which already exist in the conventional approach. We wish now to make some further observations about self-dual gauge fields. These form the starting point for an analysis of self-dual gauge fields by methods of algebraic geometry.

Consider a special set of complex bi-spinors [4-vectors] $x_{AA'}$ which can be written as $\ell_A z_{A'}$ where ℓ_A is fixed and $z_{A'}$ varies. It is clear that all points described by such coordinates are light-like with respect to each other: $x_{AA'}\tilde{x}^{AA'} = \ell_A z_{A'} \ell^A \tilde{z}^{A'} = 0$. This set of points for fixed ℓ defines a light-like plane. Next let us project the gauge field onto such a light-like plane.

$$F^{\ell}_{UV} = x^{AA'}\tilde{x}^{BB'} F_{AA',BB';UV} = \ell^A z^{A'} \ell^B \tilde{z}^{B'} F_{AA',BB';UV} \tag{3.13}$$

But a self-dual field takes the form

$$F_{AA',BB';UV} = \frac{1}{2} \varepsilon_{AB} \overset{+}{F}_{A'B';UV} \tag{3.14}$$

so for self-dual configurations the projection (3.13) vanishes. We thus come to the important conclusion that on arbitrary light-planes the gauge potential, for self-dual fields, is integrable.

$$\ell^A A_{AA'} = g_\ell^{-1} \ell^A \partial_{AA'} g_\ell \quad . \tag{3.15}$$

The program of reconstructing $A_{AA'}$ from the above by methods of algebraic topology is being pursued actively, but we shall not discuss this topic further.[17]

3.3. Dirac Equations in Spinorial Formalism

An important feature of pseudoparticle configurations is that they produce zero-eigenvalue modes in the Dirac equation for a [Euclidean] fermion in the pseudoparticle field; that is one can solve

$$i\gamma^\mu(\partial_\mu + A_\mu)\psi = 0 \tag{3.16}$$

with several normalizable functions. Here, ψ has 2T + 1 components which transform according to some definite, irreducible representation of SU(2)

$$\delta\psi = iT^a\psi\theta_a$$

$$[T^a, T^b] = i\,\varepsilon_{abc}T^c \tag{3.17}$$

and A^μ is the Yang-Mills potential in an anti-hermitian matrix representation: $iA^\mu = A^\mu_a T^a$. In general A^μ need not solve the Yang-Mills equations, but is always taken to be sufficiently well-behaved that the Yang-Mills action is finite; consequently the gauge configuration is characterized by an integer valued Pontryagin index.

When the gauge potential is the conformal, self-dual configuration (3.12), the spinorial formalism may be used to simplify Eq. (3.16) considerably. We now present this analysis and solve Eq. (3.17) completely for iso-spin ½ and 1.

The Dirac matrices in (3.16), satisfy Euclidean anti-commutation relations $\{\gamma^\mu, \gamma^\nu\} = 2\delta^{\mu\nu}$ which can be realized in a fashion such that γ_5 is diagonal.

$$\gamma^\mu = \begin{pmatrix} 0 & \alpha^\mu \\ \bar{\alpha}^\mu & 0 \end{pmatrix}$$

$$\gamma_5 = \gamma_1\gamma_2\gamma_3\gamma_4 = \begin{pmatrix} I & 0 \\ 0 & -I \end{pmatrix} \quad . \tag{3.18}$$

In this representation Eq. (3.16) decouples into two separate equations for two-component spinors of definite chirality.

$$\psi = \begin{pmatrix} \psi^+ \\ \psi^- \end{pmatrix}$$

$$i\,\alpha^\mu(\partial_\mu + A_\mu)\psi^- = 0 \tag{3.19a}$$

$$i\,\bar{\alpha}^\mu(\partial_\mu + A_\mu)\psi^+ = 0 \quad . \tag{3.19b}$$

When we discuss the Atiyah-Singer index theorem in the next sub-Section, we shall show that (3.19b) has no normalizable solutions, and, therefore, only (3.19a) need be considered.

In the spinorial formalism, (3.19a) transcribes into

$$\partial_{AA'}\psi^{A'}{}_{;U_1\ldots U_{2T}} + A_{AA';U_1V}\psi^{A'\ V}{}_{;\ U_2\ldots U_{2T}} \quad . \tag{3.20a}$$

Here the spinor carries the index A', describing two spatial components; it is entirely symmetric in the 2T indices U_1 which refer to the 2T + 1 components of iso-spin. Substitution of the conformal solution into the above gives the equation we analyze.

$$\frac{1}{2T}\,\partial_{AA'}\psi^{A'}{}_{;U_1\ldots U_{2T}} + \frac{1}{2}\,\varepsilon_{AU_1}(\partial_{BB'}\log\rho)\psi^{B'\ B}{}_{;\ U_2\ldots U_{2T}}$$

$$-\frac{1}{2}\,(\partial_{U_1A'}\log\rho)\psi^{A'}{}_{;A\ldots U_{2T}} + \text{permutations} = 0 \quad . \tag{3.20b}$$

A straightforward but lengthy sequence of manipulations of indices, which among other things involves separating the above into symmetric and anti-symmetric parts in (A, U_1), yields the result that a normalizable solution necessarily has the form[16]

$$\psi_{A';U_1\ldots U_{2T}} = \rho^{T}\partial_{U_1A'}\rho^{-2T}\chi_{U_2\ldots U_{2T}} + \text{permutations} \tag{3.21a}$$

where χ satisfies

$$\partial_{U_2B'}\rho^{-2T+1}\partial^{CB'}\chi_{CU_3\ldots U_{2T}} + \text{permutations} = 0 \quad . \tag{3.21b}$$

We now specialize to iso-spin ½ and 1. For the former

$$\psi_{A';U} = \rho^{\frac{1}{2}}\,\partial_{UA'}\rho^{-1}\chi$$

$$\partial_{UB'}\partial^{CB'}\chi = 0$$

or

$$\Box\,\chi = 0 \quad . \tag{3.22}$$

Of course only singular functions solve the harmonic equation; however we can tolerate singularities, provided they are absent from the gauge-invariant norm density $\psi^{A';U}\psi_{A';U}$, so that the spinor is normalizable. Therefore we can allow in χ only poles which are already present in ρ. In this way we get n+1 solutions for χ.

$$\chi^{(i)} = \frac{\lambda_i^2}{(x - y_i)^2}$$

$$i = 1, \ldots, n+1 \quad . \tag{3.23}$$

Of these only n $\psi^{(i)}$'s are independent since $\sum_i \chi^{(i)} = \rho$, and $\sum_i \psi^{(i)} = 0.$[18]

For iso-spin 1

$$\psi_{A';U_1U_2} = \rho\underbrace{\partial_{U_1A'}\rho^{-2}}\chi_{U_2}$$

$$\partial_{U_2B'}\rho^{-1}\partial^{CB'}\chi_C = 0 \quad . \tag{3.24}$$

Solutions are conveniently exhibited by setting $\chi_C = M_{CC'}u^{C'}$, where $u^{C'}$ is a constant spinor, with two arbitrary components which provide two solutions for each matrix M. Eq. (3.24) is solved by the following n+1 expressions for M.

$$M^{(1)(i)}_{CC} = \frac{\lambda_i^2}{(x - y_i)^4}(x - y_i)_{CC'}$$

$$i = 1, \ldots, n+1 \quad . \tag{3.25a}$$

An additional n+1 forms are

$$M^{(2)(i)}_{CC'} = -\rho\frac{\lambda_i^2}{(x - y_i)^2}\varepsilon_{CC'}$$

$$-\sum_{\substack{j=1\\ j\neq i}}^{n+1}\frac{\lambda_i^2\lambda_j^2}{(y_i - y_j)^2}\left[\frac{(x - y_i)_{CC''} + (x - y_j)_{CC''}}{(x - y_i)^2(x - y_j)^2}\right](y_i - y_j)_{C'}{}^{C''}$$

$$i = 1, \ldots, n+1 \quad . \tag{3.25b}$$

However of these 2n + 2 matrices, only 2n are linearly independent, since the following relationships are readily established.

$$\sum_{i=1}^{n+1} M_{CC'}^{(2)(i)} = -\rho^2 \varepsilon_{CC'} \tag{3.26a}$$

$$\sum_{i=1}^{n+1} \lambda_i^2 M_{CC'}^{(1)(i)} + \sum_{i}^{n+1} M_{CC''}^{(2)(i)} (y_i)^{C''}{}_{C'} = \rho^2 x_{CC'} \quad . \tag{3.26b}$$

One finds all the solutions to be normalizable; thus there are 4n zero-eigenvalue modes for iso-spin 1 Fermi fields.

We conclude this discussion of the solutions to the Dirac equation by noting that, since the eigenfunctions have definite chirality, bilinears $\psi^\dagger \Gamma \psi$ vanish for a vectorial Dirac matrix Γ. In particular $\psi^\dagger T^a \gamma^\mu \psi$ is zero; hence the functions A^μ and ψ are also solutions to the coupled Yang-Mills fermion equations, when A^μ has definite duality and ψ is a chiral eigenstate solution of the Dirac equation. [We show below that such solutions exist not only for iso-spin ½ and 1, but also for arbitrary iso-spin T.]

3.4. Atiyah-Singer Index Theorem

In the two examples discussed earlier - iso-spin ½ and 1 Fermi fields moving in a self-dual Yang-Mills potential - we found a number of zero-eigenvalue modes of definite chirality. The existence of these modes has far-reaching physical consequences; moreover it is related to the anomaly of the axial vector current[19] and to topological properties of the gauge fields. This unexpected connection between physics and mathematics is best understood with the help of the "Atiyah-Singer index theorem", which we now explain.

Consider a linear differential operator L and its adjoint $L^\dagger$; further suppose that the number of normalizable zero-eigenvalue modes of L is n_- and that of $L^\dagger$ is n_+. The "index" is the quantity $n_- - n_+$, and the index theorem evaluates this object in terms of the properties of L. In order to make these considerations relevant to our Dirac equation (3.16), let us write it in block form, using the γ matrices in the representation (3.18), which diagonalizes chirality.

$$\begin{bmatrix} 0 & L \\ L^{\dagger} & 0 \end{bmatrix} \begin{pmatrix} \psi^{+} \\ \psi^{-} \end{pmatrix} = 0 \tag{3.27a}$$

$$L\psi^{-} = 0 \tag{3.27b}$$

$$L^{\dagger}\psi^{+} = 0 \tag{3.27c}$$

$$L = i\alpha^{\mu}(\partial_{\mu} + A_{\mu}) \tag{3.27d}$$

$$L^{\dagger} = i\bar{\alpha}^{\mu}(\partial_{\mu} + A_{\mu}) \tag{3.27e}$$

Thus we see that $n_+(n_-)$ is the number of positive (negative) chirality zero-eigenvalue solutions of the Dirac equation. The index theorem, which we derive below, when applied to (3.27) states

$$\begin{aligned} n_- - n_+ &= -\frac{1}{16\pi^2}\int d^{\mu}x\ \mathrm{tr}{}^{*}F^{\mu\nu}F_{\mu\nu} \\ &= \frac{1}{16\pi^2}\ \mathrm{tr}\ T_a T_b \int d^{\mu}x{}^{*}F_a^{\mu\nu}F_{b\mu\nu} \end{aligned} \tag{3.28a}$$

For fermions with total iso-spin T and gauge fields with Pontryagin index n, the above is evaluated with the help of

$$\mathrm{tr}T_aT_b = \frac{1}{3}T(T+1)(2T+1)\delta_{ab} \tag{3.28b}$$

and we find that the index is

$$n_- - n_+ = \frac{2}{3}\ T(T+1)(2T+1)n \tag{3.28c}$$

We have assumed that the gauge potential leads to finite action and that it carries Pontryagin index n; in all other respects, it is arbitrary. However when A^{μ} is self-dual or anti-self-dual, the index theorem may be strengthened by showing that only n_- or n_+ is non-zero: apply $L^{\dagger}$ to (3.27b) and L to (3.27c) to get

$$\left[(\partial_{\mu} + A_{\mu})^2 + 2i\bar{\sigma}_{\mu\nu}F^{\mu\nu}\right]\psi^{+} = 0 \tag{3.29a}$$

$$\left[(\partial_\mu + A_\mu)^2 + 2i\sigma_{\mu\nu}F^{\mu\nu}\right]\psi^- = 0 \tag{3.29b}$$

The duality properties of $\bar{\sigma}^{\mu\nu}(\sigma^{\mu\nu})$ [see (2.19) and (2.28)] assure that $\bar{\sigma}_{\mu\nu}F^{\mu\nu}(\sigma_{\mu\nu}F^{\mu\nu})$ vanishes for self-dual (anti-self-dual) gauge fields. Since $(\partial_\mu + A_\mu)^2$ is a positive definite operator, the differential equation without the gauge-field term does not have normalizable solutions. All known Yang-Mills solutions with finite action are self-dual or anti-self-dual; hence for these potentials there are precisely 2/3 T(T+1)(2T+1)n zero-eigenvalue modes with chirality determined by the gauge field's duality properties. Of course this general result reproduces, for T=1/2 and 1, the numbers found before: n and 4n.

When n_+ or n_- vanishes, a "vanishing theorem" is said to hold. We have seen that such a theorem can be always established when the gauge field is self-dual or anti-self-dual; however, it is not yet known whether the vanishing theorem is valid for more general field configurations.[20]

We now derive the index theorem, by a method which makes reference to the anomaly of the axial-vector current.[19] First a local version of (3.28a) is obtained; upon integration over all space, (3.28a) is regained. The derivation begins with a consideration of the full eigenvalue problem for the Dirac operator.

$$i\gamma^\mu(\partial_\mu + A_\mu)\psi_E = E\psi_E \tag{3.30}$$

It is clear that γ_5, which anti-commutes with the left-hand-side of (3.30), takes eigenfunctions ψ_E into ψ_{-E}, while the zero-eigenvalue modes can be chosen to be eigenstates of γ_5.

$$\int d^4x\,\psi_E^\dagger(x)\gamma_5\psi_E(x) = 0 \qquad E \neq 0 \tag{3.31a}$$

$$\int d^4x\,\psi_0^\dagger(x)\gamma_5\psi_0(x) = (\pm)1 \qquad \binom{\text{positive}}{\text{negative}}\text{ chirality} \tag{3.31b}$$

To proceed, we construct the resolvent of the differential operator in (3.30).

$$R(x,y;\mu) = \sum_E \frac{\psi_E(x)\psi_E^\dagger(y)}{E+i\mu} \tag{3.32a}$$

$$\left[i\gamma^\mu(\partial_\mu + A_\mu) + i\mu\right]R(x,y;\mu) = \delta^4(x-y) \tag{3.32b}$$

We shall want to take x and y coincident, which may produce infinities and ambiguities that must be regulated. A convenient, gauge invariant regularization is the Pauli-Villars scheme; from (3.32a) the same expression is subtracted with μ replaced by M, and at the end of the calculation M is passed to infinity. [It happens that one regulator mass is sufficient for the problem at hand.]

$$R_{Reg}(x,y;\mu) = \lim_{M\to\infty} [R(x,y;\mu) - R(x,y;M)] \tag{3.32c}$$

Next we form an axial-vector projection of the resolvent - which we call the "axial vector current" - and also its divergence.

$$J^\mu_5(x) = \mathrm{tr}\ i\gamma^\mu\gamma_5\ R_{reg}(x,x;\mu) \tag{3.33}$$

A simple calculation, based on (3.30) gives

$$\partial_\mu J^\mu_5(x) = 2i\mu \sum_E \frac{\psi^\dagger_E(x)\gamma_5\psi_E(x)}{E+i\mu} - \lim_{M\to\infty} \left(2iM \sum_E \frac{\psi^\dagger_E(x)\gamma_5\psi_E(x)}{E+iM}\right) \tag{3.34}$$

To complete the calculation we need to evaluate the limit. [Formally it is given by the ambiguous expression $2i\sum_E\psi^\dagger_E(x)\gamma_5\psi_E(x) = 2i\ \mathrm{tr}\gamma_5 \sum_E \psi_E(x)\psi^\dagger_E(x) = 2i(0)\delta^4(0)$.] It is here that we can use the results about the anomaly of the axial-vector current operator constructed from quantum Fermi fields which interact with an external classical vector field. Of course in the above we are not dealing with a quantum field theory; rather, we are studying differential equations in Euclidean space. Nevertheless, the objects we have constructed are recognized to be precisely the [Wick rotated] quantal amplitudes. Thus the resolvent R is exactly the [Wick rotated] propagator for a massive Fermi field in an external gauge potential, and the axial-vector current J^μ_5 is the [Wick rotated] vacuum expectation value of the axial-vector current operator for that theory. Hence we arrive at[19]

$$\partial_\mu J^\mu_5(x) = 2i\mu J_5(x) - \frac{1}{8\pi^2}\mathrm{tr}{}^*F^{\mu\nu}F_{\mu\nu}$$
$$J_5(x) = \sum_E \frac{\psi^\dagger_E(x)\gamma_5\psi_E(x)}{E+i\mu} \tag{3.35}$$

This anomalous divergence of the axial-vector current is also the local form of the index theorem.

To derive the global relation, Eq. (3.35) is integrated over all x.[21]

$$\int d^4x \partial_\mu J_5^\mu(x) = 2i\mu \int d^4x \sum_E \frac{\psi_E^\dagger(x)\gamma_5\psi_E(x)}{E+i\mu}$$

$$- \frac{1}{8\pi^2} \int d^4x \; tr^*F^{\mu\nu}F_{\mu\nu} \qquad (3.36)$$

When it is assumed that the integral on the left-hand-side produces no surface terms, and that the integral on the right-hand-side can be evaluated term-by-term with the help of (3.31), Eq. (3.28a) is regained.

The above derivation also exposes circumstances which may modify the simple, integrated expression (3.28a). The surface term for the integral of $\partial_\mu J_5^\mu$ need not vanish; the term-by-term integration may be illegitimate. In that case an additional contribution is present in (3.28a); it is called the "signature defect". We expect that such pathologies occur when long-range potentials are present in the Dirac equation. In our example the gauge potential can be long-range, but gauge-invariant quantities see only the short-range gauge field, and the simple result (3.28a) is expected to hold, as is indeed the case in the explicit computations for iso-spin 1/2 and 1. A more mathematical formulation states that the index theorem should be applied only to compact manifolds without boundaries. In that case there obviously is no surface term on the left-hand-side of (3.36); the summation over eigenvalues on the right-hand-side is truly a sum over discrete eigenvalues, and the term-by-term integration may be justified. In our example, we are on the non-compact manifold of Euclidean 4-space. However, as explained in Section 2, the conformal invariance of the theory and the assumption that the gauge fields decrease rapidly at infinity allow our problem to be mapped onto the surface of a 4-dimensional hypersphere, and a signature defect is not expected. [This consideration introduces the following subtlety: The normalizability condition for the Dirac equation in an O(5) covariant formulation requires only that $\int \frac{d^4x}{1+x^2} \psi^\dagger(x)\psi(x)$ converge, while $\int d^4x\psi^\dagger(x)\psi(x)$ may diverge. However, we have not encountered a situation in 4 dimensions where this distinction makes a difference.]

Even though the signature defect is absent in the present application of index theory, it will play a role in other physical situations. We have encountered Dirac equations in an odd number of Euclidean dimensions, where no anomaly exists, yet there are zero-eigenvalue modes.[22] These examples involve soliton-monopole potentials which include a long-range Higgs field and provide physically interesting applications of the signature defect.[23]

To illustrate the utility of the index theorem, we derive once more the result that the self-duality equation $F^{\mu\nu}=*F^{\mu\nu}$ has 8n-3 gauge invariant deformations. An infinitesimal variation of this equation, caused by an arbitrary variation δA^{μ} about a self-dual gauge potential A^{μ}, is [see (2.100) and (2.103)],

$$\delta F^{\mu\nu} - \delta * F^{\mu\nu} = 0$$

$$\delta F_{\mu\nu} = D^{A}_{\mu}\delta A_{\nu} - D^{A}_{\nu}\delta A_{\mu} \tag{3.37a}$$

Since the matrix $\bar{\sigma}^{\mu\nu}$ is anti-symmetric and anti-self-dual, the above is entirely equivalent to

$$\bar{\sigma}^{\mu\nu}D^{A}_{\mu}\delta A_{\nu} = 0 \tag{3.37b}$$

Next we write $\bar{\sigma}^{\mu\nu}$ as $\frac{1}{2i}(\alpha^{\mu}\bar{\alpha}^{\nu} - \delta^{\mu\nu})$ and impose the background gauge condition on the small variations.

$$D^{A}_{\mu}\delta A^{\mu} = 0 \tag{3.37c}$$

Hence the equation one is left to solve is

$$\alpha^{\mu}D^{A}_{\mu}(\bar{\alpha}^{\nu}\delta A_{\nu}) = 0 \tag{3.37d}$$

But we recognize (3.37d) to be two decoupled Dirac equations for two Dirac two-component spinors in the adjoint (T=1) representation - these two spinors make up the two columns of the matrix $\bar{\alpha}^{\nu}\delta A_{\nu}$ and move in the external potential A^{μ}. In other words, the above demonstrates that if ψ_a, a=1,2,3, solves

$$i\alpha^{\mu}(\partial_{\mu} + A_{\mu})\psi = 0 \tag{3.38}$$

with self-dual A^{μ}_{a}, then $A^{\mu}_{a} + u^{\dagger}\alpha^{\mu}\psi_{a}$ is self-dual to first order, with u being a constant arbitrary two-component spinor. The index theorem states that there are 4n solutions to (3.38); and, by the construction, 8n small deformations are found. One shows that they can be arranged into exactly 8n linearly independent, real combinations, and one further finds that 3 of them are infinitesimal gauge transformations $D^{A}_{\mu}\theta$. Hence the number of infinitesimal deformations is 8n-3, in agreement with our previous computation.[24] The explicit solution of the Dirac equation for iso-vector fermions, presented earlier, provides therefore explicit formulas for the small deformations of a self-dual gauge potential, now in the familiar background gauge, rather than in the somewhat obscure gauge employed previously.

The startling relationship between solutions of the Dirac equation for iso-vector fermions and small deformations of the self-dual gauge potential goes even further. The following fact is easily established by reducing products of gamma matrices. If $F^{\mu\nu}_a$ solves the Yang-Mills equation, then $\psi_a = F^{\mu\nu}_a \gamma_\mu \gamma_\nu u$ and $\psi^a_a = F^{\mu\nu}_a \gamma_\mu \gamma_\nu \gamma \cdot xu$ solve the iso-spin 1 Dirac equation where u is an arbitrary constant 4-component spinor. [Notice that when $F^{\mu\nu}_a$ is self-dual (anti-self-dual) only the negative (positive) chiral components of these spinors are non-vanishing.] These curious connections between gauge fields and Fermi fields are related to the super-symmetry properties of iso-vector fermions.[25]

4. MINKOWSKI SPACE SOLUTIONS

In this last Section we shall discuss some solutions to the Yang-Mills equations in Minkowski space which have recently been found. At the present time it is not clear what information about the quantum theory is contained in these classical field configurations; towards the end of the presentation we shall describe some tentative ideas that we have about this question.

4.1. O(4)×O(2) Formalism

Rather than recording the solutions straight-away, we first develop a kinematical framework in which their elegance and significance is manifest. The Yang-Mills theory possesses the O(4,2) conformal group of invariances. Under conformal transformations the coordinates x^μ transform non-linearly. But as is well known, one may introduce a light-like six-vector ξ^A, A=1...6, $\xi^2=\xi_1^2+\xi_2^2+\xi_3^2-\xi_4^2+\xi_5^2-\xi_6^2=0$, which has the property that (pseudo) rotations of ξ^A correspond to conformal transformations of $\xi^\mu/(\xi_5+\xi_6)$, $\mu=1,2,3,4$. The action of the conformal group is thus linearized on this null-cone, and it becomes convenient to use the ξ^A's as coordinates, rather than the conventional x^μ. The relationship between the two coordinate systems contains of course a large amount of ambiguity. For example, one possible mapping is

$$\xi^i = \frac{2x^i\lambda}{\lambda^2-x^2}\, f\,, \quad i=1,2,3$$

$$\xi^4 = \frac{2t\lambda}{\lambda^2-x^2}\, f$$

$$\xi^5 = \frac{\lambda^2+x^2}{\lambda^2-x^2}\, f$$

$$\xi^6 = f \tag{4.1}$$

where $x^2 = t^2 - \vec{x}^2$, λ is an arbitrary scale and f is an arbitrary function of x which parametrizes the ambiguity. Conventionally the ambiguity is removed by setting homogeneity conditions on all interesting objects of the theory. An alternate way to remove the ambiguity is to fix the value of $\vec{\xi}^2+\xi_5^2=\xi_4^2+\xi_6^2$. This we do here; we set that quantity to unity, which forces ξ^A to lie on a six-dimensional hypertorus. Thus the mapping introduces two Euclidean vectors; a 4-component $\hat{R}^\mu$ and a 2-component $\hat{r}^a$ of unit magnitude. Explicitly one has[26]

$$\hat{R}^i = \frac{2x^i\lambda}{w} \qquad i=1,2,3$$

$$\hat{R}_4 = \frac{\lambda^2+x^2}{w}$$

$$\hat{r}_1 = \frac{2t\lambda}{w}$$

$$\hat{r}_2 = \frac{\lambda^2-x^2}{w}$$

$$w^2 = (\lambda^2-x^2)^2 + 4t^2\lambda^2 \tag{4.2}$$

and all of Minkowski space is mapped, two-to-one, onto the hypertorus $\hat{r}^2 = \hat{R}^2 = 1$. The action of the $O(4)\times O(2)$ subgroup of the conformal group is then represented by independent rotations of $\hat{R}^\mu$ and $\hat{r}^a$, while the remaining conformal transformations mix the $\hat{R}^\mu$'s with the $\hat{r}^a$'s. [The metric of the $\hat{R}^\mu$ coordinates, as well as that of the $\hat{r}^a$ coordinates is Euclidean.]

The ordinary derivatives $\frac{\partial}{\partial x^\mu}$ and gauge potentials A^μ are mapped into derivatives and gauge potentials tangential to the surface of the torus. Thus we have

$$\hat{\partial}_\mu = \frac{\partial}{\partial \hat{R}^\mu} - \hat{R}_\mu(\hat{R}^\nu \frac{\partial}{\partial \hat{R}^\nu})$$

$$\hat{\Delta}_a = \frac{\partial}{\partial \hat{r}^a} - \hat{r}_a(\hat{r}^b \frac{\partial}{\partial \hat{r}^b})$$

$$\frac{\partial}{\partial x^\mu} = \frac{\partial \hat{R}_\nu}{\partial x^\mu}\frac{\partial}{\partial \hat{R}^\nu} + \frac{\partial \hat{r}_a}{\partial x^\mu}\frac{\partial}{\partial \hat{r}^a} \tag{4.3}$$

$$A^\mu(x)dx_\mu = \hat{A}^\mu(\hat{R},\hat{r})d\hat{R}_\mu + \hat{a}^a(\hat{R},\hat{r})d\hat{r}_a$$

$$\hat{R}_\mu\hat{A}^\mu = \hat{r}_a\hat{a}^a = 0 \tag{4.4}$$

[Compare with the similar mapping discussed in Section 2.4 for the O(5) formalism in Euclidean space.]

From $\hat{A}^\mu$ and $\hat{a}^a$ one constructs an "electric" field

$$\hat{E}_{a\mu} = \hat{\Delta}_a \hat{A}_\mu - \partial_\mu \hat{a}_a + [\hat{a}_a, \hat{A}_\mu] \tag{4.5a}$$

and a "magnetic" field.

$$\hat{H}_{\mu\nu} = \partial_\mu \hat{A}_\nu - \partial_\nu \hat{A}_\mu - \hat{R}_\mu \hat{A}_\nu + \hat{R}_\nu \hat{A}_\mu + [\hat{A}_\mu, \hat{A}_\nu] \tag{4.5b}$$

Both are tangential, i.e., $\hat{R}^\nu \hat{H}_{\mu\nu}=0, \hat{R}^\nu \hat{E}_{a\nu}=0, \hat{r}^a \hat{E}_{a\mu}=0$. The nomenclature is derived from the fact that near the origin $\hat{E}_{2i}$ and $\hat{H}_{ij}$ are proportional to the electric F_{oi} and magnetic F_{ij} components of $F_{\mu\nu}$.

It is convenient to parametrize the two-dimensional vector by an angle.

$$\hat{r}_1 = \cos\tau \qquad \hat{r}_2 = \sin\tau \tag{4.6}$$

Then $\hat{\Delta}_a = -\varepsilon_{ab}\, \hat{r}^b \frac{\partial}{\partial\tau}$, and we may similarly set

$$\begin{aligned} \hat{a}_a &= -\varepsilon_{ab}\, \hat{r}^b \hat{a} \\ \hat{E}_{a\mu} &= -\varepsilon_{ab}\, \hat{r}^b \hat{E}_\mu \\ \hat{E}_\mu &= \dot{\hat{A}}_\mu - \hat{\partial}_\mu \hat{a} + [\hat{a}, \hat{A}_\mu] \end{aligned} \tag{4.7}$$

where the dot refers to differentiation with respect to τ. The gauge potentials can be modified by a gauge transformation.

$$\begin{aligned} \hat{A}_\mu &\to g^{-1}\hat{A}_\mu g + g^{-1}\hat{\partial}_\mu g \\ \hat{a} &\to g^{-1} a g + g^{-1}\dot{g} \end{aligned} \tag{4.8}$$

In particular, one may gauge transform $\hat{a}$ to zero.

The Minkowski-space Yang-Mills action

$$I = \frac{1}{2} \int d^4x \; \mathrm{tr}\, F^{\mu\nu} F_{\mu\nu} \tag{4.9a}$$

becomes in terms of new variables

$$I = -\frac{1}{2} \int d\tau d\Omega \mathrm{tr}[\hat{E}^\mu \hat{E}_\mu - \frac{1}{2} \hat{H}^{\mu\nu} \hat{H}_{\mu\nu}] \tag{4.9b}$$

The range of the τ integration is from 0 to 2π: the remaining integration is over the surface of the sphere $\hat{R}^2 = 1$. Since the mapping $x^\mu \to \{\tau, \Omega\}$ is two-to-one there appears an additional factor of 1/2 in (4.9b). For the same reason, there is no periodicity requirement in τ. Note that the range of integration is compact, hence finiteness of the action is guaranteed when the fields are non-singular on the τ circle and on the Ω sphere. The Yang-Mills field equations of motion, which follow from varying the fields in (4.9b), are

$$\hat{\partial}^\mu \hat{H}_{\mu\nu} - \dot{\hat{E}}_\nu + [\hat{A}^\mu, \hat{H}_{\mu\nu}] - [\hat{a}, \hat{E}_\nu] = 0$$
$$\hat{\partial}^\mu \hat{E}_\mu + [\hat{A}^\mu, \hat{E}_\mu] = 0 \qquad (4.10)$$

4.2. Invariant Solutions

Having developed this hypertoroidial formalism, which makes explicit the $O(4)\times O(2)$ group of symmetries of the problem, we may look for solutions which are themselves invariant under interesting subgroups of $O(4)\times O(2)$. Specifically $O(4)$ invariant field configurations are obtained by setting

$$\hat{A}^\mu = i\sigma^{\mu\nu}\hat{R}_\nu f(\tau)$$
$$a = 0 \qquad (4.11)$$

The Yang-Mills equations then reduce to

$$\ddot{f} + 2f(f+1)(f+2) = 0 \qquad (4.12)$$

which is identical to the equation of motion of a point particle in a two-minimum potential, symmetric about $f=-1$.

The solutions are obvious. A first integral is immediate.

$$\frac{1}{2}\dot{f}^2 + \frac{1}{2}((f+1)^2-1)^2 = \varepsilon \qquad (4.13)$$

There are τ independent solutions: $f=0,-1,-2$; these lead to $O(4)\times O(2)$ invariant Yang-Mills potentials. $f=0$ gives the trivial, vanishing potential; $f=-2$ is a pure gauge, $f=-1$ is gauge-equivalent to the solution found by deAlfaro, Fubini and Furlan.[27] In the mechanical analog problem, $f=0$ and -2 correspond to the particle sitting at the minima of the potential; $f=-1$ is in unstable equilibrium at the maximum. The τ dependent solutions have been found by Lüscher and Schechter.[26] These are periodic functions of $\tau-\tau_o$, where τ_o is another integration constant. Two different types of solution are seen: for $\varepsilon<1/2$ there are separate oscillations about each of the two minima, for $\varepsilon>1/2$ the oscillations range widely across the central hump. The analytic expression, which we do not record here, involves Jacobi elliptic functions. For $\varepsilon=1/2$ a simple formula holds

$$f = -1 \pm \frac{\sqrt{2}}{\cosh\sqrt{2}(\tau - \tau_o)} \tag{4.14}$$

It is clear that the action of these solutions is finite.

By using the formulas (4.2) and (4.4) the Yang-Mills fields may be given in conventional variables. We do not carry out the projection, since the resulting configurations do not exhibit any noteworthy features; the fields have finite energy, but dissipate in time in accordance to general theorems which tell us that no soliton solutions exist in the pure Yang-Mills theory.[2] Rather we prefer to remain with the hypertorodial coordinates where the solutions are constant or periodic, not in time, to be sure, but in the new variable τ.

4.3. Alternate Quantization of Yang-Mills Theory

Let us recall that the evolution of a dynamical system need not necessarily be described by time evolution. Other combinations of t and $\vec{x}$ are possible evolution variables, provided all space-time is covered. In a Lorentz invariant theory one may use any time-like vector for describing evolution of initial data specified on a space-like surface. In a conformally invariant theory there are further possibilities, and in particular one can use τ to describe evolution of data specified on the Ω surface. The generator of τ translations is easily determined; it is $R = 1/2(\frac{1}{\lambda} K^o + \lambda P^o)$.

When these considerations are brought to bear on a quantum theory they lead to the well-known conclusion that in a Lorentz invariant quantum theory there are alternate methods of quantization which do not rely on a Hamiltonian evolving the quantal system in time. Indeed some years ago, light-cone quantization was profitably employed to analyze deep-inelastic scattering processes.[28] Similarly, alternate quantization methods have been suggested for conformally invariant quantum theories in Euclidean space.[29] In the present context it appears very interesting to take τ as the quantization variable. Correspondingly the Hamiltonian is R, which we now shall call H_τ, and the states which diagonalize H_τ become the "static" basis for the Hilbert space [rather than the energy eigenstates of conventional quantum theory]. Precisely this alternative for conformally invariant theories has been advocated by Fubini.[30] It is emphasized that a new quantum theory is not being developed, rather the conventional theory is discussed in terms of a new set of basis states. Indeed it was explicitly demonstrated by deAlfaro, Fubini and Furlan,[31] in the simple example of the conformally invariant quantum mechanics of a point particle in a $1/r^2$ potential, that the new approach is entirely equivalent to the conventional one.

Here we consider quantizing Yang-Mills theory with τ as the evolution variable and H_τ as the Hamiltonian.

$$H_\tau = -\frac{1}{2}\int d\Omega \mathrm{tr}[\hat{E}^\mu \hat{E}_\mu + \frac{1}{2}\hat{H}^{\mu\nu}\hat{H}_{\mu\nu}] \tag{4.15}$$

The canonical quantization procedure is entirely straightforward. The gauge $\hat{a}=0$ is very convenient, the canonical coordinates are then $\hat{A}^\mu$ with conjugate momenta $\hat{E}^\mu = \dot{\hat{A}}^\mu$. Then Eq. (4.10), $\hat{D}_\mu \hat{E}^\mu = 0$, becomes Gauss' law which has to be imposed as a condition on the physical states. Equal-τ commutators involve delta functions on the Ω surface and in the non-interacting case H_τ can be explicitly diagonalized. One important result emerges: the spectrum is discrete, the eigenvalues are dimensionless integers. This of course is a consequence of the fact that the space is compact; infra-red divergence has been tamed.

4.4. Semi-classical Quantization of Solutions

We shall not go into the obvious details of the canonical approach. Rather we want to inquire to what extent the solutions which we have previously discussed can be used to perform a semi-classical analysis of the quantum theory. Clearly the constant solutions f=0 and -2, which are pure gauges, correspond to the vacuum; the latter being a non-vanishing, pure gauge potential.

$$\begin{aligned} \hat{A}^\mu &= -2i\sigma^{\mu\nu}\hat{R}_\nu \\ &= g^{-1}\hat{\partial}^\mu g \\ g &= \alpha\cdot R \end{aligned} \tag{4.16a}$$

This gauge has unit winding number, and obviously describes one of the many classically degenerate vacua of Yang-Mills theory. The other vacua are O(3) invariant configurations.

$$\hat{A}^\mu_n = g^{-n}\hat{\partial}^\mu g^n \tag{4.16b}$$

Of course there is tunnelling between the vacua; the pseudoparticle solution which exists for imaginary τ insures this. [The pseudoparticle is just the kink solution of (4.12) with $\ddot{f}$ replaced by $-\ddot{f}$.]

The f=-1 solution is seen to correspond to an unstable vacuum, hence no quantum state is associated with it. Nevertheless we can use this solution to compute the height of the barrier which separates the two minima; it is $\frac{3\pi^2}{2}$. [Another curious feature of this solution can be noted. If the pseudoparticle is continued from imaginary τ to real τ, we obtain a complex self-dual gauge potential. Since the equations are non-linear one does not expect, a priori, that the real and imaginary parts of this complex configuration satisfy the Yang-Mills equation. Nevertheless, the real part of the self-dual complex potential is just the f=-1 solution.[32]]

Of course it is the periodic solutions that offer the most interesting probe into the quantum theory since it is possible to quantize them by the Bohr-Sommerfeld method, thus obtaining the semi-classical spectrum of H_T. Before proceeding, let us review the Bohr-Sommerfeld method as applied to field theory.

For quantum mechanics of a point particle in one-dimensional motion, the WKB quantization condition reads

$$(n+\tfrac{1}{2})\pi = \int_{q_1}^{q_2} dq\ p(q) \tag{4.17a}$$

where p(q) is the local momentum $\sqrt{2E-2V(q)}$ and the q_i are the turning points of the bound motion. The quantity $n\pi$ on the left-hand-side arises from the correspondence principle; the quantity $1/2\pi$ is derived from the details of one-dimensional motion, and is specific to that problem. The approximation is presumed accurate for large n. When one drops 1/2 compared with n, one is left with the Bohr-Sommerfeld quantization, which after a change of variable from q to t may also be written as

$$n\pi = \int dt\ \dot{q}(t)p(t) \tag{4.17b}$$

where the integration is now over a semi-period.

Although the full WKB condition may also be derived for many degrees of freedom,[33] as well as for a field theory with infinite degrees of freedom,[34] we remain with the simpler Bohr-Sommerfeld condition. For several degrees of freedom (4.17b) generalizes to

$$n\pi = \int dt \sum_m p_m(t)\dot{q}_m(t) \tag{4.18a}$$

while for field theory we take

$$n\pi = \int dt \int dV\Pi(t,v)\dot{\Phi}(t,v) \tag{4.18b}$$

Eq. (4.18b) instructs us to find a periodic solution Φ, multiply its time derivative by the canonical momentum H and integrate over all variables V, save the evolution parameter t. The resulting quantity, $\int dV\Pi\dot{\Phi}$, depending on t as well as various constants of integration, is then integrated over a semi-period of t and set equal to $n\pi$, thus achieving one quantization condition on the constants of motion.

For the Yang-Mills theory governed by the action (4.9b), the Bohr-Sommerfeld condition reads

$$n\pi = -2 \int d\tau \int d\Omega \, \mathrm{tr}\{\frac{\delta\mathcal{L}}{\delta\dot{\hat{A}}^{\mu}} \dot{\hat{A}}^{\mu} + \frac{\delta\mathcal{L}}{\delta\dot{\hat{a}}} \dot{\hat{a}}\} \tag{4.19a}$$

Since

$$\frac{\delta\mathcal{L}}{\delta\dot{\hat{a}}} = 0, \quad \frac{\delta\mathcal{L}}{\delta\dot{\hat{A}}} = \frac{1}{2} \hat{E}_{\mu} \tag{4.19b}$$

(4.19a) becomes

$$n\pi = - \int d\tau \int d\Omega \, \mathrm{tr}\hat{E}^{\mu}\dot{\hat{A}}_{\mu}$$

$$= - \int d\tau \int d\Omega \, \mathrm{tr}\{\hat{E}^{\mu}\hat{E}_{\mu} + \hat{E}^{\mu}(\hat{\partial}_{\mu}\hat{a} - [\hat{a},\hat{A}^{\mu}])\} \tag{4.19c}$$

An integration by parts and use of the equation for $\hat{E}^{\mu}$ shows that the second term in the curly brackets can be set to zero, and we are left with the gauge-invariant quantization condition.

$$n\pi = - \int d\tau \int d\Omega \, \mathrm{tr}\hat{E}^{\mu}\hat{\mathbf{E}}_{\mu} \tag{4.19d}$$

We insert into this formula the known periodic solutions, for which

$$\hat{A}^{\mu} = i\sigma^{\mu\nu}\hat{R}_{\nu} f(\tau)$$

$$\hat{E}^{\mu} = i\sigma^{\mu\nu}\hat{R}_{\nu} \dot{f}(\tau) \tag{4.20a}$$

with f satisfying

$$\dot{f}^2 = 2\varepsilon - f^2(f+2)^2 \tag{4.20b}$$

and get

$$n\pi = 6\pi \int d\tau \dot{f}^2(\tau) = 6\pi \int df \sqrt{2\varepsilon - f^2(f+2)^2} \tag{4.21}$$

where the f integration ranges between the two turning points of the classical motion. The meaning of ε is clear. If we evaluate the τ-Hamiltonian, Eq. (4.15), for the solution (4.20), we find $3\pi^2\varepsilon$. Evidently our semi-classical procedure provides the semi-classical eigenvalues of H_τ, which we call $E_\tau = 3\pi^2\varepsilon$.

To determine the dependence of E_τ on n, the f integration in (4.21) must be performed. The formulas involve the complete elliptic integrals of the first and second kind. Two distinct expressions emerge depending whether ε is less than or greater than 1/2. We record here only the asymptotic forms. For small n, degenerate levels are found with

$$E_\tau = n \tag{4.22}$$

For large n, there is no degeneracy and

$$E_\tau \propto n^{4/3} \tag{4.23}$$

[When small oscillations about any of the vacuum configurations are canonically quantized one also obtains a linear spectrum.] The quantal significance of the periodic solution may be given: when it is expanded in a Fourier series, the Fourier coefficients provide a semi-classical approximation to the matrix elements of the quantum field $\hat{A}^\mu$ between successive bound states.[35]

What corrections to these semi-classical results are envisioned? It is clear that tunnelling removes the degeneracy; this is clear and causes no conceptual problems. Much more problematical are the questions which arise if one confronts this entire program with the realities of Yang-Mills perturbation theory. The problem is of course that the well-known anomalies prevent the theory from being conformally invariant - the renormalization procedure introduces conformal symmetry breaking.[36] In other words it is not obvious how to relate results of the alternate quantization method to the physically relevant, Poincaré covariant theory.

One of two approaches is possible. The theory is regulated in a conventional way; H_τ acquires a τ dependence; bound states disappear but perhaps some kind of adiabatic perturbation theory can be used to study further properties of the spectrum.[37] Alternatively, a non-conventional regularization scheme may be adopted such that H_τ remains a constant of motion. The theory loses translation covariance, since translation generators acquire a τ dependence. Let us suppose however that even in the renormalized theory it is true that $P^\mu = \lim_{\lambda\to\infty} \frac{2}{\lambda} R^\mu$. Then information about the translationally covariant theory could be regained in the limit.

4.5. Reduced Yang-Mills Theories

To conclude these considerations, we put aside the serious obstacles which still exist in assessing the physical relevance of our alternate proposals, and proceed to another suggestion for obtaining results about Yang-Mills theory. It is apparent that at the present time the model appears too complicated for a complete solution, either classically or quantum mechanically. Nevertheless it has been possible to obtain complete classical solutions which respect a symmetry. A suggestion for the analysis of the quantum theory is to reduce the full degrees of freedom to those that are invariant under a subgroup of the conformal group. In this way we obtain quantum systems which are considerably simpler than the full Yang-Mills theory, yet may retain some of the physical properties of the complete theory.

A very simple model is obtained if we freeze out all but the O(4) invariant degrees of freedom. The Ansatz (4.11) leads to the action

$$I = 6\pi^2 \int d\tau \left[\frac{1}{2}\dot{f}^2 - \frac{1}{2} f^2(f+2)^2\right] \tag{4.24}$$

The quantum mechanics, though trivial, already exhibits some of the features of the complete theory: classical degeneracy, vacuum tunnelling, an anharmonic oscillator bound state spectrum.

Richer is the model where only O(3) symmetry is imposed. The most general O(3) Ansatz is

$$\begin{aligned} \hat{A}^\mu &= i\sigma^{\mu\nu}\hat{R}_\nu f_1 + iP^{\mu\alpha}\sigma_{\alpha\beta}\hat{C}^\beta f_2 \\ &\quad + i\, P^{\mu\nu}\hat{C}_\nu\hat{C}_\alpha\sigma^{\alpha\beta}\hat{R}_\beta f_3 \\ \hat{a} &= i\,\hat{R}_\alpha\sigma^{\alpha\beta}\hat{C}_\beta f_4 \\ P^{\mu\nu} &= g^{\mu\nu} - \hat{R}^\mu\hat{R}^\nu \end{aligned} \tag{4.25}$$

Here $\hat{C}$ is a unit 4-vector which picks out the direction of O(4) breakdown to O(3). We set $R\cdot C=\cos\theta$, $0\leqslant\theta\leqslant\pi$, and the f_i's depend on τ and θ. With the redefinitions

$$\phi_1 = 1 + \sin^2\theta f_1$$
$$\phi_2 = \sin\theta(\cos\theta f_1 + f_2)$$
$$A^\theta = f_1 + \cos\theta f_2 + \sin^2\theta f_3$$
$$A^\tau = \sin\theta f_4 \qquad (4.26)$$

the action becomes

$$I = 2\pi \int d\tau d\theta\{|(\partial_\mu + iA_\mu)\Phi|^2 - \frac{1}{4}\sin^2\theta F^{\mu\nu}F_{\mu\nu} - \frac{1}{2\sin^2\theta}(1 - |\Phi|^2)^2\} \qquad (4.27)$$

where $F^{\mu\nu} = \partial^\mu A^\nu - \partial^\nu A^\mu$, $\Phi = \phi_1 + i\phi_2$, and the metric is $\begin{pmatrix}1 & 0\\ 0 & -1\end{pmatrix}$ with the first coordinate being τ, the second θ. The above is an Abelian Higgs model in a 2-dimensional space of constant curvature, with a 3-parameter O(2,1) conformal invariance group of coordinate transformations.

$$\delta\xi^\alpha : \begin{array}{l}\delta\tau = a + b\sin\tau\cos\theta + c\cos\tau\cos\theta\\ \delta\theta = \phantom{a + {}} b\cos\tau\sin\theta - c\sin\tau\sin\theta\end{array}$$

$$\delta\Phi = (\delta\xi^\alpha)\partial_\alpha\Phi$$
$$\delta A^\mu = (\delta\xi^\alpha)\partial_\alpha A^\mu + (\partial^\mu\delta\xi^\alpha)A_\alpha \qquad (4.28)$$

The gauge theory possesses an infinite set of topologically inequivalent classical vacua, just as does the complete theory.

$$\Phi = e^{2in\theta}$$
$$A^\tau = 0$$
$$A^\theta = 2n \qquad (4.29)$$

The imaginary time $[\tau \to i\tau]$ version of the theory has n-pseudoparticle solutions. [The Euclidean model is a conformal transformation of Witten's Lagrangian.[14]] Thus the simplified theory gives rise to the multiple vacua and tunnelling of the complete theory.

At present nothing more is known about the reduced classical or quantum mechanical theory (4.27), beyond of course its O(4) invariant "sub-theory". There is a fine tradition in theoretical physics of studying 2-dimensional models for clues to realistic

4-dimensional ones. We hope that an understanding of our 2-dimensional model (4.27) will help unravel the complexities of the full Yang-Mills theory from which, after all, it was obtained.

This material was presented at various conferences in the summer and early fall of 1977, and formed the content of lectures given at Berlin, Kiev and at the Brookhaven National Laboratory. We are grateful to I. Biyalinicki-Birula, E. Caianiello, J. Iliopoulos, A. Kamal, H. Kleinert and O. Parasiuk for giving us the opportunity to discuss this work, and to K. Yoshida, who provided an early version of the manuscript. Also we acknowledge with thanks the assistance of M. Ansourian, F. Ore and B. Schechter in the preparation of our Review.

REFERENCES

* This work is supported in part through funds provided by ERDA under Contract EY-76-C-02-3069.*000

† Material based on lectures presented at:

Workshop on Theoretical Problems in Quantum Chromodynamics
Crete, Greece, 20-30 June, 1977

Workshop on Solitons
Salerno, Italy, 27 June - 23 July, 1977

Banff Summer Institute on Particles and Fields
Banff, Canada, 26 August - 3 September,1977

Mystery of the Soliton
Warsaw, Poland, 26-30 September, 1977

1) For previous reviews see R. Rajaraman, Phys. Rep. 21C, 227 (1975); R. Jackiw, Acta Physica Pol. B6, 919 (1975); S. Coleman, Erice Lectures (1975); J.-L. Gervais and A. Neveu, Phys. Rep. 23C, 237 (1976); R. Jackiw, Rev. Mod. Phys. 49, 681 (1977); S. Coleman, Erice Lectures (1977).

2) S. Coleman, Erice Lectures (1975); H. Pagels, Phys. Lett. 68B, 466 (1977); S. Coleman, Commun. Math. Phys. 55, 113 (1977).

3) A. Belavin, A. Polyakov, A. Schwartz and Y. Tyupkin, Phys. Lett. 59B, 85 (1975).

4) G. 't Hooft, these Proceedings.

5) R. Jackiw and C. Rebbi, Phys. Rev. D14, 517 (1976).

6) S. Adler, Phys. Rev. D6, 3445 (1972); S. Adler, Phys. Rev. D8, 2400 (1973).

7) F. Ore, Phys. Rev. D15, 470 (1977).

8) F. Ore, Phys. Rev. D16, 1041 (1977); Phys. Rev. D (in press).

9) S. Chadha, A. D'Adda, P. DiVecchia and F. Nicodemi, Phys. Lett. 67B, 103 (1977).

10) G. 't Hooft, Phys. Rev. D14, 3432 (1976).

11) F. Wilczek in Quark Confinement and Field Theory, D. Stump and D. Weingarten, eds., (Wiley, New York, 1977); F. Corrigan and D. Fairlie, Phys. Lett. 67B, 69 (1977).

12) R. Jackiw, C. Nohl and C. Rebbi, Phys. Rev. D15, 1642 (1977).

13) A less general version of this result was obtained by G. 't Hooft, Coral Gables proceedings, 1977.

14) The first construction of multi-pseudoparticle configurations with arbitrary Pontryagin index was achieved, with a method different from the one described here, by E. Witten, Phys. Rev. Lett. 38, 121 (1976). His pseudoparticles are distributed in an O(3) symmetric configuration.

15) R. Jackiw and C. Rebbi, Phys. Lett. 67B, 189 (1977).

16) R. Jackiw and C. Rebbi, Phys. Rev. D16, 1052 (1977).

17) R. Ward, Phys. Lett 61A, 81 (1977); C. N. Yang, Phys. Rev. Lett. 38, 1377 (1977); M. Atiyah and R. Ward, Commun. Math. Phys. 55, 117 (1977).

18) B. Grossman, Phys. Lett. 61A, 86 (1977).

19) S. Adler in Lectures on Elementary Particles and Quantum Field Theory, Vol. 1, S. Deser, M. Grisaru and H. Pendleton, eds. (MIT Press, Canbridge, 1970); S. Treiman, R. Jackiw, D. Gross, Lectures on Current Algebra and Its Applications, (Princeton University Press, Princeton, 1972).

20) An example of the Dirac equation in a potential with no definite duality properties has been analyzed by L. Dolan, Harvard University preprint. No zero-eigenvalue modes are found.

21) S. Coleman (unpublished) suggested that the index theorem may be derived from the axial-vector anomaly. For further discussion see L. Brown, R. Carlitz and C. Lee, Phys. Rev. D16, 417 (1977); as well as Jackiw and Rebbi, Ref. 16.

22) R. Jackiw and C. Rebbi, Phys. Rev. D13, 3398 (1976).

23) For further discussion of the index theorem, signature defect, etc., see J. Kiskis, Phys. Rev. D15, 2329 (1977); M. Ansourian, Phys. Lett. (in press); B. Schroer and K. Nielsen, Nordita preprint. For a discussion from the mathematical point of view see M. Atiyah, V. Patodi, and I. Singer, Math. Proc. Camb. Phil. Soc. 77, 43 (1975); 78, 405 (1975); 79, 71 (1976).

24) This application of the index theory is due to Brown, Carlitz and Lee, Ref. 21. It is also possible to apply index theory directly to the small deformation equation, without reference to fermions; see A. Schwarz, Phys. Lett. 67B, 172 (1977); M. Atiyah, N. Hitchin and I. Singer, Proc. Nat. Acad. Sci. USA 74, 2662 (1977).

25) D. Freedman and D. Gross, unpublished; Chadha, D'Adda, DiVecchia and Nicodemi, Ref. 9; B. Zumino, Phys. Lett. 69B, 369 (1977).

26) M. Lüscher, DESY preprint; B. Schechter, Phys. Rev. D (in press).

27) V. deAlfaro, S. Fubini and G. Furlan, Phys. Lett. 65B, 163 (1976); see also M. Cervero, L. Jacobs and C. Nohl, Phys. Lett. B (in press); W. Bernreuther, MIT preprint.

28) For a review see R. Jackiw, Springer Tracts in Modern Physics, Vol. 62, G. Höhler, ed., (Springer-Verlag, Berlin, 1972).

29) S. Fubini, A. Hanson and R. Jackiw, Phys. Rev. D7, 1732 (1973).

30) S. Fubini, Nuovo Cim. 34A, 521 (1976).

31) V. deAlfaro, S. Fubini, G. Furlan, Nuovo Cim. 34A, 569 (1976).

32) C. Rebbi, Phys. Rev. D, in press.

33) V. Maslov, Teor. Mat. Fiz. 2, 30 (1970) [Theor. Math. Phys. 2, 21 (1970)]; M. Gutzwiller, J. Math. Phys. 12, 343 (1971).

34) R. Dashen, B. Hasslacher and A. Neveu, Phys. Rev. D10, 4114 1974); 10, 4130 (1974); V. Korepin and L. Faddeev, Teor. Mat. Fiz. 25, 147 (1975) [Theor. Math. Phys. 25, 1039 (1976)].

35) C. Nohl, Ann. Phys. 96, 234 (1976); A. Klein and F. Krejs, Phys. Rev. D13, 3282 (1976).

36) S. Coleman and R. Jackiw, Ann. Phys. 67, 552 (1971).

37) C. Lovelace, Nucl. Phys. B99, 109 (1975).

TRANSVERSE MOMENTUM DISTRIBUTION OF PARTONS IN QUANTUM CHROMODYNAMICS

C. S. Lam

McGill University

Montreal, Canada

1. INTRODUCTION

The material discussed below is based on the work[1] I did recently with Tung-Mow Yan of Cornell University.

Dr. Brown (C.N. Brown of Fermilab gave a seminar on the observation of Υ) mentioned two nights ago that the μ-pair mass distribution calculated from the Drell-Yan model depends on the parton transverse momentum distributions. Furthermore, the knowledge of parton transverse momentum is absolutely necessary for the calculation of the transverse momentum distribution of μ-pairs. Similar remarks can be made about many other experimental quantities, viz., all those quantities whose calculations depend on part but not all of the parton momenta.

Naively, one might expect the root mean square average transverse momentum of partons, $\langle k_T^2\rangle^{\frac{1}{2}}$, to be of the order of 300 MeV, reflecting the average transverse momenta of secondary hadrons in production processes. However, this cannot be strictly correct for otherwise νW_2 in deep inelastic lepton scattering would scale[2], contrary to recent experimental results[3]. Also, the transverse momenta of μ-pairs are larger than what can be obtained from $\langle k_T^2\rangle^{\frac{1}{2}} \simeq$ 300 MeV. The question is what should k_T be? This is obviously a dynamical question, involving strong interaction dynamics, and as such is difficult to answer even when a theory is specified. This is similar to the difficulty encountered in trying to calculate the parton longitudinal momentum distribution from first principles. However, the non-scaling of deep inelastic scattering tells us that the parton distribution depends on the momentum transfer Q. Moreover, this variation with Q for large enough Q is calculable in quantum chromodynamics (QCD) because of its asymptotic freedom nature. We may therefore hope that the variation of k_T with Q is also calculable in QCD, and this is what we are going to do below. For simplicity, I will assume throughout a theory with three colors and four flavors, though formulas with any number of flavors can be similarly worked out[1].

I will review in section 2 what QCD has to say about the parton distribution function f(x,Q), where x stands for the longitudinal momentum fraction of the parton in an infinite momentum frame. In section 3, I will discuss the prediction of QCD for $f(x,k_T,Q)$. Finally, the result of section 3 is applied in section 4 to explain the transverse momentum distribution of μ-pairs.

2. DISTRIBUTION FUNCTION f(x,Q)

2.1. Results from Renormalization Group and Operator Product Expansion[4]

In an asymptotically free theory, it is convenient to use the variable $t = \log(Q^2/\Lambda^2)$ instead of Q, where Λ is some scale parameter. Introducing the x-moments by the definition,

$$f_n(t) = \int_0^1 x^{n-1} f(x,t)\,dx \tag{1}$$

the t-dependence of all flavor-non-singlet quarks is given by

$$q_n^{NS}(t) = C_n t^{A_n/2\pi b} \quad . \tag{2}$$

The flavor-singlet quark q^S, on the other hand, may mix with gluon G. Their t-dependences are given by two terms

$$\left.\begin{aligned} q_n^S(t) &= \alpha_n(k_n t^{\lambda_n/2\pi b}) + \alpha_n'(k_n' t^{\lambda_n'/2\pi b}) \\ G_n(t) &= \beta_n(k_n t^{\lambda_n/2\pi b}) + \beta_n'(k_n' t^{\lambda_n'/2\pi b}) \end{aligned}\right\} \tag{3}$$

where

$$\alpha_n^2 + \beta_n^2 = 1 \qquad \alpha_n'^2 + \beta_n'^2 = 1 \quad . \tag{4}$$

In Eqs. (2) and (3), the constants C_n, k_n, k_n' are free parameters reflecting the initial distributions of quarks and gluons and $2\pi b = 25/6$. The powers A_n, λ_n, λ_n' which govern the dependence on t, as well as the coefficients α_n, β_n, α_n', β_n', are computable and are given numerically in Table 1 and graphically in Fig. 1. We see from the Table and Figure the following general features: (i) A_n, λ_n, λ_n' are all negative, with the single exception of λ_2, which is zero; (ii) $|\lambda_n'| > |A_n| > |\lambda_n|$, and they are all increasing functions of n; (iii) $|\lambda_n|$ gets closer and closer to $|A_n|$ as n increases; correspondingly $|\alpha_n|$ and $|\beta_n|$ get closer to unity. As a matter of fact they are already quite close for n = 4. Actually, for large n, $|\beta_n|/|\alpha_n| = O(1/n \log n)$.

The corresponding variation of f(x,t) with t is shown schematically in Fig. 2. The dotted curve represents the behavior at a larger t than that of the solid curve. Since all $f_n(t)$ $(n > 2)$ decrease with t, and large n probes large x, the solid curve has to drop to get to the dotted curve in the large x region. On the other hand, for sufficiently large t, q_2^s and G_2 are constants, implying that the areas under the curves xf(x,t) (as a function of x) to be approximately independent of t. This can be fulfilled only when the solid curve rises to the dotted curves at small x, as sketched in Fig. 2.

2.2. The Physical Picture of Kogut and Susskind (KS)

If partons were free, deep inelastic structure functions would scale. In a very interesting paper, Kogut and Susskind[5] discussed how parton interactions lead to scaling violations of the type sketched in Fig. 2.

Consider a quark (solid line) moving very fast in an infinite momentum frame as shown in Fig. 3. It emits gluons (dotted lines),

Table 1. Powers and coefficients in Eqs. (2) and (3) for a QCD system with three colors and four flavors.

n	A_n	λ_n	α_n	β_n	λ_n'	α_n'	β_n'
2	-1.8	0.0	0.60	0.80	-3.1	0.71	-0.71
3	-2.8	-2.5	0.97	0.25	-5.8	0.30	-0.96
4	-3.5	-3.4	0.99	0.11	-7.7	0.17	-0.99

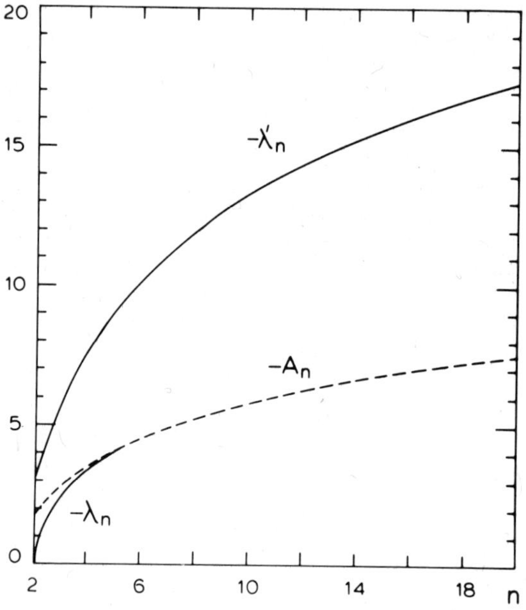

Fig. 1. Powers of t governing the variation of the nth moments of parton distributions. These powers are defined in Eqs. (2) and (3).

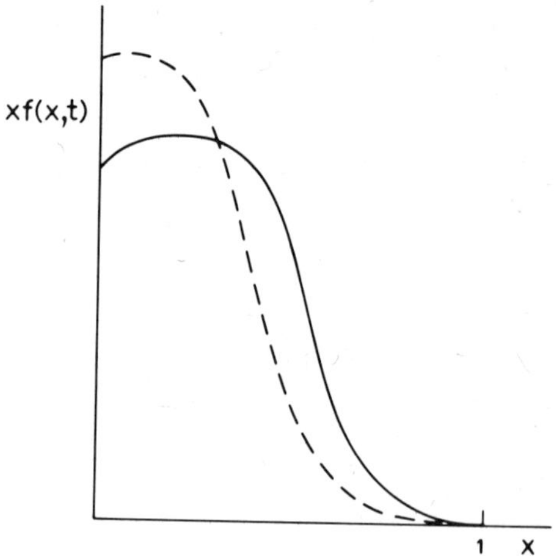

Fig. 2. Qualitative picture of how parton distributions change with Q. The dotted curve has a larger Q.

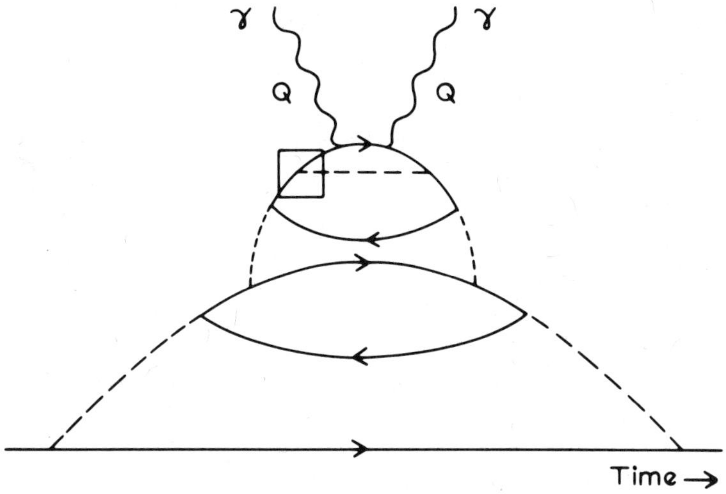

Fig. 3. Different parton layers in the Kogut-Susskind picture of interacting partons.

and the gluons can be transformed into pairs, etc. In this frame, it costs far too much energy for partons to move backwards, thus each branching in the diagram must reduce the momentum of the branched parton. Moreover, due to transverse motion of these newly created partons, a "cloud" is formed around the original parton by these branching processes. As we go further and further down the branching chain, the size of the cloud will be expected to become smaller and smaller than the previous one, since almost by definition the cloud of the previous one includes the cloud of the present one. Also, the longitudinal momentum would have been split into smaller and smaller fractions. Now a virtual photon of mass $-Q^2$ will hit a charged parton plus its surrounding cloud out to a spatial extent of order Q^{-1}. As Q increases, the photon is effectively looking into smaller spatial regions, thus further down the chain in Fig. 3. Hence it will see more partons of smaller longitudinal momentum, and less partons with large longitudinal momentum, which is precisely the behavior shown in Fig. 2.

2.3. QCD and Parton Model

In a very interesting paper, Altarelli and Parisi[6] were able to apply the Kogut-Susskind (KS) picture to derive the quantitative QCD result outlined in section 2.1. Although the results are the same as section 2.1, this approach has the distinct advantage of being intuitive, thus making it easier to be generalized.

As was discussed in the last section, we go deeper and deeper into the layers of Fig. 3 by increasing t. We pass into a deeper layer through a vertex which splits a parton into two, e.g., the one enclosed in a box in Fig. 3. The manner in which a parton distribution changes with t is therefore governed by the probability functions calculable from these vertices using the QCD Feynman rules. This is essentially the content of the "master equations" of Altarelli and Parisi[6]. Quantitatively, these equations are as follows:

$$\frac{dq^i(x,t)}{dt} = \int_0^1 dy\, dz\, \delta(x-yz)\left(\frac{\alpha(t)}{2\pi}\right)[P_{qq}(z)q^i(y,t) + P_{qG}(z)G(y,t)]$$

$$\frac{dG(x,t)}{dt} = \int_0^1 dy\, dz\, \delta(x-yz)\left(\frac{\alpha(t)}{2\pi}\right)[P_{Gq}(z)\sum_{i=1}^{2f} q^i(y,t) + P_{GG}(z)G(y,t)]. \tag{5}$$

Here f = 4 is the number of flavors. q^i (i = 1,...,8) describes the quark and anti-quark distributions, and G describes the gluon distribution. $\alpha(t) = (bt)^{-1}$ is the running fine structure constant of QCD, where b is given below Eq. (4). The transition probability density of converting a parton A into a parton B carrying a fraction z of the longitudinal momentum of A is described by the function $(\alpha(t)/2\pi)P_{BA}(z)$, where A and B can be either a quark or a gluon. These functions can be calculated[6] from the diagrams of Fig. 4. (The boxed vertex in Fig. 3 is depicted in Fig. 4(a)). They turn out to be

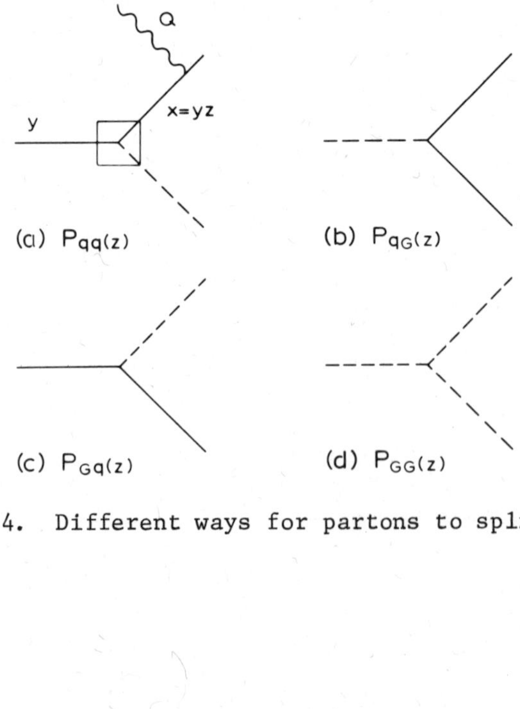

Fig. 4. Different ways for partons to split up.

$$
\begin{aligned}
P_{qq}(z) &= (4/3)(1+z^2)/(1-z)_+ + 2\delta(1-z) \\
P_{Gq}(z) &= (4/3)[1 + (1-z)^2]/z \\
P_{qG}(z) &= (1/2)[z^2 + (1-z)^2] \\
P_{GG}(z) &= 6[z/(1-z)_+ + (1-z)/z + z(1-z) + (25/36)\delta(1-z)] \;.
\end{aligned} \tag{6}
$$

The 2f+1= 9 equations in (5) can be decoupled in the following way. In terms of the singlet quark distribution function

$$q^S(x,t) = \sum_{i=1}^{2f} q^i(x,t) \tag{7}$$

and the non-singlet quark distribution functions

$$q^{NS}(x,t) = q^i(x,t) - q^j(x,t) \;, \qquad i \neq j \tag{8}$$

Eq. (5) decouples as follows,

$$\frac{dq^{NS}(x,t)}{dt} = \int_0^1 dy\, dz\, \delta(x-yz)\left(\frac{\alpha(t)}{2\pi}\right)P_{qq}(z)q^{NS}(y,t) \tag{9}$$

$$\frac{df^S(x,t)}{dt} = \int_0^1 dy\, dz\, \delta(x-yz)\left(\frac{\alpha(t)}{2\pi}\right)P(z)f^S(y,t) \tag{10}$$

where

$$f^S(x,t) \equiv \begin{pmatrix} q^S(x,t) \\ G(x,t) \end{pmatrix}$$

and

$$P(z) \equiv \begin{pmatrix} P_{qq}(z) & 2fP_{qG}(z) \\ P_{Gq}(z) & P_{GG}(z) \end{pmatrix} \tag{11}$$

Taking x-moments of (9) and (10), they are reduced to

$$\frac{dq_n^{NS}(t)}{dt} = \left(\frac{\alpha(t)}{2\pi}\right)(P_{qq})_n q_n^{NS}(t) \tag{12}$$

$$\frac{df_n^S(t)}{dt} = \left(\frac{\alpha(t)}{2\pi}\right) P_n f_n^S(t) \quad . \tag{13}$$

Equation (2) is immediately obtained from (12) if we identify A_n with $(P_{qq})_n$. Similarly, Eqs. (3) and (4) will result from the integral of (13) after diagonalization if we note that λ_n and λ_n' are the eigenvalues of P_n.

3. DISTRIBUTION FUNCTION $f(x,k_T,Q)$

We will discuss in this section the variation of the parton distribution with Q when considered as a function of longitudinal and transverse momenta. In particular, we will consider how the average transverse momentum, as a function of x, varies with Q.

3.1. Intuitive Expectations

KS[5] argued sometime ago that $\langle k_T^2 \rangle \sim Q^2/\log Q^2$. The $\log Q^2$ in the denominator comes from asymptotic freedom effects, whereas the Q^2 on top may be thought of as resulting from dimensional analysis. In terms of the picture of section 2.2, we may further expect $\langle k_T^2 \rangle$ to be a decreasing function of x. This is because when we go into deeper layers of the split, the spatial extent gets smaller,

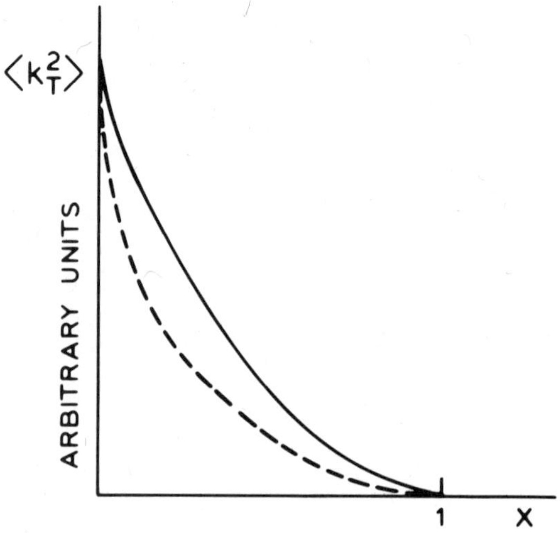

Fig. 5. x-dependence of parton transverse momentum at large Q. The dotted line corresponds to a larger Q.

resulting in a larger transverse momentum fluctuation $\langle k_T^2 \rangle$. In the meantime, the split has decreased x. Therefore large $\langle k_T^2 \rangle$ corresponds to small x, and vice versa, as sketched in Fig. 5.

3.2. Master Equation

If transverse momentum is taken into account, the master equation (5) may be generalized to read as follows:

$$\frac{dq^i(x,k_T,t)}{dt} = \int_0^1 dy\, dz\, \delta(x-yz) \int d^2k_{1T} d^2k_{2T}\, \delta^2(\vec{k}_T - z\vec{k}_{1T} - \vec{k}_{2T}) \cdot$$

$$\cdot \left(\frac{\alpha(t)}{2\pi}\right)\left[P_{qq}(z,k_{2T},t) q^i(y,k_{1T},t) + P_{qG}(z,k_{2T},t) G(y,k_{1T},t)\right]$$

$$\frac{dG(x,k_T,t)}{dt} = \int_0^1 dy\, dz\, \delta(x-yz) \int d^2k_{1T} d^2k_{2T}\, \delta^2(\vec{k}_T - z\vec{k}_{1T} - \vec{k}_{2T}) \cdot \tag{14}$$

$$\cdot \left(\frac{\alpha(t)}{2\pi}\right)\left[P_{Gq}(z,k_{2T},t) \sum_{i=1}^{2f} q^i(y,k_{1T},t) + P_{GG}(z,k_{2T},t) G(y,k_{1T},t)\right].$$

Here $P_{BA}(z,k_{2T},t)\, dz\, d^2k_{2T}\, dt$ is the probability of parton A changing into parton B, as depicted in Fig. 6. (y,k_{1T}) and (x,k_T) are the longitudinal momentum fractions and the transverse momenta of

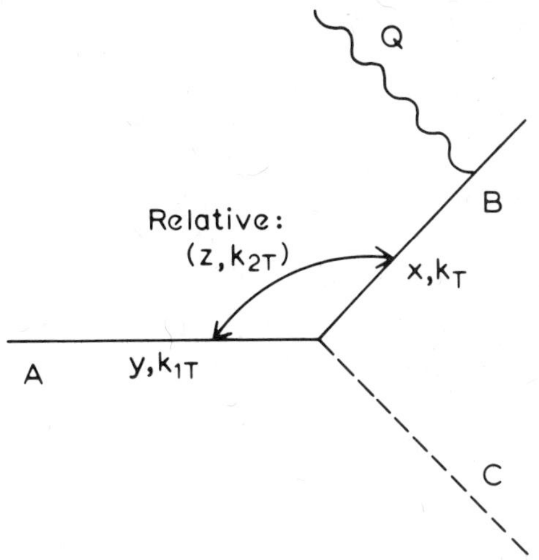

Fig. 6. Kinematics for parton decomposition used in the master equation of section 3.2.

partons A and B, respectively. (z,k_{2T}) are the relative longitudinal momentum fraction and the transverse momentum between B and A. The δ-functions in (14) simply reflect kinematical relationships between these variables.

Imagine Fig. 6 to be a part of a much larger diagram, say for forward Compton scattering. Then the large space-like momentum Q from the incoming photon will have to flow through the adjacent lines to reach the outgoing photon. Therefore the invariant mass squared of line B, $M_B^2 \equiv M_{AC}^2$, is expected to be of order $-Q^2$ (say $-\lambda(z)Q^2$). Since

$$M_B^2 \equiv M_{AC}^2 = -k_{2T}^2/(1-z) \tag{15}$$

we obtain the relationship

$$k_{2T}^2 = (1-z)\,Q^2\,\lambda(z) \tag{16}$$

determining k_{2T} as a function of z and Q^2. Hence

$$P_{BA}(z,k_{2T},t) = P_{BA}(z,k_{2T})\,\delta(k_{2T}^2 - \lambda(z)(1-z)Q^2)/\pi \tag{17}$$

for some $P_{BA}(z,k_{2T})$. Now Eq. (14) reduces to Eq. (5) upon integration of all transverse momenta, since

$$\int P_{BA}(z,k_{2T},t)\,d^2k_{2T} = P_{BA}(z) \quad , \tag{18}$$

$$\int q^i(x,k_T,t)\,d^2k_T = q^i(x,t) \quad , \text{ etc.}$$

Thus $P_{BA}(z,k_{2T})$ is independent of k_{2T} and is equal to the functions in Eq. (6).

At this point we do not have an easy way to determine what the function $\lambda(z)$ is (however, see the remarks in section 3.4). Actually, the arguments leading to Eqs. (15)-(17) do not even say that $\lambda(z)$ cannot be a fuzzy function of z; viz., M_B^2 in Eq.(15) may take a range of values even for a given z and Q. In that case, the δ-function in (17) is effectively replaced by a smeared δ-function.

Equation (14) is difficult to solve in its full generality, but its k_T-moments, $\langle k_T^2\rangle$, $\langle k_T^4\rangle$, ..., are easier to solve. In this talk, we will discuss only $\langle k_T^2\rangle$.

Let

$$h^i(x,t) \equiv \int q^i(x,k_T,t)k_T^2 d^2k_T \equiv \langle k_T^2\rangle_i \, q^i(x,t)$$
$$H(x,t) \equiv \int G(x,k_T,t)k_T^2 d^2k_T \equiv \langle k_T^2\rangle_G \, G(x,t) \quad . \tag{19}$$

Then (14) gives rise to the following equations for h^i and H:

$$\frac{dh^i(x,t)}{dt} = \int_0^1 dy\, dz\, \delta(x-yz)\left(\frac{\alpha(t)}{2\pi}\right)\{z^2[P_{qq}(z)h^i(y,t)+P_{qG}(z)H(y,t)]$$
$$+ \Lambda^2 e^t \lambda(z)(1-z)[P_{qq}(z)q^i(y,t) + P_{qG}(z)G(y,t)]\}$$
$$\frac{dH(x,t)}{dt} = \int_0^1 dy\, dz\, \delta(x-yz)\left(\frac{\alpha(t)}{2\pi}\right)\{z^2[P_{Gq}(z)\sum_{i=1}^{2f} h^i(y,t)+P_{GG}(z)H(y,t)]$$
$$+ \Lambda^2 e^t \lambda(z)(1-z)[P_{Gq}(z)\sum_{i=1}^{2f} q^i(y,t) + P_{GG}(z)G(y,t)]\} \tag{20}$$

where Λ is a scale parameter introduced in section 2.1.

As in the treatment of Eq. (5), Eq. (20) may be considerably decoupled by considering the flavor singlet and non-singlet parts:

$$h^{NS}(x,t) \equiv h^i(x,t) - h^j(x,t) \; , \qquad i \neq j$$
$$h^S(x,t) \equiv \sum_{i=1}^{2f} h^i(x,t) \quad . \tag{21}$$

Then ordinary differential equations in the t-variable may be written down for the x-moments, defined in a way analogous to Eq. (1). The equations for h^{NS} are all identical, and not coupled; the equations for h^S and H are coupled. The solutions of these equations are as follows[1]:

$$h_n^{NS}(t) = (t/t_o)^{A_{n+2}/2\pi b}\,[h_n^{NS}(t_o) + (\Lambda^2 B_n/2\pi b)\int_{t_o}^{t}(dt'/t')e^{t'} \cdot$$
$$\cdot (t'/t_o)^{(A_n - A_{n+2})/2\pi b}\, q_n^{NS}(t_o)] \tag{22}$$

$$\begin{pmatrix} h_n^S(t) \\ H_n(t) \end{pmatrix} = (t/t_o)^{P_{n+2}/2\pi b} \left\{ \begin{pmatrix} h_n^S(t_o) \\ H_n(t_o) \end{pmatrix} + (\Lambda^2/2\pi b) \int_{t_o}^{t} (dt'/t') e^{t'} \right.$$

$$\left. (t'/t_o)^{-P_{n+2}/2\pi b} \; \bar{B}_n (t'/t_o)^{P_n/2\pi b} \begin{pmatrix} q_n^S(t_o) \\ G_n(t_o) \end{pmatrix} \right\} \tag{23}$$

Here $A_n \equiv (P_{qq})_n$, and

$$B_n \equiv \int_0^1 dz\; z^{n-1} \lambda(z)(1-z) P_{qq}(z) \quad . \tag{24}$$

The 2×2 matrix P_n is the same as the one in Eqs. (11) and (13), whereas the matrix $\bar{B}_n$ is defined by replacing $P_{qq}(z)$ in (24) with the matrix $P(z)$ in (11).

3.3. Behavior at Large t

The first term in Eqs. (22) and (23) reflect the initial transverse momentum distributions of the partons. As remarked in the introduction, this can be calculated reliably only when we know more about strong interactions. However, these first terms decrease with t, whereas the second terms increase exponentially with t. Thus at large enough t, the initial distributions have been forgotten and a universal behavior emerges. Since

$$\int^t e^{t'} (t')^{\alpha}\, dt' = e^t t^{\alpha}[1 + 0(t^{-1})] \quad , \tag{25}$$

to leading order in t, the integrals in (22) and (23) may be replaced by their integrands evaluated at the upper limits. This gives

$$h_n^{NS}(t) \simeq \xi(t)\, B_n\, q_n^{NS}(t)$$

$$\begin{pmatrix} h_n^S(t) \\ H_n(t) \end{pmatrix} \simeq \xi(t)\, \bar{B}_n \begin{pmatrix} q_n^S(t) \\ G_n(t) \end{pmatrix} \tag{26}$$

where

$$\xi(t) = \Lambda^2 e^t/2\pi bt = Q^2 \alpha(t)/2\pi \quad . \tag{27}$$

Finally, we note that Eq. (26) may also be written in component form as

$$h_n^i(t) \simeq \xi(t)[(B_{qq})_n\, q_n^i(t) + (B_{qG})_n\, G_n(t)]$$
$$H_n(t) \simeq \xi(t)[(B_{Gq})_n \sum_{i=1}^{2f} q_n^i(t) + (B_{GG})_n\, G_n(t)] \tag{28}$$

where

$$(B_{ij})_n \equiv \int_0^1 dz\; z^{n-1}\, \lambda(z)(1-z)P_{ij}(z) \equiv \int_0^1 dz\; z^{n-1}\, B_{ij}(z). \tag{29}$$

Asymptotically, $q_2^i(t)$ and $G_2(t)$ are independent of t, hence it follows that $h_2^i(t)/\xi(t)$ and $H_2(t)/\xi(t)$ are also independent of t.

What is usually required for applications is the behavior of $\langle k_T^2\rangle$ as a function of x and t, but not of n and t. Let us therefore perform the inverse Mellin transform on (28). Using the convolution theorem, which says that if

$$A_n = B_n\, C_n$$

then

$$A(x) = \int_0^1 dy\; dz\; \delta(x-yz)B(y)C(z)$$
$$= \int_x^1 (dy/y)B(y)C(x/y) = \int_x^1 (dz/z)B(x/z)C(z) \tag{30}$$

we obtain

$$h^i(x,t) \simeq \xi(t) \int_x^1 (dz/z)[B_{qq}(x/z)q^i(z,t) + B_{qG}(x/z)G(z,t)]$$
$$H(x,t) \simeq \xi(t) \int_x^1 (dz/z)[B_{Gq}(x/z) \sum_{i=1}^{2f} q^i(z,t) + B_{GG}(x/z)G(z,t)] \;, \tag{31}$$

where

$$B_{ij}(y) = \lambda(y)(1-y)P_{ij}(y) \quad . \tag{32}$$

More specifically, from (6) and (32) we get

$$\begin{aligned} B_{qq}(y) &= (4/3)\lambda(y)(1+y^2) \\ B_{Gq}(y) &= (4/3)\lambda(y)(1-y)[1+(1-y^2)]/y \\ B_{qG}(y) &= (1/2)\lambda(y)(1-y)[y^2+(1-y)^2] \\ B_{GG}(y) &= 6\lambda(y)[y+(1-y)^2/y+y(1-y)^2] \ . \end{aligned} \tag{33}$$

We may now obtain $\langle k_T^2 \rangle$ as a function of x and t from (19), (31), and (33).

(A) Quarks. Let the value of $z\,q^i(z,t)$ and $z\,G(z,t)$ at $z = 0$ be $a(t)$ and $b(t)$ respectively. Then for $x \to 0$,

$$\begin{aligned} h^i(x,t) &\to \xi(t) \int_x^1 (dz/z^2)[a(t)B_{qq}(x/z) + b(t)B_{qG}(x/z)] \\ &\to (\xi(t)/x)[a(t) \int_0^1 B_{qq}(y)dy + b(t) \int_0^1 B_{qG}(y)dy] \ . \end{aligned} \tag{34}$$

Thus

$$\langle k_T^2 \rangle_i (x=0,t) = \xi(t) \int_0^1 [B_{qq}(y) + b(t)/a(t))B_{qG}(y)]dy \ . \tag{35}$$

Now for $x \to 1$, Assume

$$\begin{aligned} q^i(z,t) &\to C_i(t)(1-z)^{\eta_i(t)} \\ G(z,t) &\to C_G(t)(1-z)^{\eta_G(t)} \end{aligned} \tag{36}$$

when $z \to 1$. Since both the present data and the asymptotic formulas from QCD suggest that $\eta_G > \eta_i$, and since from (33), $B_{qG}(y)$ vanishes at $y = 1$, only the first term of (31) contributes to h^i as $x \to 1$. The result is

$$\langle k_T^2 \rangle_i (x\to 1,t) = \xi(t)(8/3)\lambda(1)(1-x)/(\eta_i(t) + 1) \ . \tag{37}$$

We therefore expect $\langle k_T^2 \rangle_i$ to behave like the solid curve in Fig. 5. Moreover, $b(t)/a(t)$ probably approaches a constant asymptotically since $G_n(t)/q_n^S(t) = \beta_n/\alpha_n$ (see Eq. (3)) are independent of t. Thus if the arbitrary unit in Fig. 5 is $\xi(t)$, then the value of $\langle k_T^2\rangle_i$ at $x = 0$ is unchanged for a larger t, but $\eta_i(t)$ is expected to increase. Thus as t increases, the solid curve would drop to the dotted curve.

(B) Gluons. For $x \to 0$, we obtain from Eq. (31) a formula analogous to Eq. (35):

$$\langle k_T^2\rangle_G\ (x=0,t) = \xi(t)\int_0^1 [B_{Gq}(y)(a(t)/b(t)) + B_{GG}(y)]dy\ . \qquad (38)$$

Referring to the explicit form of B_{Gq} and B_{GG} in Eq. (33), both functions have a (1/y) singularity as $y \to 0$. Hence if (38) is to remain finite, we must have a y factor in $\lambda(y)$. This factor is to be expected also on other grounds. If we go through the same arguments leading to Eqs. (15) and (16), but applied this time to M_C^2 rather than M_B^2, then Eq. (16) will be replaced by $k_{2T}^2 = z\bar{\lambda}(z)Q^2$, suggesting that $\lambda(z)$ must have a factor of z, and $\bar{\lambda}(z)$ must contain a factor of (1-z).

It is less straightforward to obtain the $x \to 1$ behavior for $\langle k_T^2\rangle_G$. This difficulty stems from the fact that $\eta_G > \eta_i + 2$ at present t if i represents up or down quarks in nucleons. On the other hand, according to Eqs. (2) and (3), and the fact that $|\beta_n|/|\alpha_n| = O(1/n \log n)$, asymptotically all η_i will become equal and η_G will only be infinitesimally larger than $\eta_i + 1$. The same kind of analysis that led to Eq. (37) suggest that $\langle k_T^2\rangle_G$ would behave like the $\eta_i + 2 - \eta_G$ power of (1-x). Thus at very large t, it behaves qualitatively in the way depicted in Fig. 5, but at presently available t, it may not behave like that at all.

3.4. Another Approach

When partons do not interact, deep inelastic structure functions scale. In that model, Feynman[7] gave the following formula for the ratio $R = \sigma_L/\sigma_T$ of longitudinal and transverse cross sections:

$$R = 4(\langle k_T^2\rangle + m^2)/Q^2\ . \qquad (39)$$

Here m is the quark mass. Politzer and Kogut and Shigemitsu[8] have applied the diagrammatic calculation of R using Fig. 7 to obtain[9] $\langle k_T^2\rangle_n = \alpha(t)Q^2/3\pi(n+1)$. This gives rise to a linear dependence near $x = 1$ as in our asymptotic behavior (37). In fact, if in the spirit of (27) and (28), we identify $(B_{qq})_n$ with 2/3(n+1), then (29) and (33) imply that

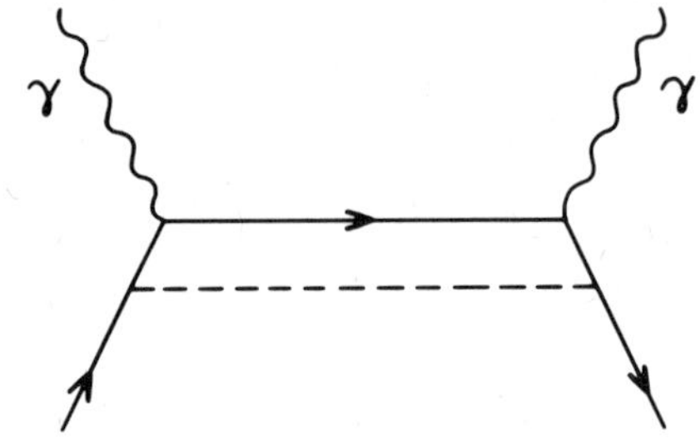

Fig. 7. Diagram used to calculate R.

$$\lambda(z) = \frac{1}{2}\,\frac{z}{1+z^2} \quad . \tag{40}$$

Note the appearance of the factor z discussed below Eq. (38).

This approach does not include the renormalization group interactions contained in the master equations of section 3.2. It also makes use of Feynman's formula (39). This is a kinematical formula derived in the free parton model assuming $\langle k_T^2\rangle/Q^2 \ll 1$, which is not the case in QCD for small x. However, it is very interesting that very similar results are obtained by these different approaches.

4. μ-PAIRS FROM HADRONIC COLLISIONS

The transverse momentum of μ-pairs produced by hadronic collisions reveals almost directly the transverse momenta of the partons, if the Drell-Yan[10] mechanism is assumed in the production process. In this mechanism, a quark from one hadron annihilates with an anti-quark from the other hadron to form a virtual photon which subsequently decays into a μ-pair. Thus $\langle p_T^2\rangle$, the average of the square of the transverse momentum of the μ-pair, is just the sum of the corresponding quantities for the colliding partons:

$$\langle p_T^2\rangle(\mu\text{-pair}) = \langle k_T^2\rangle_q(x_1,t) + \langle k_T^2\rangle_{\bar{q}}(x_2,t) \quad . \tag{41}$$

If s is the c.m. energy squared of the colliding hadrons, and Q the mass of the μ-pair, then $x_1x_2 = Q^2/s$, or $x_1 = x_2 = Q/\sqrt{s}$ if the μ-pair has no longitudinal momentum. Assuming further that what is derived for space-like photon may be used for time-like photon at the same $|Q^2|$, then we can calculate Eq. (41) from the formulas in the previous section using $t = \log(Q^2/\Lambda^2)$.

To get an idea as to what p_T^2 might be as a function of Q, let us assume the quark distribution q(x,t) to be $(1-x)^3/\sqrt{x}$, and the anti-quark distribution to be $(1-x)^6/x$. We use Eqs. (19) and (31) to compute $\langle k_T^2 \rangle$, ignoring the smaller term G(z,t) in (31), and replacing $B_{qq}(x/z)$ by its value at unit argument. For $\lambda(1)$, we use the value 1/4 given by Eq. (40). For Λ, we use 0.7 GeV. Finally, Eq. (31) contains the Q factor in $\xi(t)$, making $\langle k_T^2 \rangle$ vanish at Q = 0, whereas we expect $\langle k_T^2 \rangle \simeq (.3\ \text{GeV})^2$ even at zero Q. This discrepancy owes its origin to the asymptotic nature of Eq. (31), wherein the initial condition term $h^i(t_0)$ had been dropped from Eqs. (22) and (23). We therefore restore it by hand by adding to $\langle p_T^2 \rangle^{1/2}$ calculated from (31) the additional constant $\sqrt{2}$x(.3) GeV. The result is the solid curve in Fig. 8. The date (preliminary) are those shown by Dr. Brown two nights ago. Shown for comparison are also two dotted lines depicting linear variations with Q, one tangential to the solid curve at Q = 0, the other going through the CP II points. The relationship between $\langle p_T^2 \rangle^{1/2}$ and $\langle p_T \rangle$ is $\langle p_T^2 \rangle^{1/2} = 4\langle p_T \rangle/\pi$, and is derived from the distribution $(1 + p_T^2/(1.5\ \text{GeV})^2)^{-3}$ in p_T^2 observed by the CFS group.

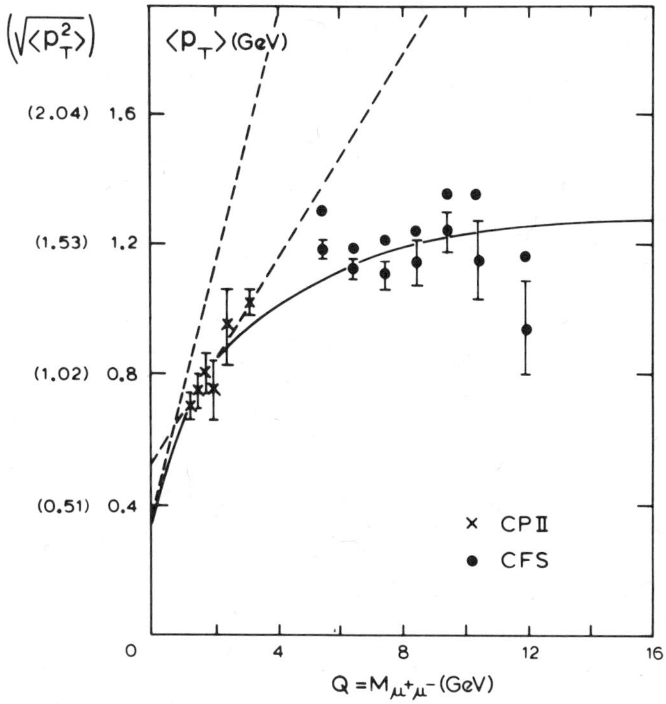

Fig. 8. μ-pair transverse momentum distribution. The solid curve is the theoretical curve and the dotted curves are straight lines shown for the purpose of comparison only.

We see that the solid curve gives a fair representation of the data. Calculations are now underway using the exact formulas (22) and (23), and a much more realistic fit to quark and antiquark distribution functions. We are also applying these k_T distributions to calculate other parton processes, but we have no result at this stage.

REFERENCES

1) C.S. Lam and T.M. Yan, Phys. Lett. 71B, 173 (1977).

2) S.D. Drell, D.J. Levy and T.M. Yan, Phys. Rev. Lett. 22, 744 (1969); Phys. Rev. 187, 2159 (1969); Phys. Rev. D1, 1035 (1970).

3) H.L. Anderson et al., Phys. Rev. Lett. 37, 4 (1976); 38, 1450 (1977); E.M. Riordan et al., SLAC-PUB-1634 (1975); C. Chang et al., Phys. Rev. Lett. 35, 901 (1975).

4) H.D. Politzer, Phys. Rev. Lett. 30, 1346 (1973); D. Gross and F. Wilczek, Phys. Rev. Lett. 30, 1343 (1973); Phys. Rev. D8, 3633 (1973); Phys. Rev. D9, 980 (1974); G. Altarelli, talk given at the XI Rencontres de Moriond, Flaine, France, 1976.

5) J. Kogut and L. Susskind, Phys. Rev. D9, 697, 3391 (1974); J.B. Kogut, Phys. Lett. 65B, 281 (1971).

6) G. Altarelli and G. Parisi, Ecole Normale Supérieure preprint, March 1977; Nucl. Phys. B126, 298 (1977).

7) R.P. Feynman, Photon-Hadron Interactions (Benjamin, New York, 1972). See Lecture 28.

8) H.D. Politzer, Harvard preprints; J. Kogut and J. Shigemitsu, Cornell preprint.

9) A. Zee, F. Wilczek and S.B. Treiman, Phys. Rev. D10, 2881 (1974).

10) S.D. Drell and T.M. Yan, Phys. Rev. Lett. 25, 316 (1970); Ann. Phys. (N.Y.) 66, 578 (1971).

TRIMUONS

R.J.N. Phillips

Rutherford Laboratory

Chilton, Oxon, England

"There is something fascinating about Science. One gets such wholesome returns of conjecture out of such a trifling investment of fact".[1]

This is a brief review of mechanisms suggested for neutrino trimuon events[2-5] (see lectures by A.K. Mann). The first data[2,3] suggested a high threshold, probably unique charge signature $\nu N \to \mu^-\mu^-\mu^+ X$, and rate $\mu^-\mu^-\mu^+/\mu^- \sim 5.10^{-4}$. The secondary muons were rather energetic, unlike those in dimuon events, suggesting that they did not come from charm decays. Later data[4,5] include events with slower secondary muons, and at least one with different charge signature $\nu \to \mu^-\mu^+\mu^+$. Several mechanisms may be at work. It is too soon to decide what is responsible for trimuons, but we can at least examine and compare the suggestions.

It is generally easier to get fast muons from the lepton rather than the hadron vertex; the energy of a struck quark is degraded by a fragmentation function as it turns into a hadron, before it can decay to a muon, whereas a produced heavy lepton can decay directly. Hence many suggestions include heavy lepton production. Most of the mechanisms fall into three classes, illustrated in Fig. 1.

a) Lepton cascade class[6-11]. All three muons come from the cascade decay of a heavy lepton, produced at the neutrino vertex.

b) Lepton-hadron class[8,11-12]. Two muons come from heavy lepton decay and one from the hadron vertex, e.g. from heavy quark decay.

c) Hadron class[11,13-15]. One muon comes from a normal $\nu \rightarrow \mu$ transition, while the others come from the hadron vertex, e.g. from heavy quark cascade decay.

LEPTON CASCADE MECHANISMS

At least two new heavy leptons are needed. A single-stage $M^- \rightarrow \mu^- \mu^- \mu^+$ decay would give unique trimuon invariant mass, which is not found. The observed invariant masses require M^- to be very heavy, $m(M^-) \sim 7$ GeV or more, so that M^- production is kinematically suppressed relative to μ^- by an order of magnitude at 100 GeV. Since the trimuon mechanism also relies on two muonic branching ratios each of order 0.2, the coupling εG at the $\nu \rightarrow M^-$ vertex must be a substantial fraction of the full Fermi coupling strength G to achieve a $3\mu/\mu$ rate of order 5.10^{-4}; in fact $\varepsilon^2 \sim 0.1$ is needed[6].

There are generally many competing cascade modes. For instance, if there is only one charged W-boson (as in $SU(2) \times U(1)$ theories), the couplings appearing in Fig. 1 also allow the decays in Fig. 2, offering an alternative route to trimuons and even pentamuons. The decay $M^- \rightarrow \mu^- \mu^- \mu^+$ gives very energetic trimuons if an off-diagonal $M^- \rightarrow \mu^-$ neutral current coupling exists; this once seemed a possible explanation of the exceptional event No. 119, but now a similar event No. 281 has been found, with a different 3μ invariant mass[5]. There are lots of e, τ and hadron modes too. The relative rates depend on the masses and coupling strengths; in fact Fig. 1a could well be the dominant trimuon mode[6].

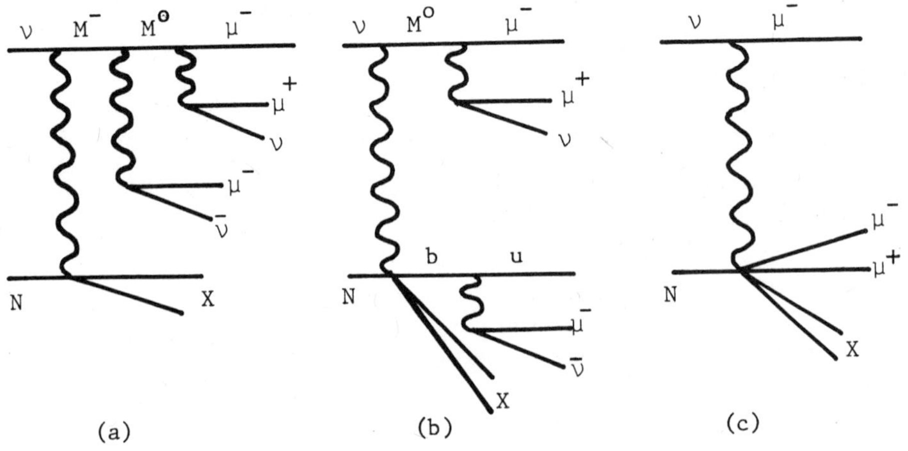

Fig. 1. Three classes of trimuon mechanisms.

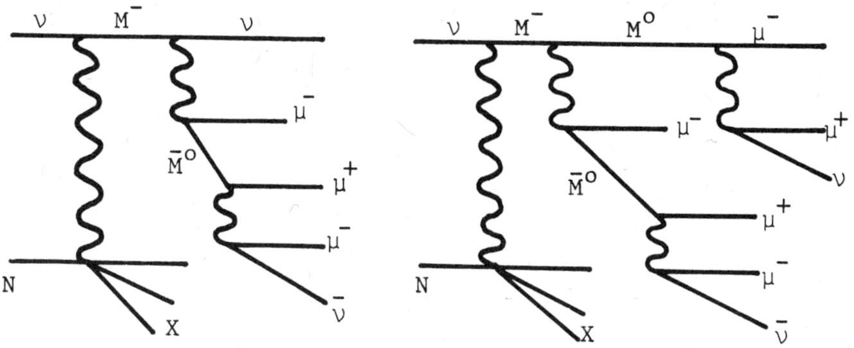

Fig. 2. Other cascade decay possibilities.

In any model one can sum the decay modes to estimate mean lifetimes τ and also mean track lengths $\bar{\ell}$ for production in a particular neutrino beam. Typical results are[6]

$$\tau(M^-) \sim 10^{-15}\,\text{sec} \qquad \bar{\ell}(M^-) = 2\ \mu\text{m}$$

$$\tau(M^o) \sim 10^{-12}\,\text{sec} \qquad \bar{\ell}(M^0) = 4\ \text{mm}$$

with masses 7 GeV, 2.5 Gev for M^-, M^0; $\bar{\ell}$ refers to the neutrino beam of ref. 3. M^- track lengths are thus on the borderline for detection in emulsions: M^0 might be seen in bubble chambers.

Lepton cascades can fit most of the early HPWF data[3], see e.g. Figs. 3 and 4. The muon energy distributions are about right, except for event 119 that has rather high E for the fast μ^-. The scale of invariant mass distributions m_{--+}, m_{--} and m_{-+} is set by the M^- mass: you can see why it has to be big. m_{-+} has two contributions from each event, for the two $\mu^-\mu^+$ pairings; we get a useful constraint by taking the lower value only in each event, which gives a lower limit on the M^0 mass. Fig. 4 shows that $m(M^0) \gtrsim 2$ GeV is needed (this applies also to Lepton-Hadron mechanisms). Transverse momentum distributions, not shown here, are also sensitive to heavy lepton masses.

In an SU(2) × U(1) gauge theory, it is natural to put (M^0, M^-) into a left-handed doublet like (ν_μ, μ^-). Then a moderate mixing angle ε between μ^- and M^- provides all the $\nu \to M^-$, $M^- \to M^0$, $M^0 \to \mu^-$ couplings we need. (Right handed doublets are optional; we do not need them here). The trimuon rate requires $\varepsilon^2 \sim 0.1$, comparable to the Cabibbo angle $\theta_c^2 \sim 0.05$, but this strength of mixing significantly disturbs the observed universality of couplings to the (ν_μ, μ^-), (ν_e, e^-), (u, d_c) doublets. One remedy[6,16] is to mix

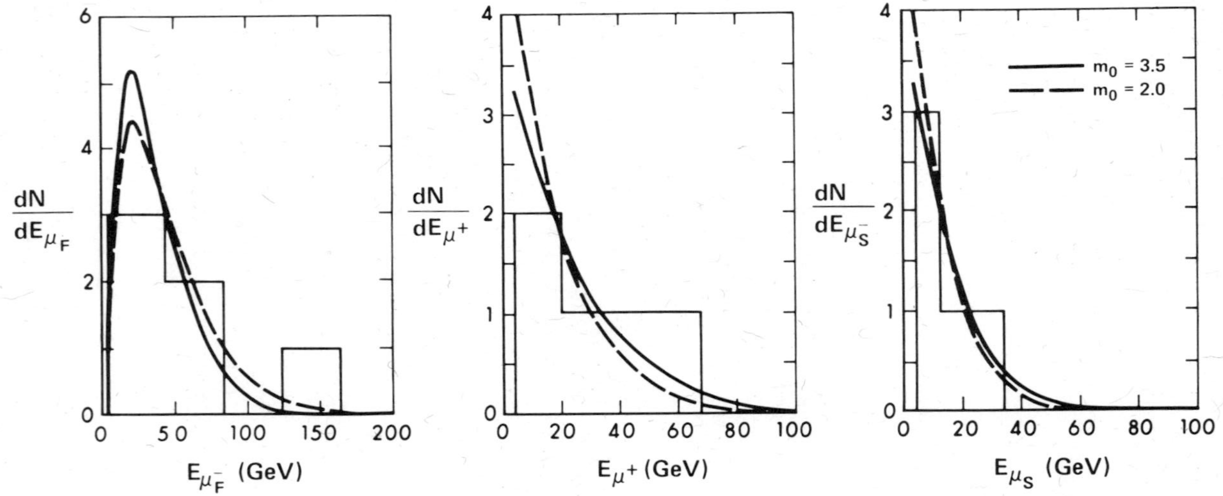

Fig. 3. Fast μ^-, μ^+, slow μ^- E-distribution data[3] compared with lepton cascade calculations[6].

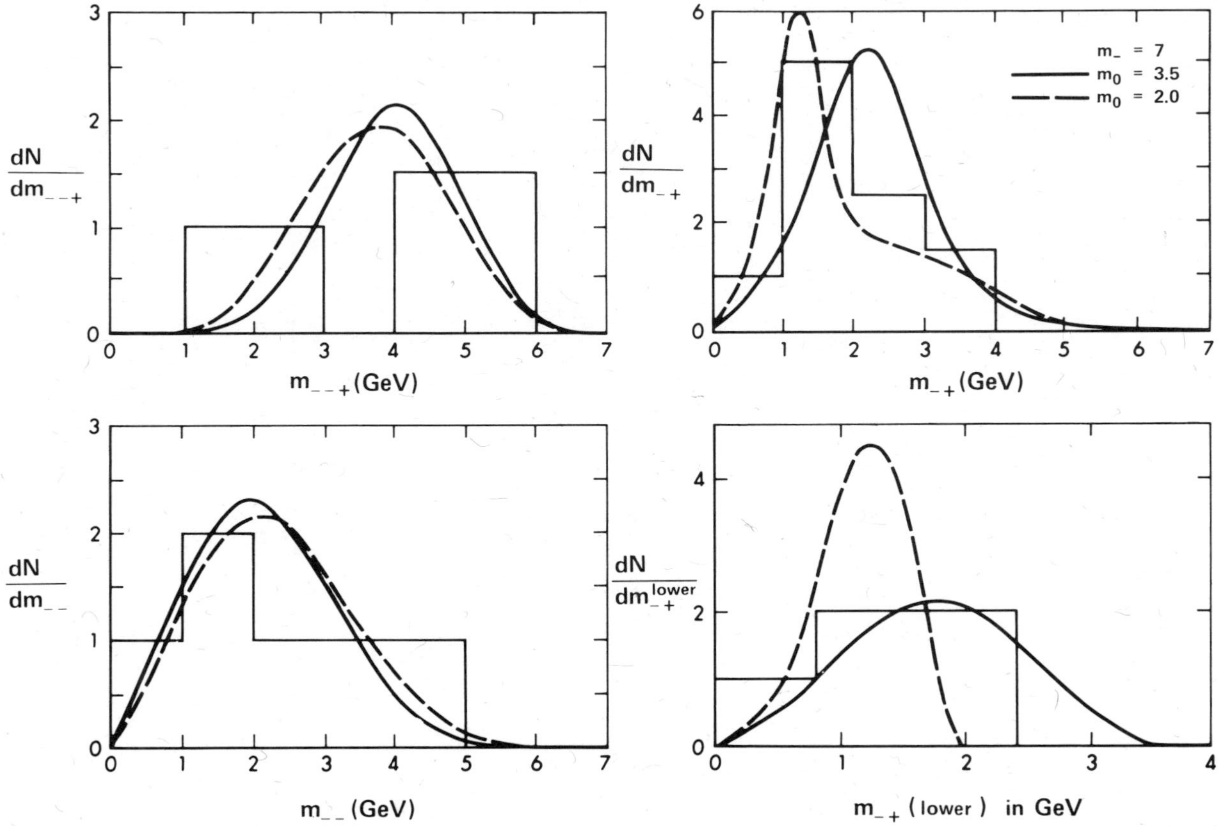

Fig. 4. Trimuon invariant mass distributions[3] and a lepton cascade model.

other heavy leptons and quarks with e^-, d_c to restore the symmetry, which may appear unaesthetic; another is to seek greater freedom in higher symmetry groups.

In SU(3) × U(1) models there are 9 gauge bosons. If for example[8,9] we put the relevant leptons in triplets as in Fig. 5, the $\nu \to M^-$ transition is handled by a W' boson distinct from the W boson of $\nu \to \mu^-$ transitions. The strength of M^- production depends on the W' mass. With a little μ^--M^- mixing in the right-handed triplet, we get all the cascade couplings needed without upsetting conventional results. This is one of many possibilities. Reference 10 suggests a gauge group SU(2) × SU(2) × U(1) with lepton representations including a doubly charged M^{--}.

Some higher symmetry schemes[9] essentially introduce a new weak quantum number, produced in association by the new W' boson exchange. M^- production is then accompanied by new quark production, leading to a higher threshold in the recoil hadron energy[17].

LEPTON-HADRON MECHANISMS

Here we have at least one new lepton M^0 and probably one new quark. To get the observed trimuon rate we need substantial flavour-changing neutral currents at both the lepton and quark vertices, which are virtually ruled out in SU(2) × U(1). We need a higher symmetry from the outset.

Fig. 6 shows a suggested SU(3) × U(1) assignment[8], with quark triplets and lepton antitriplets for the relevant states. (Similarly e^- appears with heavy leptons E_1^0, E_2^0; c and s with a heavy quark b'). Incidentally this gives μ^- and e^- pure vector neutral current couplings, desirable these days, but also a right-handed $\bar{\nu}u \to \mu^+ b$ transition guaranteed to cause a high-y anomaly. This illustrates a general point, that every gauge model has dozens of constraints to satisfy, from the measured muon g-factor to the

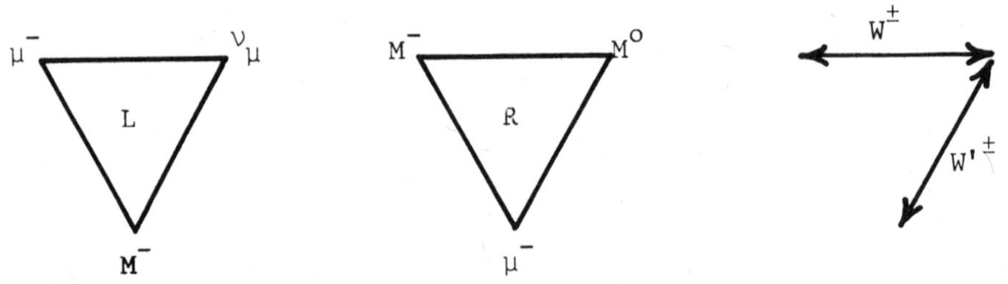

Fig. 5. Possible SU(3) × U(1) lepton triplets[8,9]; charged intermediate boson coupling routes are also shown.

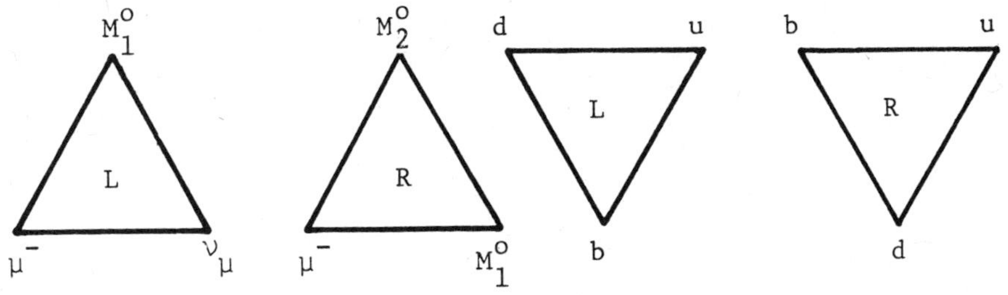

Fig. 6. Possible SU(3) × U(1) quark triplets and lepton anti-triplets, for a lepton-hadron mechanism[8].

limits on τ decays: there is less freedom than at first appears.

In detailed calculations[11], lepton-hadron mechanisms give fits to data comparable to the lepton cascade examples shown above. However, some of their predictions differ strikingly. While both give the same signature $\nu N \to \mu^- \mu^- \mu^+ X$ for neutrinos, their antineutrino predictions are different,

$$\bar{\nu}N \longrightarrow \mu^+\mu^+\mu^- X \qquad \text{(Lepton cascade)}$$

$$\longrightarrow \mu^+\mu^-\mu^- X \qquad \text{(Lepton hadron)} \quad .$$

In each case, the whole lepton vertex is charge conjugated. However, the lepton-hadron process has a neutral current, and the muon from the hadron vertex stays the same.

Both mechanisms predict same-sign dileptons. If in Figs. 1a, 1b the M^o decay is replaced by $M^o \to \mu^- +$ hadrons, we get $\nu N \to \mu^- \mu^- X$ with three or more times the trimuon rate. Exchanging a $\mu^-\nu$ vertex for $e^-\nu$ gives $\nu N \to \mu^- e^- X$ at the same rate. See Fig. 7. There

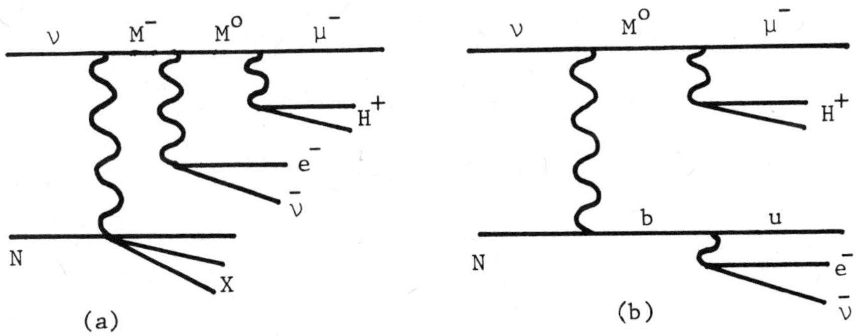

Fig. 7. $\nu N \to \mu^- e^- X$ in lepton cascade and lepton hadron mechanisms.

are striking differences between the predicted energy distributions, however, especially for $\mu^- e^-$ events. In lepton cascades, the e^- comes from the first stage and tends to have more energy than μ^-. In lepton-hadron models, however, the e^- comes from hadronic decay and has less energy. For an experiment like the BNL-Columbia heavy neon exposure at Fermilab, typical model calculations[18] give

$$\langle E(\mu^-)\rangle / \langle E(e^-)\rangle = 0.7 - 0.8 \qquad \text{lepton cascade}$$

$$= 2.7 - 3.6 \qquad \text{lepton hadron .}$$

This brings out an important point. The physics of same-sign dileptons may be closely related to that of trimuons, and they should be studied together.

Finally it is interesting that both classes of heavy lepton mechanism predict dimuons from $M^o \to \mu^- \mu^+ \nu$ decay, with distributions differing distinctively[19] from charm-decay dimuons. The latter tend not to populate a symmetric region of the $E(\mu^-) - E(\mu^+)$ scatter plot. It is not yet established whether there is a significant excess of these "symmetric dimuons": see lectures by A.K. Mann.

HADRON MECHANISMS

We may expect $\mu^+\mu^-$ pairs to be produced in the recoil hadron shower by the same mechanisms (meson decay, associated charm production or whatever) that produce dileptons in hadron-hadron collisions. To judge from the latter case[20], however, the rate would be only $\sim 10^{-4}$ (with muon acceptance cuts) and the $\mu^+\mu^-$ invariant mass would almost always be less than 1 GeV, unlike many trimuon events.

In fact the intermediate W-boson is pointlike, and does not interact simply like a hadron. We can therefore look also for hard-scattering mechanisms, that are not included in the mainly soft hadron-hadron-like category above. Some suggestions are:

i) New quark production and cascade decay[11,13], such as the $d \to t \to b \to u$ sequence in Fig. 8a. Calculations in ref. 11 show the decay muons to be rather slow. The degradation of energy through fragmentation functions may be minimized if heavy quark fragmentation is concentrated near $z = 1$, unlike light quarks; this question is being debated[21]. To get a big enough trimuon rate, the $d \to t$ coupling has to be substantial, and we meet the same problems as in lepton cascades. In $SU(2) \times U(1)$ models with left-handed doublets, we need sizable u-t mixing which will upset μ-e-

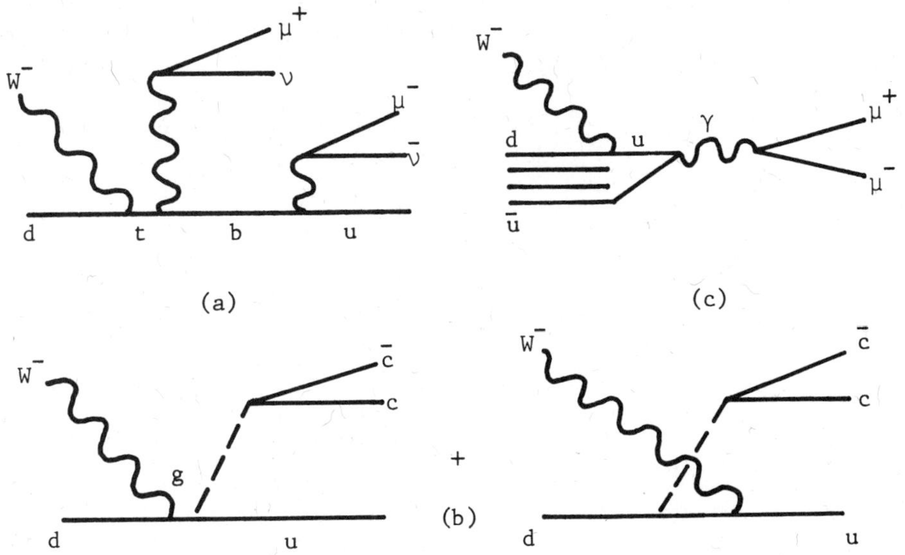

Fig. 8. Various hard-scattering hadron mechanisms.

quark universality unless we introduce extra mixings in the other multiplets, which may seem unaesthetic. (For more discussion of heavy quark cascades in $SU(2) \times U(1)$ models, see lectures given by F. Gilman and ref. 22). Higher gauge groups may be more attractive. If $b \to c \to d$ instead of $b \to u$, we can get 4μ states.

ii) Charm-anticharm creation by a gluon, radiated by the struck quark: see Fig. 8b. The $3\mu/\mu$ rate is calculated[14] to be of order 10^{-5} at 300 GeV, rather small for present purposes.

iii) Pair production by a photon, radiated by the struck quark: in Fig. 8b replace $\bar{c}c$ by $\mu^-\mu^+$ and g by γ. The rate is roughly $\alpha^2/\pi \sim 10^{-5}$.

iv) Internal Drell-Yan pair production: see Fig. 8c. I first heard of this second-order quark process at the present meeting[15]. This process is remarkable - and unique among the processes discussed, in all classes - in that it channels almost all the incident neutrino energy into the three muons. Can it be the explanation of the exceptional events 119 and 281? The rate has not yet been calculated carefully.

OTHER MECHANISMS

Some outside the classes above are:

i) Muon tridents: $\mu^-(\mu^+)$ is produced normally by $\nu(\bar{\nu})$, and radiates a $\mu^-\mu^+$ pair. I understand this has been calculated by Bjorken (unpublished), and the rate for massive pairs is too small.

ii) Production via charged Higgs bosons[23]. The trimuon rate falls very rapidly as the Higgs mass increases: $M(H^\pm) < 10$ GeV is required.

iii) Production via neutral Higgs bosons: H^o couples to the intermediate $W^\pm$ in normal charged-current scattering, and decays via heavy leptons and quarks to multimuon states[24]. Predictions are not very specific yet.

CONCLUSION

Clearly there are far too many models, but most of them will go away when the dust settles and the experimental facts emerge. Meanwhile there is still some work to do in sharpening the distinctions between different mechanisms.

I am indebted to V. Barger, D. Cline, T. Gottschalk, D. Nanopoulos, D. Reeder, D. Scott and T. Walsh for collaborations and discussions on the whole trimuon business, and to C. Goebel for introducing me to ref. 1.

REFERENCES

1) M. Twain, "Life on the Mississippi", (c. 1882).

2) B.C. Barish et al., Phys. Rev. Lett. 38, 577 (1977).

3) A. Benvenuti et al., Phys. Rev. Lett. 38, 1110, 1183 (1977).

4) J. Steinberger, report at Budapest Conference (1977).

5) A. Benvenuti et al., report at Hamburg Conference (1977).

6) V. Barger et al., Phys. Rev. Lett. 38, 1190 (1977) and Wisconsin preprints COO-596, 597, 598 (1977).

7) C. Albright et al., Phys. Rev. Lett. 38, 1187 (1977) and Stony Brook preprint ITP-SB-77-32 (1977).

8) P. Langacker and G. Segrè, Phys. Rev. Lett. 39, 259 (1977) and Pennsylvania preprint UPR-0073T.

9) B.W. Lee and S. Weinberg, Phys. Rev. Lett. 38, 1237 (1977).

10) A. Zee, F. Wilczek and S.B. Treiman, Phys. Lett. 68B, 369 (1977).

11) R.M. Barnett and L.M. Chang, SLAC-PUB-1932 (1977).

12) D. Horn and G.G. Ross, Phys. Lett. 69B, 364 (1977).

13) F. Bletzacker et al., Phys. Rev. Lett. 38, 1241 (1977) and Stony Brook preprint ITP-SB-77-34 (1977).

14) T. Walsh and B.M. Young, to be published.

15) D.R. Winn, private communication.

16) M. Suzuki, Phys. Rev. Lett. 35, 1553 (1975); V.K. Cung and C.W. Kim, Phys. Rev. D14, 1376 (1976); Y. Abe et al., Phys. Lett. 62B, 207 (1976).

17) V. Barger et al., Wisconsin preprint COO-600 (1977).

18) V. Barger et al., Wisconsin preprint COO-602 (1977).

19) V. Barger et al., Wisconsin preprint COO-603 (1977).

20) K.J. Anderson et al., Phys. Rev. Lett. 37, 799 (1976); J.G. Branson et al., Phys. Rev. Lett. 38, 1334 (1977).

21) M. Suzuki, LBL-6173; V. Barger et al., Wisconsin preprint COO-601 (1977); R. Odorico, CERN-TH.2360 (1977); J.D. Bjorken, SLAC preprint.

22) J. Ellis et al., CERN-TH.2346 (1977).

23) Y. Tomozawa, Michigan preprint UM-HE-77-20 (1977).

24) P.D. Morley and L. Pilachowski, Iowa State preprint.

AN APPROACH TO MEASUREMENT IN QUANTUM MECHANICS*

E.C.G. Sudarshan, T.N. Sherry and S.R. Gautam

University of Texas

Austin, Texas 78712

ABSTRACT

We consider an unconventional approach to the measurement problem in quantum mechanics - we treat the apparatus as a classical system, belonging to the macro-world.

In order to have a measurement the apparatus must interact with the quantum system. As a first step, we embed the classical apparatus into a larger quantum mechanical structure, making use of Superselection rules. We can project back to the classical system. We now couple the apparatus and system such that the apparatus remains classical (principle of integrity), and unambiguous information of the values of a quantum observable are transferred to the variables of the apparatus. Finally, we project back to the classical formulation.

Further measurement of the classical apparatus can be done, causing no problems of principle. Thus interactions causing pointers to move (which we do not treat) can be added.

We examine the restrictions placed by the principle of integrity on the form of the interaction between classical and quantum systems.

I. INTRODUCTION

The work we are going to discuss concerns a rather unusual and unconventional approach to the description of what we mean by measurement.[1,2] We do not claim to have worked out a complete theory which realistically describes a measurement process - rather we have concentrated on points of principle, illustrating by means of some simple examples.

The unusual feature of our work is that we treat the apparatus as a system belonging to the regime of classical mechanics. Our reasons for considering such an unusual possibility are philosophical since a piece of apparatus belongs to the macro-physical world it would seem more reasonable to treat it in a macro-physical way. This is not of its own accord too convincing; nevertheless, we feel that it is worthwhile to pursue such thoughts. Useful ideas could be unconventional.

We shall not be dealing directly with the problem of pointer readings or irreversible processes. Rather, we will consider a measurement complete if unambiguous information has been transferred from the quantum system being examined to the classical apparatus system. We can list the chief requirements for what we mean by measurement as follows:

1. the apparatus is classical,
2. the apparatus interacts with the quantum system,
3. the interaction is such that the apparatus retains its classical identity - principle of integrity,
4. the interaction results in unambiguous information concerning the values of the relevant system observables being transferred to the observables of the classical system.

An important point to notice about the above outlined procedure is the following: in classical mechanics the measurement of dynamical variables is possible <u>in principle</u> without causing a disturbance on the system. (The technological problem is another matter.) Once we have successfully transferred the information into the classical system, the transferring of this information to an observer, or anther system, is straightforward.

Clearly, the first problem we must examine is how to cause classical and quantum systems to interact. To achieve this we introduce an unorthodox embedding of classical mechanical systems into quantum mechanical systems. This procedure makes use of a superselection principle.[3] Once this is understood, we can go on to make use of it to couple together the classical and quantum mechanical systems. The requirements we listed above will place certain restrictions on the allowed couplings - and this we shall examine in some detail.

The proof of any pudding is in the eating. So, we shall then discuss some questions relating to an example. This example is a form of the Stern-Gerlach[4] experiment.

Many of the ideas we will present are at the level at which

quantum mecahnics is usually discussed. In particular we refer to the discussion of Superselection, where manipulations valid for bounded operators are assumed to hold in some sense for unbounded self-adjoint operators on infinite dimensional Hilbert space. Also we assume that unbounded operators can be treated as having eigenvectors corresponding to points belonging to their continuous spectra. We are presently carrying out a more careful analysis to see if these assumptions can be made more precise.

SCHEMATIC DIAGRAM

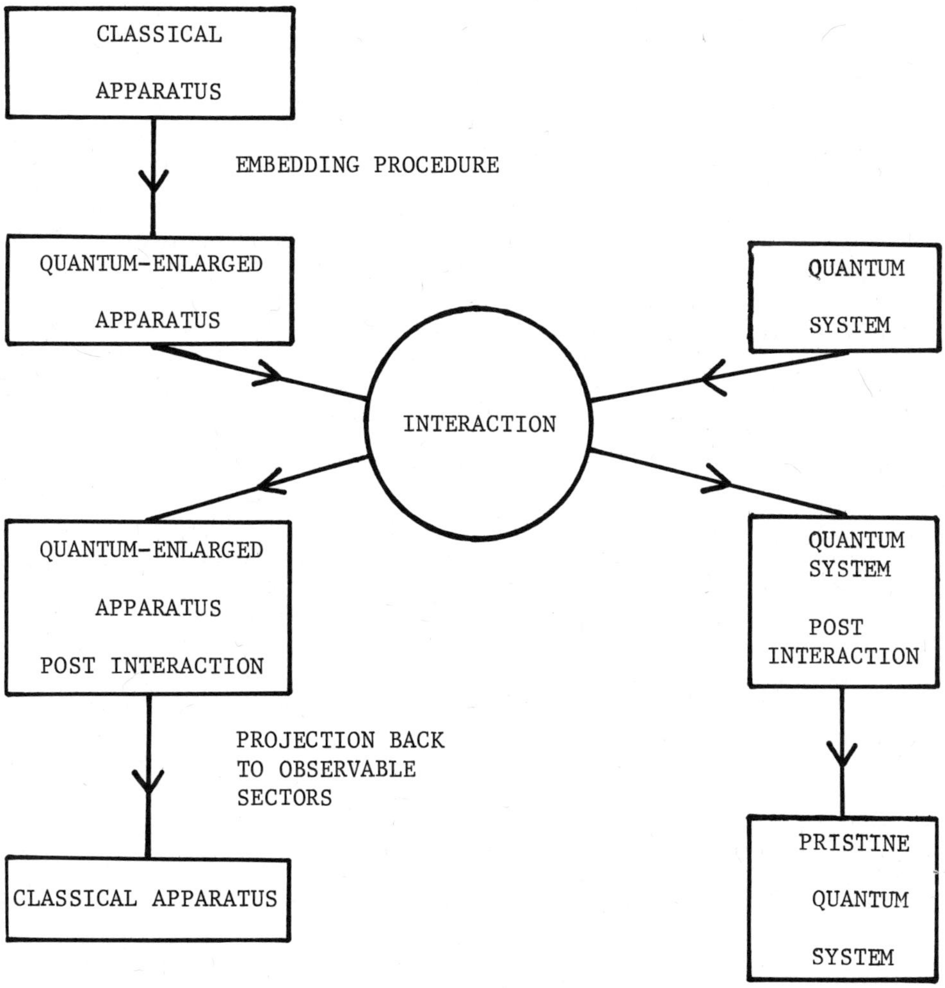

II. SUPERSELECTION AND NON-OBSERVABLES

Within the context of a Hilbert space of states (to which all physical states should belong), there is defined the concept of a selection rule. This has to do with the dynamics of an isolated quantum system. Given two distinct subspaces of the total Hilbert space, a selection rule operates between them if the vectors of one remain orthogonal to the vectors of the other in time - no spontaneous transitions can occur if the system is left to evolve on its own.

The concept of a Superselection rule[4] is a generalization of this concept - a superselection rule operates if a selection rule operates and further, no observable can connect states belonging to the different subspaces.

In many cases the selection rule can be understood to arise as a result of a symmetry of the dynamics of the system. The dynamical symmetry gives rise to invariant charge operators. It is between the eigenspaces of the charge operators that the selection rule operates. Similarly, the superselection rule can be understood to arise from a symmetry, not restricted just to the dynamics however. There is an associated charge operator - the superselecting operator - and it is between its eigenspaces that the superselection rule operates.

The existence of a superselection rule has certain consequences. The selection rule means that a state which belongs to a particular eigenspace of charge operators, $\mathcal{h}_i$, initially, will always belong to $\mathcal{h}_i$. The superselection part has the following consequences: in a superposition over distinct subspaces of the superselecting operators the relative phases are not measurable. No observable can connect the subspaces, and so the relative phases cannot be measured by any observable. Such a state we call 'an effective mixture state'. The subspaces $\mathcal{h}_i$ of the Hilbert space are then said to be superselected. Furthermore, those operators which fail to commute with the superselecting operators are not observable.

Within this conventional scheme the set of all observables is included within the commutant of the set of superselecting operators, and those operators which fail to commute with the superselecting operators form a subset of the set of non-observable operators.

We found it necessary in the course of our investigations to use a different notion of a superselecting operator.[2] The essential properties listed above are:

1. it commutes with the time evolution operator (selection rule)
2. it commutes with all observables

3. it is an observable.

In conventional quantum mechanical systems the time evolution operator is observable, and so the first condition is a special case of the second. We actually gain in generality by dropping 1 in its explicit form. We furthermore extend 2 to the status of an 'if and only if' condition and delete 3 , but these are not relevant for our present considerations.

As motivation for these changes, we see that now an operator will be unobservable if and only if it fails to commute with some superselecting operator - a reasonable definition. Thus given an arbitrary algebra of operators one can check the consistency of a particular choice of unobservable operators.

The extended concept clearly allows a superselecting operator which is not a constant of the motion. Then in the evolution of time states can spontaneously move from one superselected subspace to another. Thus, in a superposition over distinct subspaces $\mathcal{H}_i$, the relative phases are not observable - but may be measurable. At a later time the superposition may evolve into a state belonging to one of the subspaces $\mathcal{H}_i$. In that case the different superpositions would lead to different states, which could be distinguished. But this difference can only be seen after the state is altered. In addition, the Hamiltonian operator is not observable, since it does not commute with the relevant superselecting operator.

Clearly it will only be in a highly unconventional theory that such a concept could have a use. We will now describe such a theory.

III. THE EMBEDDING PROCEDURE

If we are to describe the apparatus in a measurement experiment as a classical system, even when measuring the attributes of a quantum system, then we must be able to couple together classical and quantum systems. It is not a priori obvious how this should be done. On the face of it, it appears that the structure of these theories are so different, that a coupling cannot be achieved. The differences result from the following:

1. in quantum physics the dynamical variables are non-commuting while in classical physics they are commuting;
2. a quantum state is specified by the eigenvalues of a complete commuting set of operators, a proper subset of the set of dynamical variables. A classical state, on the other hand, is specified by values for all the dynamical variables;
3. a linear superposition of 'pure' quantum states is also a pure quantum state, but a linear superposition of pure classical states is an incoherent state.

Despite these inherent differences, we will now describe a procedure for embedding a classical system into a quantum system. When this is done we shall use the procedure in the description of a measurement experiment.

Consider a classical mechanical system in the Hamiltonian formulation. The canonical coordinates are, eg., $(q^1, \ldots, q^n)$ and $(p^1, \ldots, p^n)$. The Hamiltonian is a function of these variables $H(q, p)$. The time development is given by

$$\dot{q} = \frac{\partial H}{\partial p}(q, p) \quad , \qquad \dot{p} = -\frac{\partial H}{\partial q}(q, p) \quad .$$

Such a system is a simple one - with no complexities to describe the action of pointers for example.

Let us now consider these variables q and p as operators acting on a Hilbert space, in much the same way as x is an operator in Schrödinger's wave mechanics. We denote them by $\omega^\mu = (q^1 \ldots q^n, p^1 \ldots p^n)$. We introduce operators which are conjugate to ω with respect to commutation,

$$[\omega^\mu, \pi^\nu] \equiv \omega^\mu \pi^\nu - \pi^\nu \omega^\mu = i \quad .$$

In the ω-representation we would have $\pi^\nu = -i\partial/\partial\omega^\nu$. Then, if we define the time evolution, or Hamiltonian, operator to be

$$H_{op} = -\frac{\partial H(\omega)}{\partial \omega^\mu} \varepsilon^{\mu\nu} \pi^\nu$$

where

$$\varepsilon^{\mu\nu} = \{\omega^\mu, \omega^\nu\}_{p.b.}$$

the Heisenberg picture equations of motion for $\omega^\mu(t)$ exactly mimic the classical equations given earlier.

The operators $\{\omega^\mu\}$ and $\{\pi^\mu\}$ have continuous spectra. Strictly speaking they do not have eigenvectors within the Hilbert space. Following conventional practice we shall include in the Hilbert space the non-normalizable states $|\omega'\rangle$ where

$$\omega|\omega'\rangle = \omega'|\omega'\rangle \quad .$$

Then, as in the Hamiltonian formulation where one could specify a precise initial value for $\omega(0) = (q(0), p(0))$, so also one can choose a precise initial state

$$|\omega^o\rangle = |q^o, p^o\rangle$$

which is an eigenvector of the operators q(0) and p(0).

We now supplement this structure by the Superselection principle.

A. ω^μ are superselecting operators, for all times, or equivalently

B. π^ν are unobservable operators for all times, in the sense that we saw earlier.

Notice here that there is no selection rule involved - the superselecting operators are not constants of the motion. Thus if one chooses as initial state an eigenvector of the superselecting operators, at later times the system will not be in the same eigenspace. Also the Hamiltonian operator is not itself an observable.

The observable sector of the quantum theory mimics exactly the classical theory. We have thus embedded the classical theory in a larger quantum mechanical framework.

The embedding procedure outlined above can be understood as a mapping from the classical phase space to a set of quantum mechanical operators and the space upon which they act. This map can be written as follows:

$$Q : \{\omega\} \rightarrow \{\omega_{op}\} \times \{\pi_{op}\} \times \hbar$$

where

$$Q(\omega) = (\omega_{op}, \pi_{op}, |\omega\rangle)$$

and

$$\omega_{op} |\omega\rangle = \omega|\omega\rangle$$

$$[\omega_{op}, \pi_{op}] = +i \quad .$$

Here we explicitly put a subscript 'op' to denote an operator. Then with the definition of H_{op} given earlier, we have at a later time t

$$Q(\omega(t)) = (\omega_{op}(t), \pi_{op}(t), |\omega\rangle)$$

where

$$\omega_{op}(t)\,|\omega\rangle = \omega(t)\,|\omega\rangle$$

$$[\omega_{op}, \pi_{op}] = i$$

and

$$\pi_{op}(t) = e^{iH_{op}t}\,\pi_{op}(0)\,e^{-iH_{op}t} \quad .$$

The mapping Q can be inverted, that is we can map back from the quantum system to the original classical system

$$Q^{-1} : \{\omega_{op}\} \times \{\pi_{op}\} \times \hbar \rightarrow \{\omega\}$$

where

$$Q^{-1}(\omega_{op}(t), \pi_{op}(t), |\omega\rangle) = \omega(t)$$

and

$$\omega_{op}(t)\,|\omega\rangle = \omega(t)\,|\omega\rangle \quad .$$

By means of these mappings we can achieve steps 1 and 3 in the schematic description of our work. We now consider step 2.

Example:

A simple example will help to illustrate these ideas. Consider a classical free particle in one dimension. Its Hamiltonian formulation is:

canonical coordinates: (q, p)

Hamiltonian: $H(q, p) = \frac{1}{2m} p^2$.

In the quantum enlarged formulation we have

canonical coordinates: $\omega = (q, p)$ and $\pi = (\pi^q, \pi^p)$

and Hamiltonian operator: $H = \frac{1}{m} p\pi^q$.

The Heisenberg equations of motion for the observables ω are

$$\dot{p}(t) = 0$$

$$\dot{q}(t) = \frac{1}{m} p(t)$$

with solutions

$$p(t) = p(0)$$

$$q(t) = q(0) + \frac{t}{m} p(0) \quad .$$

We choose as the state of the system $|q_o, p_o>$ then at the later time t

$$q(t)|q_o, p_o> = (q_o + \frac{t}{m} p_o)|q_o, p_o>$$

and

$$p(t)|q_o, p_o> = p_o|q_o, p_o> \quad .$$

IV. INTERACTION BETWEEN CLASSICAL SYSTEM AND QUANTUM SYSTEM

Since classical systems and quantum systems have such different structures, the manner in which they could interact is not clear. However, the interaction of systems of like structure is a straightforward exercise. The naturalness of the embedding procedure described suggests that this way to cause the interaction may be the correct way.

Consider a quantum mechanical system with dynamical variables $\{\xi\}$. The theory is specified when the energy operator of the isolated quantum system, $X(\eta)$, and the commutation relations among the ξ are known. The Hamiltonian operator for the uncoupled apparatus and system is

$$-\frac{\partial H(\omega)}{\partial \omega^\mu} \varepsilon^{\mu\nu}\pi^\nu + X(\eta) \quad .$$

A coupling term is $\Phi(\omega, \pi; \xi')$. This coupling will affect both classical and quantum systems.

However, the coupling is restrained by the requirement that after interacting the apparatus remain classical. The classical nature of the apparatus system is characterized by:

1. $\omega^{\mu}(t)$ observable for all t,
2. $\omega^{\mu}(t)$ and $\omega^{\nu}(t')$ compatible for all t and t', which are equivalent conditions.

Immediately these require that the coupling term be at most linear in the non-observables π. Thus we concentrate on couplings

$$\phi^{\mu}(\omega, \eta')\pi^{\mu} + h(\omega, \hat{\varepsilon}) \quad .$$

However this alone does not guarantee that 1 and 2 are satisfied for all times because the primary and secondary coupling functions depend on unspecified sets of quantum variables $\{\eta'\}$ and $\{\hat{\varepsilon}\}$ respectively.

We will now see what further restrictions can be deduced for these couplings. Let us first restrict our attention to couplings which are analytic everywhere. In this case an equivalent condition to examine is

$$3. \quad [\frac{d^m}{dt^m}\,\omega^{\mu}(t), \frac{d^n}{dt^n}\,\omega^{\nu}(t)] = 0 \qquad m, n \geqslant 0 \quad , \quad \text{all } t \quad .$$

From this condition we can derive a set of conditions which must be satisfied by the coupling functions so that the apparatus remain classical - the Integrity criteria. In the analytic case they are both necessary and sufficient, whereas in the non-analytic case they are merely necessary.

From the Hamiltonian operator

$$H_{op} = F^{\mu}(\omega, \eta')\pi^{\mu} + G(\omega, \hat{\varepsilon})$$

$$F^{\mu}(\omega, \eta') \rightarrow (\frac{\partial H}{\partial \omega}\,\varepsilon + \phi)$$

$$G(\omega, \hat{\varepsilon}) \rightarrow X(\eta) + h(\omega, \hat{\varepsilon})$$

we see that the time derivatives of ω are

$$\dot{\omega}^{\mu}(t) = - F^{\mu}(\omega, \eta')$$

$$\ddot{\omega}^{\mu}(t) = i[F^{\mu}, F^{\nu}\pi^{\nu} + G] \quad .$$

The first criterion that we have follows at once.

A. $\{\phi^{\mu}\}$ forms a commuting set.

Let $\{\rho\}$ be the maximal algebraically independent subset of $\{\phi^{\mu}\}$, allowing ω-dependent coefficients. We consider extensions of this set to a maximal commuting algebraically independent subset of the algebra of dynamical variables $\{\rho\} \cup \{\rho'\}$, and we denote the algebra generated by this set allowing ω-dependent coefficients by $\mathcal{A}_{\omega}(\rho, \rho')$.

The remaining criteria we derive by ensuring that the time derivatives of $\omega^{\mu}(t)$ satisfy the condition 3 are

B. $F^{\nu}[\rho_m, \pi^{\nu}] + [\rho_m, G] \in \mathcal{A}_{\omega}(\rho, \rho') \; \forall m$.

C. $F^{\nu}[\rho'_m, \pi^{\nu}] + [\rho'_m, G] \in \mathcal{A}_{\omega}(\rho, \rho')$

for those ρ'_m which occur in the expansions of $[\rho_m, H_{op}]$,

$[[\rho_m, H_{op}], H_{op}], \ldots$.

These criteria do not force ϕ^{μ} to depend on a commuting set of quantum variables. If the ϕ^{μ} are such, then B and C reduce to $[\rho_m, G]$ and $[\rho'_m, G]$ belonging to the algebra. These are conditions on h, the secondary coupling function. On the other hand, if the ϕ^{μ} depend in a non-trivial manner on non-commuting operators these criteria restrict ϕ^{μ} also.

The primary coupling function's dependence on non-commuting quantum variables will lead to a problem when we wish to read the information stored in the classical variables. It will be of the ϕ^{μ}, not the underlying non-commuting quantum operators. It is only when ϕ^{μ} depend on commuting quantum variables that information on the ϕ^{μ} can be seen to give us similar informations about the quantum variables.

Thus there may be certain interactions between the classical apparatus and the quantum system which allow the classical apparatus to remain classical, but fail to yield a measurement.

In the cases that a measurement can result from the interaction, there are more steps to be taken before one can consider it accomplished. The quantum-enlarged apparatus has interacted with the quantum system, and information has been transferred to the apparatus observables.

However, in its present form we cannot 'measure' the apparatus variables without disturbing the system. A measurement of the $\omega^{\mu}(t)$ will cause a disturbance to the $\pi^{\mu}(t)$. This in turn, if more interactions with the quantum system are envisaged, may have an effect on the quantum system. It is only when the apparatus is described in the classical Hamiltonian formulation that its observables can be measured at will.

Thus we must project back, by means of the map Q^{-1} to the classical phase space. Because this mapping is achieved via the eigenvalues, the classical variables will now possess the information about the relevant eigenvalues.

At the same time we must also project the quantum system back to a pristine form - projecting out all π-dependence in operators which do not determine the state of the system.

V. APPLICATION OF MEASUREMENT MODEL

We will now briefly discuss a particularly simple example to illustrate a measurement problem, and the use of the integrity criteria.

The quantum system to be examined is an inert quantum spin system. The operators are

$$S^2 = S_1^2 + S_2^2 + S_3^2 \quad \text{and} \quad S_1, S_2, S_3 \ ,$$

satisfying

$$[S_i, S_j] = i\, \varepsilon_{ijk}\, S_k$$

$$[S^2, S_i] = 0 \ .$$

The Hamiltonian of this quantum system is zero.

The apparatus we use is also the simplest we can imagine, with no claim to realism, a freely moving classical particle. In quantum enlarged form, the operators are

$$\omega = (q_i, p_i), \qquad \pi = (\pi_i^q, \pi_i^p)$$

and the Hamiltonian of the apparatus is $\frac{1}{m}\, \vec{p}\cdot\vec{\pi}^q$.

In order to describe an interaction between the apparatus and system we envisage the quantum spin system being carried along as internal degrees of freedom by the (electrically neutral) particle (classical) - it gives the magnetic moment of the classical particle.

The interaction will be induced by sending the particle through an inhomogeneous magnetic field. Thus our experiment can be considered as a Stern-Gerlach experiment - the heavy atoms are usually treated as classical.

The Hamiltonian operator will take the form

$$H_{op} = \frac{1}{m}\vec{p}\cdot\vec{\pi}^{q} - \gamma\vec{B}(q)\cdot\vec{S} - \gamma[\frac{\partial}{\partial q_j}\vec{B}(q)\cdot\vec{S}]\pi^{p}_{j} \quad .$$

Here $\vec{\mu} = \gamma\vec{S}$ is the magnetic moment caused by the internal degrees of freedom.

$\gamma\ \partial/\partial q_j\ \vec{B}(q)\cdot\vec{S} = \phi^{p}_{j}$ is chosen to give the correct classical equations of motion for a particle of magnetic moment $\vec{\mu}$ moving in an inhomogeneous magnetic field, and $\vec{\mu}\cdot\vec{B}(q)$ is the classical potential energy for this motion.

We allow for the most general (analytic) coupling between the apparatus and system by allowing $B_i(q)$ to be arbitrary analytic functions. The coupling functions ϕ^{p}_{j} will now depend in a non-trivial manner on non-commuting quantum operators.

We must check the integrity criteria, however.

A. $\{\phi^j\}$ commuting set, if and only if $\frac{\partial}{\partial q_i}\vec{B}(q) \times \frac{\partial}{\partial q_j}\vec{B}(q) = 0$

i.e. they are parallel.

B. $[\phi_i, H_{op}]$ commuted with ϕ_j, if and only if the inhomogeneous magnetic field takes the form $B(q)\ \vec{n}$, $\vec{n}$ a constant unit vector.

C. All the other criteria are then satisfied.

In this case the primary coupling functions depend on the single quantum variable $S_n \equiv \vec{S}\cdot\vec{n}$ and the interaction leads to a measurement.

If we allow non-analytic couplings, B restricts the external magnetic field to the form

$$\sum_m B^m(q)\vec{n}_m$$

where $\vec{n}_m$ are constant unit vectors, and the functions $B^m(q)$ have disjoint regions of support. In this case, however, the criteria are necessary, but not sufficient. The other criteria one must check reduce the magnetic field to the form $B(q)\ \vec{n}$.

<u>Result</u>: the only interaction which preserves the classical nature of the apparatus is one where the magnetic field is unidirectional

in regions through which the particle passes.

Let us now see how the measurement proceeds in this example. Consider a magnetic field

$$B(q) = \begin{cases} 0 & q_2 \leqslant y_0 \\ aq_3+b & y_0 < q_2 < y_1 \\ 0 & y_1 \leqslant q_2 \end{cases}$$

where $\vec{n}$ is along the q_3-direction, and $a < 0$. We shall ignore the edge effects caused by this choice of B(q). Elsewhere we have shown that the edge effects can be treated without altering substantially the solution.[5] We must specify the state of the system initially by

$$|\psi\rangle \otimes |\varphi\rangle$$

where the apparatus state is $|\hat{q}_2, \hat{p}_2; \hat{q}_3, \hat{p}_3\rangle = |\psi\rangle$

$$\hat{q}_2 < y_0 \qquad \hat{q}_3 = \hat{p}_3 = 0$$

$$\hat{p}_2 > 0$$

and the quantum state is $|\varphi\rangle = |s, S_1 = s'\rangle$. The Hamiltonian is

$$\frac{1}{m} \vec{p}\cdot\vec{\pi}^q - \gamma\vec{B}(q)\cdot\vec{S} + \phi_3\pi_3^p$$

$$\phi_3 = \begin{cases} 0 \\ -a\gamma S_3 \\ 0 \end{cases} .$$

The equations of motion are

$$\dot{p}_2(t) = 0 \qquad \dot{p}_3(t) = -\phi_3$$

$$\dot{q}_2(t) = \frac{1}{m} p_2(t) \qquad \dot{q}_3(t) = \frac{1}{m} p_3(t)$$

$$\dot{S}_3(t) = 0 .$$

The solutions in the region to the right of the magnetic field are

$$p_3(t) = p_3(0) - a\gamma S_3(0)(t_1-t_0)$$

$$q_3(t) = q_3(0) + \frac{1}{m} p_3(0)t - \frac{a\gamma}{2m} S_3(0)(t_1-t_0)^2 - \frac{a\gamma}{m} S_3(0)(t_1-t_0)(t-t_1)$$

where $t_0 = \frac{m}{\hat{p}_2}(y_0-\hat{q}_2)$, $t_1 = t_0 + (y_1-y_0)\frac{m}{\hat{p}_2}$.

Now we see that the apparatus observables $q_3(t)$ and $p_3(t)$ have become functionally dependent on the quantum operator S_3. To see that this will result in a measurement we need to examine the action of $p_3(t)$ on $|\psi> \otimes |\varphi>$

$$p_3(t)(|\psi>\otimes|\varphi>) = p_3(t)|\psi> \otimes \sum_{s''} <s,S_3 = s''|s,S_1 = s'>|s,S_3 = s''>$$

$$= p_3(t) \sum_{s''} a_{s''s'} |\psi> \otimes |s,S_3 = s''>$$

$$= \sum_{s''} a_{s''s'} [-a\gamma(t_1-t_0)s''] |\psi> \otimes |s,S_3 = s''>$$

-the single trajectory has split into (2s+1) beams of particles, each with state

$$|\psi> \otimes |s,S_3 = s''> \quad .$$

Now, in the projection back to the classical phase space, each beam will have values for $q_3(t)$ and $p_3(t)$ dictated by their action on $|\psi> \otimes |\varphi>$. Thus we see that here also the beams will be split into many trajectories. As it is now in purely classical form, a measurement of which beam the particle belongs to will yield a measurement (for the observer) of the value of S_3 for the internal quantum spin system.

VI. CONCLUSION

We have presented an unusual approach to the measurement problem in which the strictly classical apparatus interacts directly with the quantum system. We have examined this interaction, in particular whether or not the classical system might 'become quantized' as a result of it. Finally we have discussed a simple example to illustrate the procedures.

We have elsewhere[5] examined more complicated arrangements of the systems in section V, for example with crossed magnetic fields. The model is seen to be able to describe the outcomes.

We are currently examining approximate methods of solution, the rigorous treatment of superselection, and other experiments which can be described by the model.

REFERENCES

*) Work supported in part by the US Energy Research and Development Administration.

1) E.C.G. Sudarshan, Pramana 6, 117 (1976).

2) E.C.G. Sudarshan and T.N. Sherry, CPT preprint.

3) G.C. Wick, A.S. Wightman and E.P. Wigner, Phys. Rev. 88, 101 (1952).

4) See for example, D. Bohm, Quantum Theory (Prentice-Hall, New York, 1951).

5) E.C.G. Sudarshan, T.N. Sherry and S.R. Gautam, CPT preprint.

A SURVEY OF VORTICES IN GAUGE THEORIES*

Hsiung Chia Tze†

Stanford Linear Accelerator Center

Stanford, California 94305, U. S. A.

1. INTRODUCTION

The objects in my talk this evening are possibly the least extended among those discussed in the past few mornings. They are the vortices of gauge theories, the grand daddy of all subsequent extended objects of renormalizable relativistic field theories. While the idea of using vortex filaments as fundamental objects traces back to Thomson[1], Dirac[2] and Buneman[3], it has only recently flowed into the main stream of particle theory. This happened with the key suggestion of Nielsen and Olesen[4] as well as Tassie[5] that the Abrikosov flux line solution[6] of the static Higgs-Kibble (alias Landau-Ginzburg) model can be identified with the celebrated Nambu string of dual resonance model[7]. Ever since there has been much activity on the subject. A glance at the literature shows a clear leveling off of papers on this topic. So it seems timely for you to sit back and for me to attempt a survey of this field. In a one hour talk, my presentation can only aim at an overview (which is probably what most of you are up to near the end of an active day); I must be very brief. All details are to be filled in at your own pace and leisure. For that purpose and for the record I have accumulated an extensive but certainly incomplete reference list on vortices. I must remark at the outset that much of what I will be talking about is essentially familiar to people in solid or liquid state physics. As far as static solutions are concerned, the only difference with particle theory may be that our non-Abelian groups can be other than SO(3).

Admittedly colored by my own involvement in the subject[8,9] my talk will deal only with the following topics. After stating the necessary boundary conditions for the existence of vortex string solutions, we give the generalized London flux quantization law for any compact gauge group. The topology of knot configurations is briefly mentioned. Then come the questions of the classical stability of the vortices and their possible relevance to particle physics. To illustrate the last point, I will present a rather beautiful realization of a generalization of Nambu's new mechanics by non-Abelian dyons whose Wu-Yang type monopole is a point vortex. These dyons are shown to obey Nambu's equations for non-Abelian tops.

2. TOPOLOGY OF VORTICES

Briefly, in complete analogy to a type II superconductor, Nielsen and Olesen[4] argued that the Higgs vacuum of unified gauge models can be perforated by relativistic vortex filaments. Each vortex is composed of two structures, a tube of normal vacuum (of the size of the coherence length $\xi \sim m_s^{-1}$) filled up with a quantized magnetic flux bundle (of the size of the penetration length $\lambda \sim m_v^{-1}$). The magnetic field strength is such so as to undo the Higgs mechanism and locally restore gauge invariance. Of course, there is also the alternative possibility of symmetry restoration by a high temperature regime. Generally such local restoration takes the forms of vortices which can have the geometries of points, lines, sheets, etc. Here we shall deal almost exclusively with vortex strings. Since, if they exist at all, analytic vortex solutions to the field equations are rather rare, we choose to look at the topology of the solutions, an aspect of the problem which allows for exact answers and should be the first step toward finding solutions. Our starting point for looking at realistic topological solutions is the standard non-Abelian Higgs-Kibble system with the renormalizable Lagrangian

$$\mathcal{L}(x) = -\frac{1}{4}\vec{F}_{\mu\nu}\cdot\vec{F}_{\mu\nu} - \sum_{i=1}^{K}|D_\mu\Phi(x)|^2 - U(\Phi_1,\ldots\Phi_K) . \qquad (2.1)$$

K is the requisite number of Higgs multiplets for a well-defined vortex, otherwise our notation is self explanatory. For any compact gauge group G, the condition for finite energy per unit length of axisymmetric vortex solutions is

$$\begin{aligned} D_\mu\Phi &= (\partial_\mu - iea_\mu)\Phi \xrightarrow[r\to\infty]{} 0 \\ a_\mu &\equiv \vec{A}_\mu\cdot\vec{T} \\ r^2 &= x^2 + y^2 \end{aligned} \qquad (2.2)$$

i.e. the Higgs scalars must be covariant constants at radial infinity. Equation (2.2) is nothing but the Meissner mechanism of the flux confinement. Its integral expression is

$$\Phi_i(y) = S_\ell(y,x)\Phi_i(x) \tag{2.3}$$

where

$$S_\ell(y,x) = T\exp\left(-ie\int_x^y a_\mu(z)dz^\mu\right)$$

is the nonintegrable phase factor. T is the ordering operator for the matrices along the path ℓ linking the points x and y at spatial infinity. The motions of this phase factor $S_\ell(y,x)$ in the group manifold G constitute the kernel of the topological classification. This is so since $S_\ell(y,y)$ for closed paths ℓ measures the connectivity of the group space given by $\pi_1(G)$, the fundamental group of G.[10,11]

Taking a circular path and beginning with the Abelian case of $G = U(1) \approx SO(2)$, the nonintegrable phase factor is of the form

$$S(\phi) = \exp(in\phi) \quad , \tag{2.4}$$

then Eq. (2.3) means that as the order parameter Φ goes around a large circle $\phi = 0 \to 2\pi$ centered at the vortex in physical space, $S(\phi)$ winds around the complex-field space n times. Since U(1) is infinitely connected, $\pi_1(U(1)) \approx Z_\infty$ the group of the integers, there are n units of flux carried by the vortex Eq. (2.4),

$$4\pi g = i/e \oint \vec{\nabla} \log S(\phi)\cdot d\vec{x} = n$$

hence the London quantization

$$eg = n/2 \quad . \tag{2.5}$$

The essential difference between Abelian and non-Abelian vortices is seen by going to G = SU(2).[12] Here a judicious choice of representation is of essential importance to the possibility of nontrivial solutions. Indeed if the breaking is achieved by a complex doublet then

$$S(\phi) = \exp(2in\theta\tau_3) \in SU(2) \quad . \tag{2.6}$$

SU(2) is simply-connected, any closed loops in its manifold are contractible to a point, $\pi_1(SU(2)) = 0$. So any apparent vortex solution can in principle be continuously gauged away into the vacuum configuration $A_i = 0$, there being no topological stability.

On the other hand, if the breaking is performed with a pair of Higgs triplets,

$$S(\phi) = \exp(in\theta\tau_3) \in SO(3) \quad . \tag{2.7}$$

Since

$$\pi_1(SO(3)) \approx \pi_1(SU(2)/Z_2) \approx Z_2 \quad , \tag{2.8}$$

there exists two topologically distinct kinds of vortices with 0 or ±1 units of flux. More succinctly the London quantization is now

$$eg = n/2 \quad (\mathrm{mod}\ 2) \quad . \tag{2.9}$$

Similarly for G = SU(3), breaking with quark like representation leads to no vortex. While using regular or any other faithful representations imply the global group is actually $SU(3)/Z_3$, the old "triality" factor group. Since

$$\pi_1(SU(3)/Z_3) \approx Z_3 \quad , \tag{2.10}$$

we have

$$eg = n/2 \quad (\mathrm{mod}\ 3) \quad . \tag{2.11}$$

There is in fact an exact parallel between the topology of non-Abelian vortices and that of Wu-Yang-Dirac monopole which would then terminate these vortices. For $G = SU(3)/Z_3$ we then have as simplest vortex configurations those of Fig. 1. They would be the mesonic (a), baryonic (b,c) and pomeronic (d) vortices if the other quantum numbers to be carried by the monopole for stability are the flavors of the quarks. In order not to collapse, the last type of vortex (d) must be spinning.

The general pattern is now clear. Consider a gauge model with a compact Lie group G suitably broken down by Higgs fields such that only a subgroup the holonomy group H leaves the vacuum

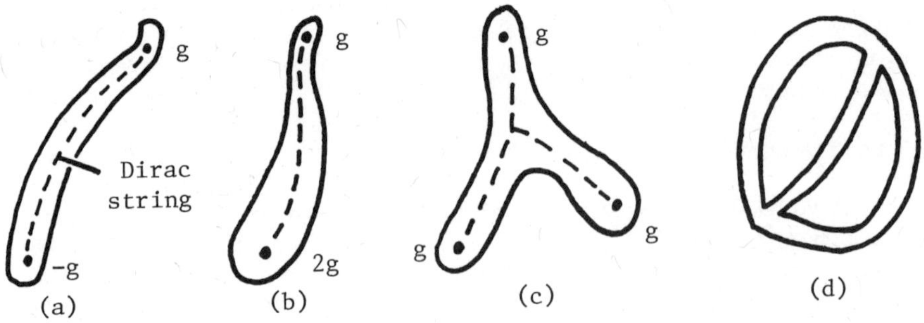

Fig. 1. Simplest vortex configurations of $G = SU(3)/Z_3$.

invariant. The condition for classical topological stability of vortices is that $\pi_1(G/H)$ be nontrivial. This means there must be topologically distinct maps from the coset space G/H onto the circle at infinity centered at the vortex core. Assuming for simplicity maximal symmetry breaking by Higgs fields taken in the adjoint representation of G then the vacuum remains invariant only under the unit matrix I of the adjoint representation, i.e., under those elements of G which are mapped into I and thus form the center of G. For SU(N) for instance the elements of the center are $I_N \exp(2\pi i\ell/N)$, $\ell = 0,1,2,\ldots,N-1$, with I_N being the $N \times N$ unit matrix taken in the fundamental representation. So taking the Higgs in the adjoint or any tensor product representation thereof means that the global group is actually $SU(N)/Z_N$. $\pi_1(SU(N)/Z_N) \approx Z_N$ and so there are then (N-1) "gauge types" of vortices, to use the terminology of Wu and Yang[12].

The case of a general compact Lie group G is now obvious. We note that the quantization of g mirrors the discreteness of $\pi_1(G)$ where by G we mean the global invariance group determined by the matter fields, the Higgs fields. The Abelian character of $\pi_1(G)$ and hence the additive combination law for the elements of Z_n is reflected in the additive law of combination of magnetic fluxes and charges defined only by modulo n, n being simply the connectivity of the global gauge group. For a non-Abelian group $\pi_1(G)$ is of finite order. Hence the varieties of distinct vortices is finite which is the key difference in the Abelian case. So to Cartan's topological classification of compact Lie groups[13] corresponds in a one to one way the topological classification of generalized Nielsen-Olesen vortices and Dirac monopoles. A detailed account of all this is to be found in Ref. 9.

3. TOPOLOGY OF KNOTTED VORTICES

As in the statistical mechanics of polymer chains[14], our vortices if sufficiently long have finite probability of forming interlocking structures and tying themselves in knots. The classification of the latter is like homotopy theory a branch of algebraic topology. For your amusement and completeness I will say a few words on the subject[15] of which I really know very little. Knot theory is associated with the English physicist and mathematician Tait. Knots are represented by their projections on some plane and are classified according to the minimum number of intersections in these projections. Those with the same number of intersections are put in a definite sequence and the order in which they appear is labelled as a subscript to the number of intersections. In Fig. 2 we display the four simplest nontrivial knots with their Alexander polynomial $\Delta(t)$, which is an invariant of the corresponding knot.

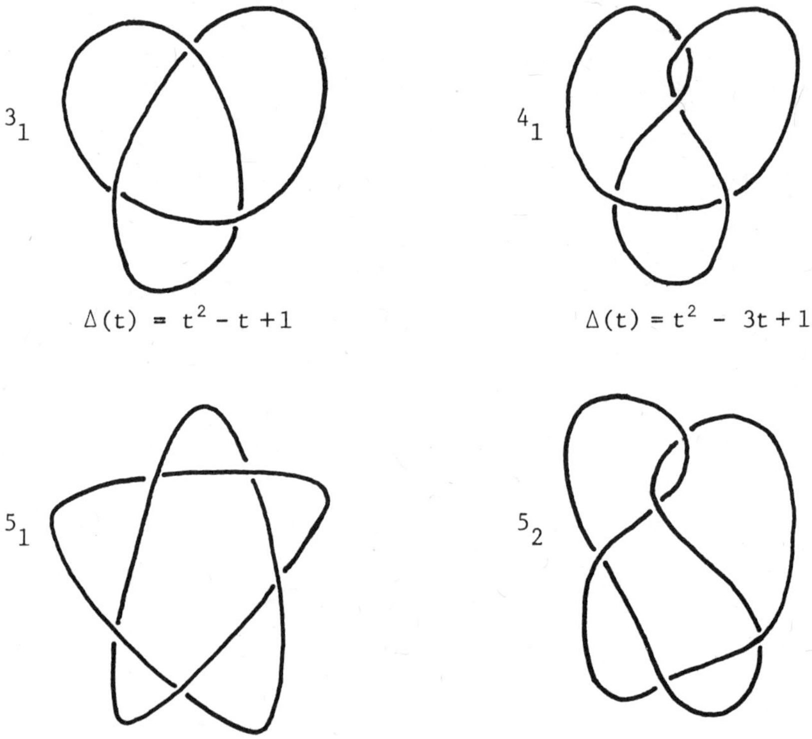

$\Delta(t) = t^2 - t + 1$ $\Delta(t) = t^2 - 3t + 1$

$\Delta(t) = t^4 - t^3 + t^2 - t + 1$ $\Delta(t) = 2t^2 - 3t + 2$

Fig. 2. Four simplest nontrivial knots.

Figure 3 gives the so-called simple knots, all the 84 knots having a minimum number of intersections less than 10. Moreover there also exists composite knots formed by various combinations of the above and whose Alexander polynomials are the products of the $\Delta(t)$'s for their constituent knots. We close this subject quickly with a reservation. The Alexander polynomial is not a complete invariant of a knot. This is so since there are for instance five of the above 84 simple knots whose $\Delta(t)$ are all $\neq 1$ yet have identical $\Delta(t)$. Furthermore there are knots whose $\Delta(t) = 1$ and is yet nontrivial. The simplest example of its kind is Fig. 4.

Finally I should mention the possibility of another new invariant associated with a flux twist[16,3] illustrated in Fig. 5 with a trivially knotted Abelian vortex. The flux twist is defined as the number of times the flux lines wind around the vortex filament.

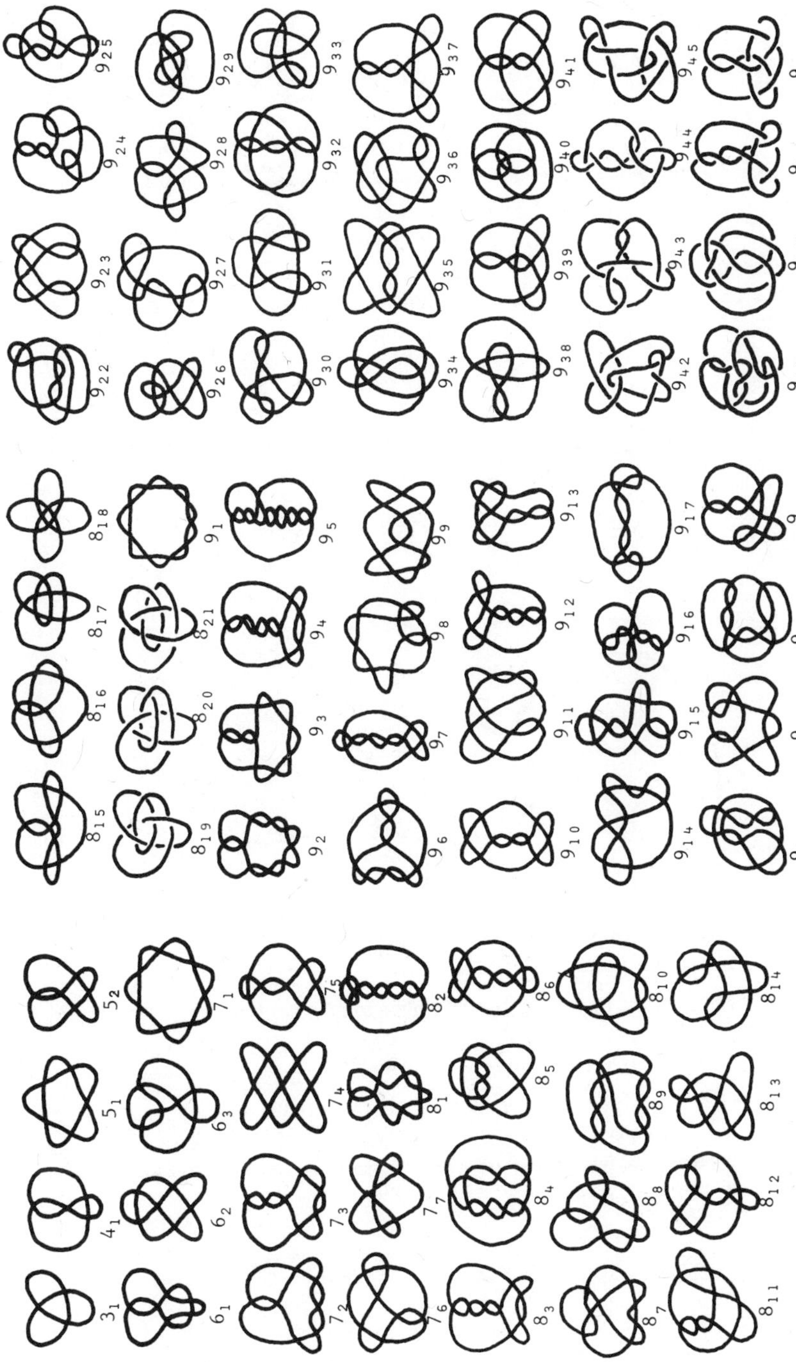

Fig. 3. Knots with degree of complexity ≤ 9. (Reproduced from Knotentheorie, Reidemeister, by kind permission of Springer-Verlag.)

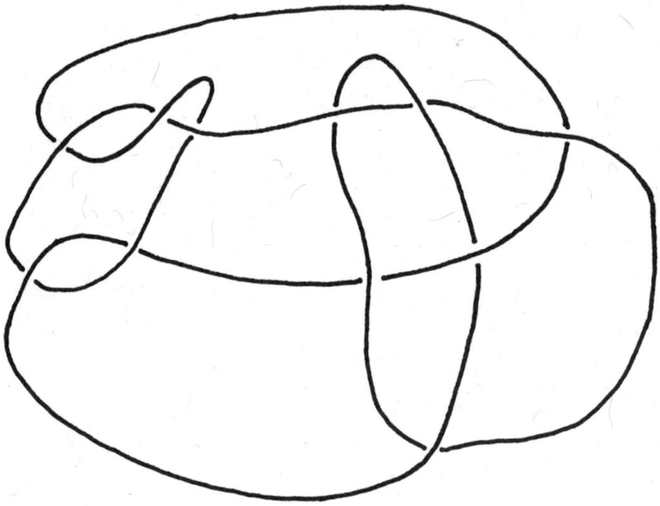

Fig. 4. Simplest nontrivial knot with $\Delta(t) = 1$.

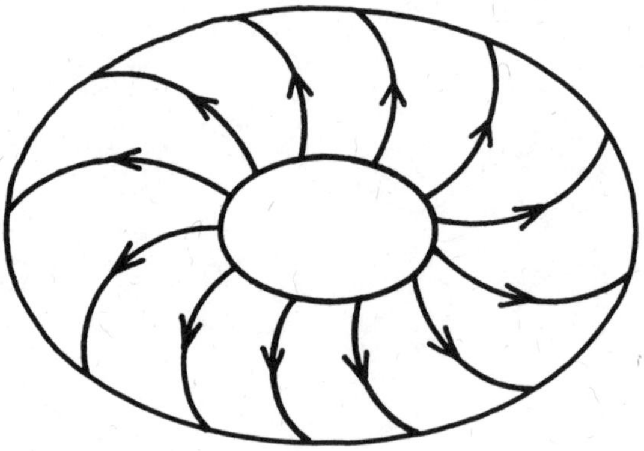

Fig. 5. Trivially knotted Abelian vortex.

All the above invariants, the flux quantum, the flux twist, the knot invariant, are tempting preys indeed to those of us who seek ever more topological conservation laws[16].

4. CLASSICAL VORTEX ENERGY AND STABILITY

For definiteness, we follow the compact analysis of Bogomol'nyi[17]. For the U(1) vortices we take $U(\Phi) = \lambda(\Phi_a\Phi_a - F^2)^2$ in Eq. (2.1) (K = 1) and write the complex field as two real fields

Φ_a (a = 1,2). The energy per unit length of a vortex of circulation n is bounded from below since we can write

$$E_n = \frac{m_v^2}{4\alpha}\,\varepsilon_n \tag{4.1}$$

$$\varepsilon_n = n + \frac{1}{2\pi}\int d^2y\Big[\frac{1}{4}\big(f_{mn} - \varepsilon_{mn}(1-Q_aQ_a)\big)^2 + \frac{1}{2}\big(\varepsilon_{mn}D_nQ_a + \varepsilon_{ab}D_mQ_b\big)^2 + \big(\kappa^2 - \frac{1}{2}\big)\big(Q_aQ_a - 1\big)^2\Big] . \tag{4.2}$$

In Eq. (4.1) use is made of the dimensionless variables

$$\alpha = e^2/4\pi\ , \qquad \kappa = \sqrt{\lambda}/e = m_s/(\sqrt{2}\,m_v)$$

$$Q_a = F^{-1}\Phi_a\ , \qquad v_m = \sqrt{2}\,F^{-1}A_m\ , \qquad f_{mn} = \nabla_{[m}v_{n]}\ , \tag{4.3}$$

and $D_mQ_a = \nabla_mQ_a + \varepsilon_{ab}v_mQ_b$.

κ is the Landau-Ginzburg parameter which distinguishes a type I ($\kappa < 1/\sqrt{2}$) from a type II ($\kappa > 1/\sqrt{2}$) superconductor. So broken gauge theories can be similarly termed as being of the first or second kind, depending on the value of κ. For $\kappa = 1/\sqrt{2}$, the minimum value of ε_n is achieved with the saturation conditions

$$f_{mn} = \varepsilon_{mn}(1 - Q_aQ_a)\ ,$$
$$\varepsilon_{mn}D_nQ_a + \varepsilon_{ab}D_mQ_b = 0 \tag{4.4}$$

which replace the field equations. For the vortex solutions of the form

$$Q_1 + iQ_2 = \exp(inS(r)\phi)\ ,$$
$$v_\phi = \frac{n}{r}\,v(r)\ , \qquad v_r = 0\ . \tag{4.5}$$

(ϕ being the azimuthal angle around the vortex). One gets the coupled system of 1st order equations which admit solutions with boundary conditions $S(r)$, $v(r) \to 1$ as $r \to \infty$ and $S(r)$, $v(r) \to 0$ as $r \to \infty$. The energy per unit length is then

$$E_n > n\sqrt{2\kappa}\ \frac{m_v^2}{4\alpha} \qquad \kappa < 1/\sqrt{2} \quad ,$$

$$E_n = n\ \frac{m_v^2}{4\alpha} \qquad \kappa = 1/\sqrt{2} \quad , \qquad (4.6)$$

$$E_n > n\ \frac{m_v^2}{4\alpha} \qquad \kappa > 1/\sqrt{2} \quad .$$

By classical instability we understand the existence of small departures from the solution which decrease its energy. Indeed the existence of a homotopic conservation law only guarantees the stability of vortices with lowest nontrivial circulation. It does not prevent those having higher flux number, n, from decaying into ones with lower n with the total flux of the system unchanged in the process. To find out we perform the usual small wave expansion about the exact solution. By suitable choice of a trial function, it was shown that for $\kappa > 1/\sqrt{2}$ strings units with flux number $n \geq 2$ are classically unstable. They break up into n vortices with unit flux. This is reflected in the forces between vortices: The latter are respectively attractive for $\kappa < 1/\sqrt{2}$, null for $\kappa = 1/\sqrt{2}$ and repulsive for $\kappa > 1/\sqrt{2}$. This is verified experimentally in type II material; by repelling one another, vortices form triangular lattices[18].

What about non-Abelian vortices? What essential differences from the Abelian case result from the remarkable modulo property of the non-Abelian flux? A similar but algebraically more intricate analysis as the above still awaits investigation. A step in this direction has been the evaluation of the critical values of non-Abelian gauge fields which lead to vacuum symmetry restoration[19].

5. THEIR RELEVANCE

We recall that vortices were originally conjectured to be the field theoretical realizations of the Nambu strings of dual models[20]. Indeed the non-Abelian configurations of our Fig. 1 are certainly suggestive of one's phenomenological intuitions about hadrons[8,9]. The quarks which presumably terminate these vortices must be Dirac monopoles as well and color symmetry must be broken down to some discrete subgroup; for SU(3) color the latter would be the group's center Z_3. For example, the Nambu DNA model[21] which has two types of strings, the λ_3- and λ_8-like strings, finds its field theoretic realization among our non-Abelian vortices. From another side, we also mention the program of Faddeev who prefers to achieve breaking of local gauge invariance by nonlinear chiral Higgs fields chosen for their aesthetic geometrical appeal[22]. In his soliton approach to hadrodynamics, the strong interactions will presumably emerge from a suitable unified theory of the electromagnetic and weak interactions. In this way, there will not be a need for quarks. Also there will be possibly vortex hadrons in these models[23]. However it must be said that in this Post-Wilsonian era of electric color confinement, solitary magnetic confinement is a lonely road travelled by a few!

Yet if we can trade an electric color supervacuum which repels magnetic flux by one magnetic vacuum which repels electric flux, vortices (particularly their zero dimensional varieties) point-like monopoles or instantons might still play a central role, in at least the continuum approaches to quark confinement[24]. Here the vacuum is envisioned as a medium whose dynamics is dominated by classical configurations of monopoles and antimonopoles seen as random fields. Various studies of confinement by such a mechanism of an electric Meisner effect have emerged recently. Time and lack of knowledge prevent me from going into any details here. Moreover you have probably heard about these topics from the morning lectures by R. Jackiw and G. 't Hooft on extended objects. Instead I would like to close by briefly presenting the main result in some work in which I have very recently been involved[25]. It points, in my opinion, to yet another direction where classical solutions in gauge theories may be of interest.

For simplicity let's consider a classical SU(2) dyon made up of a Wu-Yang monopole (i.e. a point vortex) and a test particle of mass m and isospin T_i $(i = 1,2,3)$.[26] For binding, we also include a non-electromagnetic potential V(r). Its equations of motion can be found from

$$\dot{F}(x,p,T) = \langle F,H\rangle \tag{5.1}$$

where

$$H = \frac{1}{2m}\left(\vec{p} + \frac{\vec{T}\times\vec{r}}{r^2}\right)^2 + V(r) \quad . \tag{5.2}$$

F is an arbitrary differentiable function of the dynamical variables $\vec{x}$, $\vec{p}$ and $\vec{T}$ and the generalized Poisson bracket is defined as

$$\langle A,B\rangle = \frac{\partial(A,B)}{\partial(x_i,p_i)} + \varepsilon_{abc}\,\frac{\partial A}{\partial T_a}\,\frac{\partial B}{\partial T_b}\,T_c \quad . \tag{5.3}$$

The conserved angular momentum is $\vec{J} = \vec{r}\times\vec{p} + \vec{T}$ and $\vec{J}\cdot\vec{r} = \vec{T}\cdot\hat{r}$. Thus the total angular momentum takes on integral or half-integral values depending on the tensor or spinor nature of the representation of T. We now go to a body-fixed frame defined by the orthonormal vectors

$$\vec{m}_1 = \frac{(\vec{J}\times\hat{r})\times\hat{r}}{|(\vec{J}\times\hat{r})\times\hat{r}|}\ ,\qquad \vec{m}_2 = \frac{\vec{J}\times\hat{r}}{|\vec{J}\times\hat{r}|}\ ,\qquad \vec{m}_3 = \hat{r}\ , \tag{5.4}$$

and define the components of $\vec{T}$ as $\vec{T} = \sum_{i=1}^{3} w_i \vec{m}_i$, where $\vec{w} = (w_1, w_2, w_3)$. Furthermore we define two more objects

$$\Phi = \phi - J \int^{r} \frac{dr'}{r'D(r')}\ ,$$
$$U = \int^{r} \frac{dr'}{r'D(r')}\ , \tag{5.5}$$

where ϕ is the azimuthal angle around J and $D(r) = \vec{r}\cdot\vec{p}$, the dilatation generator. U is such that $\overset{*}{U} \equiv dU/d\tau = 1$; $\tau = \phi(t)/J$ is a "time" variable.

It can be verified that the set of nine objects $\{\vec{J},U,H,\vec{w}\}$ constitutes an independent set of variables. The equations of motion for the SU(2) dyon now take the form

$$\begin{aligned}
&\overset{*}{J}_1 = \overset{*}{J}_2 = \overset{*}{J}_3 = 0\ ,\\
&\overset{*}{\Phi} = \overset{*}{H} = 0\ ,\qquad \overset{*}{U} = 1\ ,\\
&I_1\overset{*}{w}_1 = (I_3 - I_2)w_2w_3\ ,\\
&I_2\overset{*}{w}_2 = (I_1 - I_3)w_3w_1\ ,\\
&I_3\overset{*}{w}_3 = (I_2 - I_1)w_1w_2\ ,
\end{aligned} \tag{5.6}$$

where $I_3 = 2I_1 = 2I_2$. So as may be expected the $\vec{w}$ obey the Euler equations for a symmetric top. Now noting the three triplet structure of these nine dynamical variables we group them as

$$\begin{aligned}
&\vec{u} = (u_1, u_2, u_3) = (J_1, J_2, J_3)\ ,\\
&\vec{v} = (v_1, v_2, v_3) = (U, \Phi, H)\ ,\\
&\vec{w} = (w_1, w_2, w_3)\ .
\end{aligned} \tag{5.7}$$

If we then construct two "Hamiltonians"

$$H_1 = G_1 + \Phi\ ,\qquad H_2 = G_2 + H\ , \tag{5.8}$$

with

$$G_1 = \frac{1}{2}\sum_{i=1}^{3} w_2^2\ ,\qquad G_2 = \frac{1}{2} w_3^2\ ,$$

we are led to a reformulation of the dynamics of the SU(2) dyon into Nambu's generalized Hamiltonian dynamics. The corresponding Nambu equations of motion involving three triplets are

$$\overset{*}{F} = \sum_{x=\vec{u},\vec{v},\vec{w}} \varepsilon_{abc} \frac{\partial F}{\partial x_a} \frac{\partial H_1}{\partial x_b} \frac{\partial H_2}{\partial x_c} \tag{5.9}$$

$$= \{F, H_1, H_2\}, \quad \text{the Nambu bracket.}$$

For the triplet case, Eq. (5.9) is realized by the Eulerian top when $\vec{u} = (J_1, J_2, J_3)$ and $H_1 = \frac{1}{2}\sum_{i=1}^{3} J_i^2/I_i$, $H_2 = J^2/2$ which are the two manifest invariants of the problem.

Similarly we show that for the SU(3) dyon[27] (or more precisely for the SU(3)/U(2) electromagnetic and isomagnetic dyon), the generalized Nambu equations are of the form

$$\overset{*}{F} = \sum_{x=u,v,w} f_{abc} \frac{\partial F}{\partial x_a} \frac{\partial H_1}{\partial x_b} \frac{\partial H_2}{\partial x_c} , \tag{5.10}$$

where f_{abc} are the structure constants of SU(3). Here there are three octets of dynamical variables. The w_i $(i = 1,2,\ldots,8)$ are the components of the unitary spin in the body-fixed frame. The u_i and v_i $(i = 1,2,3)$ are the same as for the U(1) or SU(2) dyon. The remaining variables u_i, v_i $(i = 4,5,\ldots,8)$ are regarded as frozen degrees of freedom. This fact is a reflection of the three dimensionality of physical space. Generally it is clear what the Nambu equations look like for a G dyon[28] $(G \supset SU(2))$:

$$\overset{*}{F} = \{F, H_1, H_2\} \tag{5.11}$$

with

$$\{x_a, x_b, x_c\} = g_{abc} ,$$

where g_{abc} are the structure constants of the group G having n generators, $H_1 = G_1 + \Phi$ and $H_2 = G_2 + H$ with $G_1 = \frac{1}{2}\sum_{a=1}^{n} w_a^2$, $G_2 = \frac{1}{2} w_3^2$. Again only the first three components of the $\vec{u}$ and $\vec{v}$ are not frozen out.

The bonus of any reformulation of old problems in a new language lies in new possibilities which might show up in the new context. For dyons, the forms of Eq. (5.11) naturally invite the abstract algebraic generalization in which none of the u_i's, v_i's and w_i's are frozen out. The resulting system has then all the hallmarks of a Klein-Kaluza type unified theory which encompasses both space-time and internal symmetry, treated here on the same footing by Nambu's mechanics. In his unpublished notes,

Nambu had written down just such an Equation (5.11) as a possible top-like generalization of his new mechanics[25,29]. It is very pleasing to see such a realization among non-Abelian monopole systems as we are reminded of Dirac's invitation to exploit the kinship between mathematical and physical structures in his celebrated 1931 paper on monopoles. I conclude with an observation on the psychology of discovery. From the Nambu strings, we come to vortices then to Nambu mechanics. To all of us, this probably should not come as a surprise. Indeed my talk does illustrate that certain personal elements in the evolution of our science where, as in music, the same beautiful leit-motif appears again and again under its many mathematical forms in the works of the same individual. In the music of vortex strings this motif has been spontaneous symmetry breaking introduced in quantum field theory some 17 years ago. I have taken you about a full circle; my clearly biased survey ends here. I hope that the reference list would fill the gaps left behind.

REFERENCES

*) Supported by the Energy Research and Development Administration.

†) Present address: Department of Physics, Yale University, New Haven, CT 06520, U.S.A.

1) W. Thomson, Lord Kelvin, Mathematical and Physical Papers, Vol. 14, p. 152 (Cambridge University Press, 1910).

2) P.A.M. Dirac, Proc. Roy. Soc. A212, 330 (1952).

3) O. Buneman, Proc. Roy. Soc. A215, 346 (1952); Proc. Roy. Soc. A217, 77 (1953).

4) H. Nielsen and P. Olesen, Nucl. Phys. B57, 367 (1973).

5) L.J. Tassie, Phys. Lett. 46B, 397 (1973).

6) A.A. Abrikosov, Sov. Phys. JETP 5, 1175 (1975). For some comprehensive texts see Superconductivity, Ed. R.D. Parks (Dekker, N.Y., 1969), Vol. II, Chapters 6 and 14. P.G. de Gennes, Superconductivity of Metals and Alloys (W.A. Benjamin, N.Y., 1966); D.D. Tilley and J. Tilley, Superfluidity and Superconductivity (Van Nostrand-Reinhold, N.Y., 1974).

7) Y. Nambu, Lectures at the Copenhagen Summer Symposium, 1970 (unpublished).

8) H.C. Tze and Z.F. Ezawa, Phys. Lett. 55B, 63 (1975); Z.F. Ezawa and H.C. Tze, Nucl. Phys. B96, 264 (1975); Nucl. Phys. B100, 1 (1975).

9) Z.F. Ezawa and H.C. Tze, Phys. Rev. D14, 1006 (1976); J. Math. Phys. 17, 2228 (1976).

10) For easy readings consult: D. Speiser, in Group Theoretical Concepts and Methods in Elementary Particle Physics, Ed. F. Gürsey (Gordon and Breach, N.Y., 1964); R. Gilmore, Lie Groups, Lie Algebra and Their Applications (Wiley-Interscience, N.Y., 1974).

11) S. Coleman, Lectures at the 1975 International School of Physics, Ettore Majorana, Harvard Preprint 1975.

12) S. Mandelstam, Phys. Lett. 53B, 476 (1975); A.M. Polyakov, Sov. Phys. JETP 41, 988 (1976); Tai Tsun Wu and Chen Ning Yang, Phys. Rev. D12, 3845 (1975) especially see C.N. Yang, in Sixth Hawaii Topical Conference in Particle Physics, University of Hawaii, 1975, edited by P.N. Dobson et al. (University Press of Hawaii, Honolulu, 1976).

13) See A. Borel, Bull. Am. Math. Soc. 61, 397 (1955).

14) For example, see S.F. Edwards, J. Phys. A 1, 15 (1968).

15) R.H. Crowell and R.H. Fox, Introduction to Knot Theory (Blaisdell, London, 1963).

16) I. Khan, ICTP Preprint IC/75/148.

17) E.B. Bogomol'nyi, Sov. J. Nucl. Phys. 24, 449 (1976); E.B. Bogomol'nyi and A.I. Vainshtein, Sov. J. Nucl. Phys. 23, 588 (1976).

18) D. Cribier et al., Phys. Lett. 9, 106 (1964).

19) I.V. Krive, V.M. Pyzh and E.M. Chudnovokii, Sov. J. Nucl. Phys. 23, 358 (1976).

20) Y. Nambu, Phys. Rev. D10, 4262 (1974). A sampling of papers on this subject is A. Jevicki and P. Senjanovic, Phys. Rev. D11, 860 (1975); D. Forster, Nucl. Phys. B81, 84 (1974); J.L. Gervais and B. Sakita, Phys. Rev. D11, 2943 (1973); J.L. Gervais, A. Jevicki and B. Sakita, Phys. Rev. D12, 1038 (1975); H.J. De Vega and F.A. Schaposnik, Phys. Rev. D14, 1100 (1976); Y. Nambu, "Strings, Vortices and Gauge Fields", talk presented at the Rochester Symposium on Quark Confinement, June 1976;

M.J. Duff and C.J. Isham, Nucl. Phys. B108, 130 (1976); J. Hierahinta, E. Takasugi and K. Tanaka, Phys. Rev. D13, 3313 (1976); J. Hierahinta, Phys. Rev. D15, 1137 (1977); B. Hu, Ecole Polytechnique Preprints A250, A251, A254, A255; F. Mancini, M. Matsumoto and H. Umezawa, Lecture delivered at the XIV Annual Winter School of Theoretical Physics, Ckarpacz, March 1977; P. Vinciarelli, Phys. Rev. Lett. 38, 1179 (1977).

21) Y. Nambu, talk presented at the Tokyo Symposium on High Energy Physics, 1973 (unpublished); A. Patkos, Roland Eotvas University Preprint, October 1976; M. Kamata, Tsukuba University Preprint, February 1977.

22) L.D. Faddeev, CERN Preprint TH.2188 (1976); Phys. Dolk. 18, 382 (1974).

23) L.D. Faddeev, MPI/PAE/Pth 16 June 1974.

24) A.M. Polyakov, Nucl. Phys. B120, 429 (1977); S. Mandelstam, Proc. of the Paris Conference on Extended Systems in Field Theory, Phys. Reports 23C, 245 (1976), and to be published; G.'t Hooft, "Gauge Theories with Unified Weak, Electromagnetic and Strong Interactions", Rapporteur's talk, EPS International Conference on High Energy Physics, Palermo, 1975; V.N. Gribov, Lecture at the 12th Winter School of Leningrad Nuclear Physics Institute 1977, SLAC-Trans-176.

25) M. Hirayama and H.C. Tze, SLAC Preprint 2001 (August 1977); Phys. Rev. D (to be published).

26) P. Hasenfratz and G.'t Hooft, Phys. Rev. Lett. 36, 1119 (1976); R. Jackiw and C. Rebbi, Phys. Rev. Lett. 36, 1116 (1976).

27) E. Corrigan, D.B. Fairlie, J. Nuyts and D.I. Olive, Nucl. Phys. B106, 475 (1976).

28) D. Wilkinson and A.S. Goldhaber, CUNY Preprint ITP-SB-77-30; F.A. Bais and J. Primack, UC Santa Cruz Preprint 76/115.

29) Y. Nambu, Phys. Rev. D7, 2405 (1973).

LATTICE GAUGE THEORIES

Marvin Weinstein

Stanford Linear Accelerator Center

Stanford, California 94305, U. S. A.

1. INTRODUCTION

In the past few days we have heard several beautiful lectures describing the way in which people hope to extract interesting physical information from quantum field theories by studying their semi-classical versions. Being in the mountains it seems appropriate to describe these attempts as an attack on the semi-classical face of quantum field theory. Since all mountains have more than one face, I would like to describe in my next few lectures attempts which have been made to launch a direct attack on the quantum face (Fig. 1); hence these lectures are in a sense complementary to the preceding ones.

To be precise I will first show how one can, from the very beginning, consider the problem of solving for the spectrum of states of any given continuum quantum field theory as a giant Schroedinger problem and then explain some non-perturbative methods for diagonalizing the Hamiltonian of the theory without recourse to semi-classical approximations. Along the way the notion of a lattice will appear as an artifice to handle the problems associated with the familiar infrared and ultraviolet divergencies of continuum quantum field theory and in fact, for all by gauge theories, I will show you how to go back and forth between specific lattice theories and continuum quantum field theories formulated with spatial and momentum cutoffs. This is an important thing to be able to do in principle since, it is by no means a priori clear that the situation is as shown in Fig. 1. It may be that the situation shown in Fig. 1 where two groups are attacking different faces of the same mountain is a trick of perspective and a more Olympian view

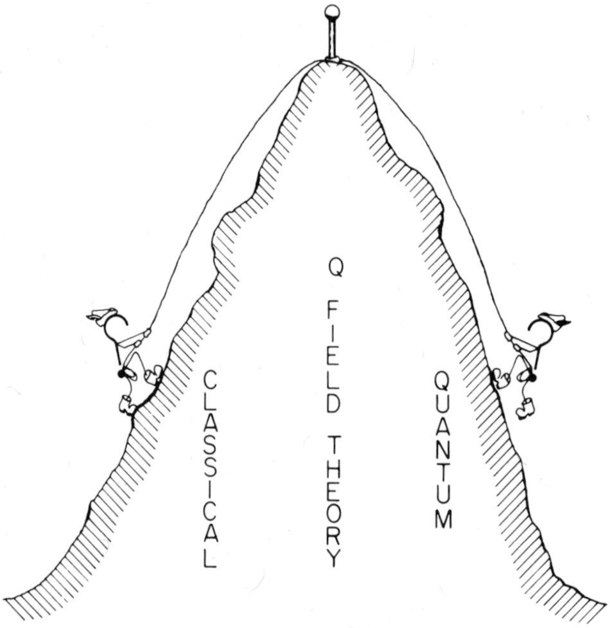

Fig. 1. Two climbers attempting to scale what appears to be the same peak.

of the situation would reveal that, contrary to our prejudice, the situation is more like Fig. 2, where we see we are in fact scaling different mountains. As this dreadful possibility could be the case in reality we must from the outset define the rules of the game and list our eventual goals so that you will understand where we are going and how we hope to get there.

First, let us address the question of goals. Here is where we get to list all the good stuff everyone has in his shopping list. We would like to understand on the basis of Lagrangian field theory:

(1) Why — as we have seen in the lecture of F. Gilman and G. Feldman — the naive quark model gives such a remarkably nice qualitative picture of hadron phenomenology. (Especially things which can be reduced to counting on our fingers kinds of questions).

(2) If the successes of the naive quark model point to the existence of bound quarks as elementary constituents of matter where are they? (i.e. why haven't they been seen in final states to date?)

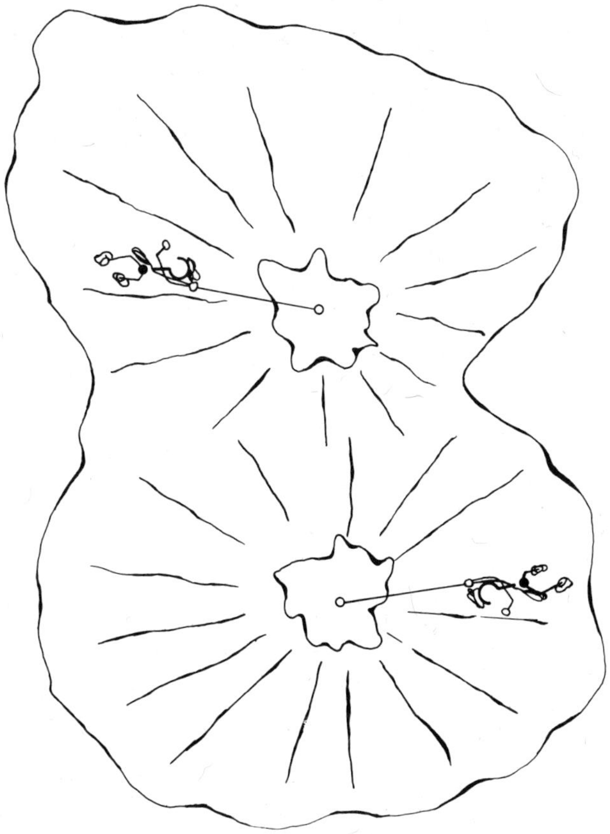

Fig. 2. A bird's eye view may reveal that the peaks are, in fact, different.

(3) If color gauge theories are in fact the right place to look for a theory of hadrons- is there any truth to the folklore that asymptotic freedom and confinement are two sides of the same coin?

Clearly, to be able to answer these questions in a really convincing way within the framework of conventional field theory it is absolutely necessary to develop techniques which are powerful enough to allow us to:

(1) Find the hadrons as bound states of the fundamental degrees of freedom.

(2) Calculate the ratio of the energy of a widely separated quark-antiquark pair to a typical bound state (e.g. the proton) in order to see if confinement does (or heresy — does not) occur in color gauge theories.

Even without specifying how we hope to develop such techniques, we can see from the fact that we are asking for the answers to questions which clearly go beyond the scope of perturbation theory that there will be certain steps which must be taken as we proceed along the way to our eventual goals. First, since we will not be able to rely on the conventional tools of Feynman graph perturbation theory we will have to find a way of formulating field theory so as to be able to discuss the problem without having to confront either infinite volume or short distance divergences. That is not to say that we will attempt to remove all divergencies from the continuum theory at the outset, but rather that the idea is to first impose sufficient cutoffs to render the theory finite. Then, in principle solve it exactly and then let all cutoffs go to infinity at the same time taking the bare parameters to the appropriate values so as to achieve a non-trivial relativistic theory with a finite spectrum of states. Obviously, if the lessons of perturbation theory hold any water for the full theory, then the fact that the multiplicative renormalization scheme can be carried out tells us that the scheme we have described must be feasible at least in principle.

Hence, I will first describe how to impose sufficient cutoffs on a given continuum theory to render all computations finite. Next I will show how to recast the resulting cutoff continuum theory in terms of a unitarily equivalent lattice theory which will enable us to better understand the quantum mechanical nature of the problem facing us. Finally, I will formulate a non-perturbative technique we propose to use to diagonalize any given lattice Hamiltonian. Clearly, from this point of view the problem of analyzing quantum field theories breaks into two distinct parts; the first being the development of a formalism which allows us to recast any cutoff continuum field theory in the form of an equivalent lattice theory and the second, being the development of techniques for solving any given lattice theory independently of how it was obtained. Except for today's lecture — whose purpose is to exhibit techniques for going back and forth between lattice theories and corresponding continuum field theories — I will focus attention on the second problem. Moreover, since the problems are not truly connected, I will study our proposed non-perturbative techniques as applied to model lattice field theories for which some exact results are known. The reason for focusing on these special models, as you will see, has nothing to do with the fact that our method is especially suited to the analysis of these 1-space-1-time dimensional models; but rather, we focus attention on these models in order to show that our methods are not producing incorrect results.

Before jumping into our technical discussion let me spend a few moments talking about the general way our approach fits into the framework of the other non-perturbative attacks on the problems of quantum field theory currently underway. Since I have been on the road so long in coming here I have to be forgiven for choosing to summarize the picture by the following road map (Fig. 3). This map — as all good maps — is essentially self explanatory and so I will limit myself to a few brief remarks. In the upper left hand corner we see the figure representing what I have labelled the lattice path-integral formalism. This, of course, stands for the program pioneered by Wilson[1] and collaborators and I haven't much

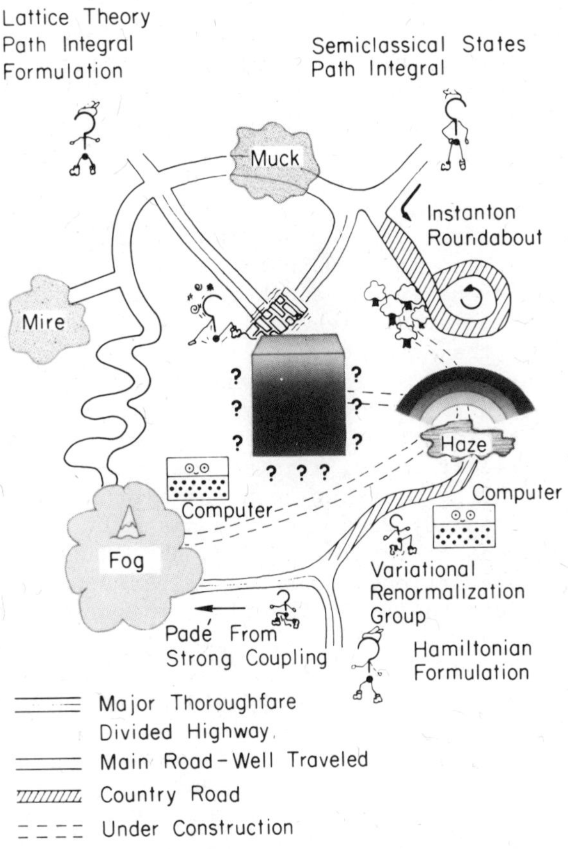

Fig. 3. A road map of some well travelled theoretical routes to the black box which presumably holds the secret of confined quarks.

to say about it. I would note that the initial notion that it would provide a super-highway which led to the mysterious black box containing the secrets of quark-confinement and the explanation of the quark model ran into a brick wall. Since then this program has followed a more tortuous path occasionally bogging down in muck and mire and at present it is obscured in a cloud of computer calculation. On the upper right hand side of the map you see the figure standing for the analysis of Euclidean path integral in terms of stationary points or semi-classical states. We have certainly heard much about this scheme in the lovely lectures by G. 't Hooft and R. Jackiw[2] — but I think it is fair to say that while the concepts one encounters are fascinating and have provided some insight into the U(1)-problem the hope that they would provide a super-highway to an understanding of quark confinement, etc., have also run into a brick wall. At present the interest in instantons, merons[3] (the darling of the Princeton group working on this problem), and other exotic beauties is based upon the <u>hope</u> that they will lead to an understanding of the physics of color-gauge theories; but I have not yet seen any compelling reason to think this has yet happened. I have chosen — with no malice intended — to signify the current state of affairs by saying that the current program is traversing the instanton roundabout but it is not yet out of the woods.

At the bottom of the map you see the figures representing attacks upon the problem based upon Hamiltonian techniques. All such methods have in one way or another made use of lattice techniques. The left hand super-highway labelled Padé from strong coupling stands for the study of lattice theories by the methods introduced by Kogut and Susskind[4] — who by the way deserve tremendous credit for initiating the program of converting the Wilson program to a Hamiltonian formalism. This program has received much attention in recent years, but the question of whether or not the method of continuing a strong coupling expansion of a lattice theory to weak coupling by Padé approximants will prove adequate to study the questions of interest is now shrouded in the fog of massive computer calculations. One can only await the results of these studies to judge their applicability to our world. Finally, I come to the much less well travelled path to which I will devote the next three lectures. This path labelled variational renormalization group approach signifies the program initiated at SLAC.[5] Our approach has been to proceed much less rapidly and study a series of simpler theories in order to achieve insight into the way our methods work, and — more importantly — how well they work. As with all other approaches we feel ours to be very promising and exciting but honesty forces me to say that we too are still lost in a haze of computation. If for some reason you notice that this path seems closer than others to the rainbow marking the "pot of gold" or in this case "black box" let me hasten to add this is probably a trick of personal perspective and as with all other theories it is the roads which are still under construction which will provide the true test of all the ideas put forward to date. As to what is in the black box, if folklore is right presumably it is the secret of confined quarks. However, one should not forget there is always the possibility that Fairbanks could be right and (Fig. 4) the box really holds a free, hungry, colored quark.

This completes the general remarks I will make and I would now like to present the plan of the discussion to follow, and then dive into the discussion of point 1.

<u>Plan of Lectures</u>

1. <u>Introduction of Basic Concepts</u>

   (a) Bosons } Lattice versions
   (b) Fermions }

2. + 3. <u>Introduce General Non-Perturbative Method for Solving Any Lattice Theory for Ground State and Spectrum of Low Lying States</u>

   (a) Ising model
   (b) Thirring model
   (c) U(1)-Goldstone model

4. <u>Discussion of Simple Gauge Theory</u>

   (a) Formulation of lattice gauge-field theory
   (b) Higgs model in 1 + 1 dimensions

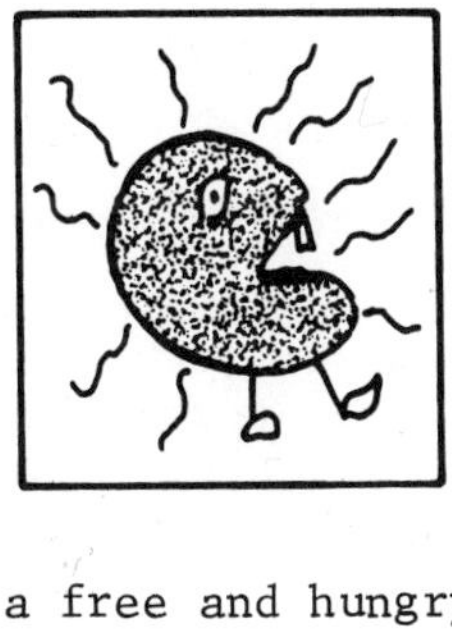

Fig. 4. Could there be a free and hungry colored quark in the black box?

Basically the plan of these lectures is as it is for obvious reasons. I wish to say a few words about why I have chosen to spend only the remainder of this first lecture introducing the notion of a lattice theory which is unitarily equivalent to a given cutoff continuum field theory, and then devoting the bulk of our time to specific lattice models. I feel this needs discussion because while I choose to focus upon the non-perturbative variational scheme we propose for studying any lattice theory, I do not wish to leave the impression that I consider the full development of the methods for relating continuum and lattice theories, as well as variations upon these methods, as unimportant. Far from it. In fact, it is my belief that when one really wishes to use the tools I will describe on the problem of analyzing the behavior of a specific continuum theory, or when one wishes to know the relationship between Feynman graph perturbation theory and calculations based upon the methods to be described, or when one wishes to ask the question of which continuum theory a given lattice theory corresponds to in the limit in which one removes all cutoffs, the fullest exploitation of the tricks I will describe only briefly in the context of free field theory will be as important as the variational techniques I will talk about for diagonalizing any given lattice Hamiltonian. I am choosing to give these questions short shrift only because of the time constraints imposed upon me by the format of this lecture series and because I wish to limit discussion to those aspects of the general program which have been most fully explored. I hope, however, that you will bear in mind that we have only begun to scratch the surface of what can be done by means of these techniques and will be encouraged to try your own hand at pushing them much further than we have done to date.

One further remark is in order, and that is that I probably will run out of time before I get to gauge theories and so probably I will only be able to make a few general remarks about the state of the art as of now — and refer you to a forthcoming series of papers on the subject[6].

By way of giving credit where credit is due, I wish to state that the work to follow has been done in collaboration with S. Drell, S. Yankielowicz, Ben Svetitsky and H. Quinn.

1.1. Free Scalar Field Theory

Let us begin our discussion of the way in which one can introduce cutoffs into a free scalar field theory and then transform it to an equivalent lattice theory. Our starting point is the usual Lagrangian

$$\mathcal{L} = \frac{1}{2}\,(\partial_\mu \phi(x))^2 - \frac{\mu^2}{2}\,\phi(x)^2 \tag{1.1}$$

and for the sake of notational simplicity alone let us specialize to $\mu = 0,1$ (i.e. a theory in 1 space + 1 time dimension). From this one forms the Hamiltonian by defining

$$\pi(x) \equiv \frac{\delta \mathcal{L}}{\delta(\partial_o \phi(x))} = \partial_o \phi(x) \tag{1.2}$$

and, assuming the theory is defined in a volume L, we obtain

$$H = \int_{-L/2}^{+L/2} dx \left(\frac{\pi(x)^2}{2} + \frac{(\partial_1 \phi(x))^2}{2} + \frac{\mu^2}{2} \phi(x)^2 \right). \tag{1.3}$$

At this point we define the quantum version of the classical theory specified in (1.1)-(1.3) by defining the equal time commutator of $\pi(x)$ and $\phi(y)$. It is here that we will choose to introduce a fundamental length in the theory and so cutoff all short distance divergences. We do this by defining the modified commutator

$$[\pi(x),\phi(y)] = -i\delta_\Lambda(x-y) \tag{1.4}$$

where

$$\delta_\Lambda(x-y) \equiv \Lambda \frac{\sin(\pi\Lambda(x-y))}{(2N+1)\sin(\frac{\pi\Lambda(x-y)}{2N+1})} \tag{1.5}$$

and where Λ is a small parameter (e.g. 10^{100} GeV) and L is defined to be

$$L = (2N+1)/\Lambda \quad . \tag{1.6}$$

Clearly, relation (1.4) implies that the fields $\pi(x)$ and $\phi(x)$ are overcomplete, and that there must exist a subset of the operators satisfying canonical commutation relations such that $\pi(x)$ and $\phi(x)$ can be written as functions of the smaller set of operators. There are many ways to see that this must be true, one way is to Fourier transform (1.4) and study the fields π_k and ϕ_k. However, the simplest way to discover the relevant independent set of variables is to observe that

$$\delta_\Lambda(x-y) = \Lambda\delta_{j_1,j_2} \tag{1.7}$$

when $x = j_1/\Lambda$ and $y = j_2/\Lambda$ for arbitrary integers j_1,j_2.

If we define

$$\phi(j) \equiv \phi(x=j/\Lambda) \tag{1.8}$$

it is then easy to show that

$$\phi(x) = \sum_{p=-N}^{+N} \frac{e^{ik_p x}}{\sqrt{2N+1}} \phi_{k_p} \tag{1.9}$$

where

$$k_p \equiv \left(\frac{2\pi\Lambda}{2N+1}\right) p \tag{1.10}$$

and where

$$\phi_{k_p} \equiv \sum_j \frac{e^{-ik_p j/\Lambda}}{\sqrt{2N+1}} \phi(j) \quad . \tag{1.11}$$

Using these formulae the Hamiltonian (1.3) can be rewritten in terms of the independent degrees of freedom as

$$H = \Lambda\left\{\sum_j \{\frac{\tilde{\pi}(j)^2}{2} + \frac{\tilde{\mu}^2}{2}\phi(j)^2\} + \sum_{j_1,j_2} \{\frac{1}{2} D(j_1-j_2)\phi(j_1)\phi(j_2)\}\right\} \tag{1.12}$$

where we have defined dimensionless fields, mass parameters, etc. by

$$\tilde{\pi}(j) = \pi(j)/\Lambda \quad , \tag{1.13}$$

$$\tilde{\mu}^2 = \mu^2/\Lambda^2 \quad , \tag{1.14}$$

and

$$D(j) \equiv \frac{1}{\Lambda^2} \sum_{p=-N}^{+N} \frac{(k_p)^2}{(2N+1)} e^{ik_p j/\Lambda} \quad . \tag{1.15}$$

N.B. the function D(j) has a particularly simple form in the limit $N \to \infty$, namely:

$$D(j) = \begin{cases} \pi^2/3 & \text{if } j = 0 \\ (-1)^j/j^2 & \text{if } j \neq 0 \end{cases} \tag{1.16}$$

Obviously, since we are dealing with a quadratic Hamiltonian (1.12) and since $D(j_1-j_2)$ is a function of only the difference of j_1 and j_2 this Hamiltonian can be diagonalized by going to k-space. If we do this we find

$$H=\Lambda\sum_{p=-N}^{N}\left(\frac{\tilde{\pi}_{-\tilde{k}_p}\tilde{\pi}_{\tilde{k}_p}}{2}+\frac{(\tilde{k}_p^2+\tilde{\mu}^2)}{2}\phi_{-\tilde{k}_p}\phi_{\tilde{k}_p}\right) \tag{1.17}$$

where

$$\tilde{k}_p \equiv \left(\frac{2\pi}{2N+1}\right)p\ . \tag{1.18}$$

If we now introduce creation and annihilation operators in the usual way, we see that

$$H = \Lambda \sum_{p=-N}^{N} \left(a^{\dagger}_{\tilde{k}_p} a_{\tilde{k}_p} + \frac{1}{2}\right)\sqrt{\tilde{k}_p^2 + \tilde{\mu}^2}\quad . \tag{1.19}$$

This form of H is quite instructive since we see that by introducing a fundamental length as in (1.4)-(1.5) we have done nothing more nor less than defining a maximum momentum cutoff on the free field theory. Our spectrum as a function of k is totally relativistic except that it terminates at $k_{max} = 2\pi N/(2N+1)$. It is worth comparing this result to the more common way of latticizing the free scalar field where one defines a lattice Hamiltonian

$$H_{nn} = \Lambda\sum_{j}\left(\frac{\tilde{\pi}_j^2}{2}+\frac{\tilde{\mu}^2}{2}\phi_j^2+\frac{1}{2}(\phi_{j+1}-\phi_j)^2\right) \tag{1.20}$$

where $\phi_{N+1} \equiv \phi_{-N}$.

The k-space form of this Hamiltonian gives

$$H_{nn} = \Lambda \sum_{p=-N}^{+N}\left(\frac{\tilde{\pi}_{-\tilde{k}_p}\tilde{\pi}_{\tilde{k}_p}}{2}+\frac{1}{2}(\tilde{\mu}^2+4\sin^2(\tilde{k}_p/2))\phi_{-\tilde{k}_p}\phi_{\tilde{k}_p}\right)$$

$$= \Lambda \sum_{p=-N}^{N}\left(a^{\dagger}_{\tilde{k}_p}a_{\tilde{k}_p}+\frac{1}{2}\right)\sqrt{\tilde{\mu}^2+4\sin^2(\tilde{k}_p/2)} \tag{1.21}$$

which approximates a relativistic spectrum only for $\tilde{k} \ll 1$. A comparison of the two spectra is shown in Fig. 5 and one sees that for momenta much smaller than the lattice mass there is no important difference between the two approaches insofar as the low momentum spectrum is concerned. There is however a huge difference in point of view since our Hamiltonian is related in a definite way to a given continuum theory and so we know how to go back and forth between the two languages at will. The first important difference between the two approaches at the level of the low

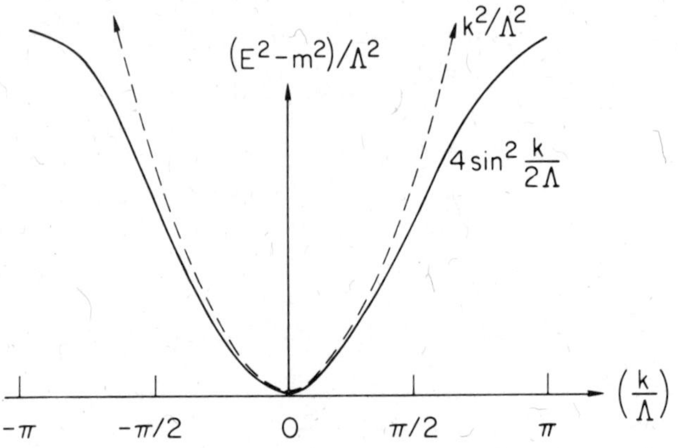

Fig. 5. Dispersion relation for cut-off free scalar field theory (solid curve) and latticized free scalar field theory (dashed curve).

energy spectrum will occur when we next study the free fermion theory.

1.2. Free Fermion Field Theory

We begin our discussion of the free fermion theory with the continuum Hamiltonian

$$H = \int_{-L/2}^{L/2} dx\{\psi^{\dagger}(x)(\frac{\alpha\partial_1}{i} + \beta M)\psi(x)\} \tag{1.22}$$

and once again modify the equal time anti-commutation relations to read

$$\{\psi^{\dagger}(x),\psi(y)\} = \delta_{\Lambda}(x-y) \quad . \tag{1.23}$$

As before, we note that this implies that the fields $\psi^{\dagger}(x)$ and $\psi(y)$ are overcomplete and we go to the independent fields

$$\tilde{\psi}(j) = (\frac{1}{\Lambda})^{\frac{1}{2}} \psi(x=j/\Lambda) \quad , \tag{1.24}$$

in terms of which one can write

$$\psi(x) = \Lambda^{\frac{1}{2}} \sum_{p=-N}^{N} e^{i\tilde{k}_p \Lambda x} \tilde{\psi}_{\tilde{k}_p} \qquad (1.25)$$

where

$$\tilde{\psi}_{\tilde{k}_p} = \sum_{j=-N}^{N} e^{-i\tilde{k}_p j} \tilde{\psi}(j) \quad . \qquad (1.26)$$

Using (1.24)-(1.26) we rewrite H as

$$H = \Lambda \left(\sum_{j_1, j_2} (\tilde{\psi}^{\dagger}(j_1) \frac{\alpha}{i} \delta'(j_1 - j_2) \tilde{\psi}(j_2)) + \sum_{j} (\tilde{\psi}^{\dagger}(j) \beta \tilde{\psi}(j)) \tilde{M} \right) \qquad (1.27)$$

where $\tilde{M} = M/\Lambda$ and

$$\delta'_{\Lambda}(j_1 - j_2) = \frac{\Lambda}{\sqrt{2N+1}} \sum_{p} i\tilde{k}_p \, e^{i\tilde{k}_p (j_1 - j_2)} \qquad (1.28)$$

which, in the limit $N \to \infty$, becomes

$$\delta'_{\Lambda}(j) = \begin{cases} 0 & \text{if} \quad j = 0 \\ (-)^j / j & \text{if} \quad j \neq 0 \end{cases} \qquad (1.29)$$

If we rewrite this in k-space we find

$$H = \Lambda \sum_{p=-N}^{+N} \tilde{\psi}^{\dagger}_{\tilde{k}_p} (\alpha \tilde{k}_p + \beta \tilde{M}) \tilde{\psi}_p \qquad (1.30)$$

which can be diagonalized to yield a theory of free fermions with an energy momentum dispersion formula given by

$$E^2(\tilde{k}_p) = \Lambda^2 (\tilde{k}_p^2 + \tilde{M}^2) \qquad (1.31)$$

which is clearly relativistic except that it cuts off at $k_{max} = 2\pi\Lambda/(2N+1)$. If we compare this to the definition of $\nabla\tilde{\psi}(j)$ given by the usual nearest neighbor formula

$$\nabla\tilde{\psi}(j) \equiv (\tilde{\psi}(j+1) - \tilde{\psi}(j)) \qquad (1.32)$$

with

$$H = \Lambda \sum_j \tilde{\psi}^\dagger(j)\left(\frac{\alpha\cdot\nabla}{i} + \beta\tilde{M}\right)\tilde{\psi}(j) \tag{1.33}$$

we find that the k-space form of (1.33) is

$$H = \Lambda \sum_{p=-N}^{+N} \tilde{\psi}^\dagger_{\tilde{k}_p}\left(\alpha \sin(\tilde{k}_p) + \beta\tilde{M}\right)\tilde{\psi}_{\tilde{k}_p} \tag{1.34}$$

which, when diagonalized yields an energy momentum dispersion formula

$$E^2_{nn}(\tilde{k}_p) = \Lambda(\sin^2(\tilde{k}_p) + \tilde{M}^2) \quad . \tag{1.35}$$

Since $|\tilde{k}_p| \leq \pi$ we see, as shown in Fig. 6, that this formulation of lattice fermions introduces a serious problem in that it leads to a doubling of the number of fermionic states having any given energy. Since this is disastrous for the free field case this is not what we wish to have happen at all; moreover in higher dimensions one gets $(2)^d$ times as many states at a given energy as the continuum theory would predict. Several methods other than the one we have described have been introduced by Wilson and Kogut and Susskind in order to avoid this problem, however they all suffer from the undesirable feature that they destroy continuous chiral symmetry of massless fermion theory so long as the cutoff (of lattice spacing $a = 1/\Lambda$) is held finite. The method we have proposed has the virtue of being simply related to the continuum theory, is as relativistic as possible and yields a chirally symmetric free massless fermion theory.

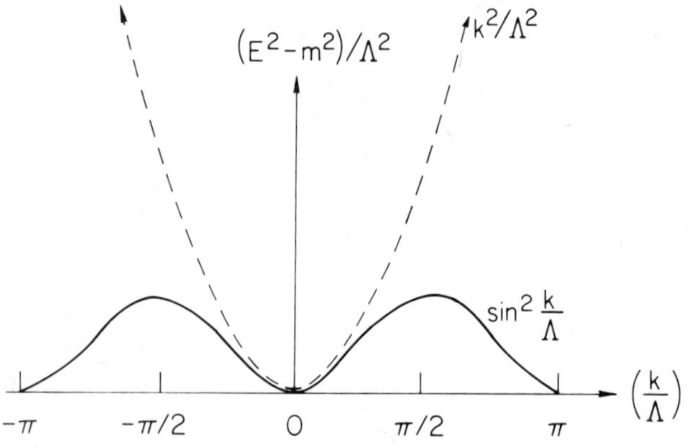

Fig. 6. The dispersion relation for the free fermion case.

1.3. Summary

Let us close this lecture with a summary of the points we have covered. First, I have shown you one way in which one can transcribe a cutoff quantum field theory as a specific lattice theory. This had the advantage that one knows precisely, for any given transcription procedure, how to smear the fields $\phi(j)$ or $\psi(j)$ so as to obtain continuum fields satisfying δ-function commutation relations in the limit $\Lambda\to\infty$. Clearly the same goes for all operators made out of polynomials in these fields and so one could, by exploiting this information, study the $\Lambda\to\infty$ limit of operator product expansions, equations of motion, etc. Secondly, we have shown that this transcription procedure handles fermions and bosons in the same way, breaks no symmetries (e.g. chiral invariance) and maintains approximate Lorentz invariance of the free field theory. Moreover, although we have not discussed this point, one could construct the operators $\Theta^{\mu\nu}_{\Lambda}(x)$ and Lorentz generators and study in an operator way how good Lorentz invariance is for low lying states. Finally, I wish to remind you that although we only carried out the transcription procedure for the free field case, there is no difficulty in carrying out the same procedure for interacting theories.

With this discussion behind us we will now forget about the question of where we got any particular lattice theory from and concentrate on the way in which one deals with any given theory without asking whether it was obtained from a continuum theory by our procedure or one of its myriad variations. We do this because, even though the quantum mechanical aspects of our problem are more easily appreciated for a lattice theory than a continuum theory, there is no reason to believe that solving a very large degree of freedom Schroedinger problem will be any more tractable than summing infinite numbers of Feynman graphs. The purpose of the next three lectures is to try and convince you that reliable non-perturbative methods for rendering the problem manageable do exist.

2. ISING MODEL IN A TRANSVERSE MAGNETIC FIELD

In this lecture and the lectures to follow I will focus on the question of how to do a variational calculation for the ground state and first few low lying states of any given lattice field theory. Although the method of analysis is quite generally applicable I will choose to develop it within the framework of specific examples for the obvious pedagogic reasons. I will begin with an analysis of a model which can be called a one dimensional Ising model in the presence of a transverse magnetic field. Before diving into calculations let us spend a few moments discussing the motivation for studying this model.

2.1. Why Study a Lattice Ising Model?

The lattice Ising model is particularly interesting for several reasons. First, it is an example of a theory which undergoes a phase transition for a critical value of coupling constant. In this case when I use the word phase transition I am not talking about temperature dependence of a system, but about a change in the symmetry properties of the ground state of the system. Thus, for the Ising model we will discuss, we will see that at a critical value of transverse field the ground state of the theory changes from being two-fold degenerate to being unique. Moreover, we will see that this corresponds to a certain discrete symmetry of the theory going from being a spontaneously broken symmetry to a normal symmetry as the coupling increases through its critical value. It is this ability to see a theory go from a spontaneously broken phase to a normal symmetry phase which makes this model so interesting, since if our present understanding of approximate hadronic symmetries has any validity we believe that the successes of the PCAC hypothesis (e.g. Adler self-consistency conditions, Adler-Weisberger relations, pion low energy theorems, $K_{\ell 3}$-predictions, etc.) point to the fact that the hadronic theory possesses a spontaneously broken or Goldstone symmetry. Since the reason why this sort of symmetry breaking takes place is a mystery, it becomes important to have a method for calculating in a theory in order to see if this sort of effect exists for any range of couplings. Having established the necessity being able to find phase transitions when they occur as a requisite for a good non-perturbative calculational scheme, we turn to the next question i.e. why study an artificial model like the Ising model in a transverse field rather than the manifestly more interesting σ-model. The answer to this is that the Ising model can be solved exactly and so we can know precisely how well our variational methods are doing. Unfortunately, the same cannot be said for the σ-model and so it would be hard to know whether or not our analysis of this theory for strong or intermediate coupling held water. Thus, we conclude that the Ising model to be studied is interesting in that it is an example of an exactly soluble model with a phase transition and hence it can serve as a benchmark against which to test our methods. Hopefully, since our methods are based on techniques which easily generalize to other theories in higher dimensions, and make no use of the special features of the theory which make it exactly soluble, the exercise is one step along the road to developing familiarity with and confidence in the variational scheme to be described. So much for motivation. Let's now go on to analyze the model in question.

2.2. The Ising Model in a Transverse Magnetic Field

The model we are interested in analyzing is defined by the Hamiltonian

$$H = \sum_j \left(\frac{\varepsilon_o}{2} \sigma_z(j) - \Delta_o \sigma_x(j) \sigma_x(j+1) \right) \tag{2.1}$$

where σ_z and σ_x are two Pauli matrices,

$$\sigma_z = \begin{pmatrix} 1 & 0 \\ 0 & -1 \end{pmatrix} \quad ; \quad \sigma_x = \begin{pmatrix} 0 & 1 \\ 1 & 0 \end{pmatrix}$$

and ε_o, Δ_o are two free parameters. Clearly, up to an overall scale factor the Hamiltonian really only depends upon the ratio $y_o = \varepsilon_o/\Delta_o$.

2.3. Exact Results

Since this model is exactly soluble it behooves us to spend a few moments summarizing those important features of the theory we will try to reproduce. First, for all values of y_o the theory is invariant under the discrete symmetry which takes $\sigma_z(j) \to +\sigma_z(j)$ and $\sigma_x(j) \to -\sigma_x(j)$ and which is generated by

$$U = e^{i\pi \sum_j \sigma_z(j)} \tag{2.2}$$

as can be seen by noting that

$$U^\dagger \sigma_z(j) U = \sigma_z(j) U^\dagger U = \sigma_z(j) \tag{2.3}$$

whereas

$$\begin{aligned} U^\dagger \sigma_x(j) U &= e^{-i\pi\sigma_z(j)} \sigma_x(j)\, e^{+i\pi\sigma_z(j)} \\ &= \sum_n \frac{(-i\pi)^n}{n!} [\sigma_z(j),[\ldots[\sigma_z(j),\sigma_x(j)]]\ldots] \\ &= \cos(\pi)\sigma_x(j) + i\sin(\pi)\sigma_y(j) \\ &= -\sigma_x(j) \quad . \end{aligned} \tag{2.4}$$

The operator U is the discrete symmetry transformation which is spontaneously broken, and since the operator $\sigma_x(j)$ has non-trivial transformation properties under U it is the fact that its vacuum expectation can become non-zero which implies the existence of a doubly degenerate ground state. To be precise, it is known that for $y_o \leq 2$,

(1) the model has a doubly degenerate ground state, or in other words there exist two states which we shall denote by $|+\rangle$ and $|-\rangle$ such that $U|+\rangle = |-\rangle$ and $U|-\rangle = |+\rangle$ for which

$$(2) \quad \langle\pm|\sigma_x(j)|\pm\rangle = \pm(1 - \frac{y_o^2}{4})^{1/8} \tag{2.5}$$

and moreover,

(3) it is known that the theory has an excitation spectrum whose energy-momentum dispersion is given by

$$E_{\tilde{k}} \cong \sqrt{(\varepsilon_o - 2\Delta_o)^2 + \tilde{k}^2} \quad . \tag{2.6}$$

Now for $y_o > 2$ it is known that the ground state, $|\psi_o\rangle$, of the theory becomes unique and is an eigenstate of U such that $U|\psi_o\rangle = |\psi_o\rangle$. The mass gap and energy-momentum dispersion relation are the same as for $y_o \leq 2$ and obviously

$$\langle\psi_o|\sigma_x|\psi_o\rangle = \langle\psi_o|U^\dagger\sigma_x U|\psi_o\rangle = - \langle\psi_o|\sigma_x|\psi_o\rangle \tag{2.7}$$

implies

$$\langle\psi_o|\sigma_x|\psi_o\rangle = 0 \ .$$

One can also calculate the ground state energy density as a function of y and the second derivative, $-\partial^2\varepsilon_g/\partial y^2$, will be of interest to us at a later point.

2.4. Some Trivial Considerations

Although I will make no attempt to explain the machinations one must go through to prove these exact results — since they involve doing a Jordan-Wigner transformation, identifying a conservation law of the theory and diagonalizing the resulting quadratic fermion Hamiltonians — I do want to spend a few moments discussing the limits $\varepsilon_o = 0$, Δ_o arbitrary and $\Delta_o = 0$, ε_o arbitrary so that the basic notions of spontaneous symmetry breaking for this model will be clear. Let us begin with the case $\varepsilon_o = 0$, i.e.

$$H_{\varepsilon_o=0} = \sum_{j=-N}^{N} [-\Delta_o\sigma_x(j)\sigma_x(j+1)] \quad . \tag{2.8}$$

Clearly, since all the operators $\sigma_x(j)$ commute with $H_{\varepsilon_o=0}$ diagonalizing $H_{\varepsilon_o=0}$ amounts to writing down all the eigenstates of

$\sigma_x(j)$. If we denote the eigenstate of $\sigma_x(j)$ of eigenvalue +1 (-1) by an arrow to the right (left) as shown in Fig. 7a, then it follows from (2.8) that the two states $|\psi_{+1}\rangle$ and $|\psi_{-1}\rangle$, shown in Fig. 7a are degenerate. Note that since U as defined in (2.2) takes σ_x to $-\sigma_x$ it maps the state $|\rightarrow\rangle_j$ into $|\leftarrow\rangle_j$ and vice versa, hence $U|\psi_{\pm 1}\rangle = |\psi_{\mp 1}\rangle$ and so the ground state of the $\varepsilon_o = 0$ system is two fold degenerate as promised. The states shown in Fig. 7b are two of the lowest lying excitations of the system.

It is essentially a matter of choice which of the two ground states of the system, or linear combination of the ground states, one chooses to base a theory on; and a definite prescription can be arrived at only from other considerations. The way one usually decides this issue is to add a small external field in σ_x direction; i.e. one adds a term

$$V(J) = -J \sum_{j=-N}^{N} \sigma_x(j) \tag{2.9}$$

to $H_{\varepsilon_o=0}$. One then studies the ground state of the system as $J \to 0$. Obviously, the energy of $|\psi_{+1}\rangle$ is given by $E_o - J(2N)$ and $|\psi_{-1}\rangle$ by $E_o + J(2N)$; hence for all $J > 0$, $|\psi_{+1}\rangle$ is the ground state of the system. Other arguments are based upon the desire to have cluster decomposition, but we will not go into that now.

Let us now consider the limit $\Delta_o = 0$, ε_o arbitrary. In this case the Hamiltonian is

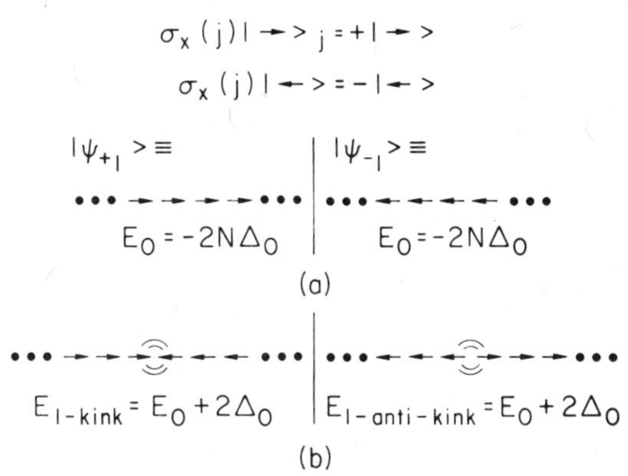

Fig. 7. Simultaneous eigenstates of $\sigma_x(j)$ and $H_{\varepsilon_o=0}$.

$$H_{\Delta_o=0} = \sum_{j=-N}^{N} \frac{\varepsilon_o}{2} \sigma_z(j) \quad . \tag{2.10}$$

As in the previous case, since $\sigma_z(j)$ commutes with $H_{\Delta_o=0}$ we can label all eigenstates of $H_{\Delta_o=0}$ by giving the eigenstates of $\sigma_z(j)$. If we let $|\uparrow\rangle_j$ and $|\downarrow\rangle_j$ denote the states such that $\sigma_z|\uparrow\rangle_j = |\uparrow\rangle_j$ and $\sigma_z|\downarrow\rangle_j = -|\downarrow\rangle_j$, then the ground state of the theory is the unique state shown in Fig. 8a and a typical lowest lying excitation is shown in Fig. 8b.

2.5. Approximate Solution by Recursive Methods

Having set the stage let us now introduce the general method by which we hope to analyze this and all other lattice field theories. As we have noted, the method we wish to use should be non-perturbative and should not rely upon any special features of the 1-dimensional problem. The method we have turned to is the Rayleigh-Ritz variational procedure, and our innovation is to devise a scheme for constructing a trial wave function for the ground state, since guessing the correct form of an infinite parameter wave function is beyond our mortal powers.

Our constructive technique is an iterative one based upon a procedure of thinning degrees of freedom. To be precise, it is based upon the observation that if one has an orthonormal set of states $|\psi_j\rangle$ then the problem of minimizing the expectation value

$$\langle\psi(\alpha)|H|\psi(\alpha)\rangle/\langle\psi(\alpha)|\psi(\alpha)\rangle \tag{2.11}$$

where

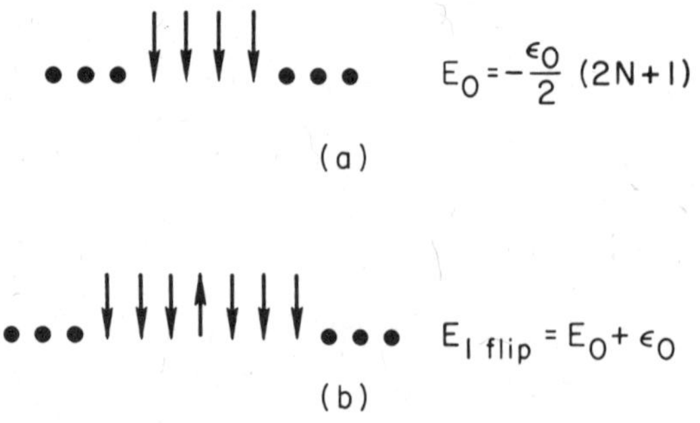

Fig. 8. Simultaneous eigenstates of $\sigma_z(j)$ and $H_{\Delta_o=0}$.

$$|\psi(\alpha)\rangle \equiv \sum_j \alpha_j |\psi_j\rangle \tag{2.12}$$

is equivalent to diagonalizing the "truncated" Hamiltonian

$$H^{TR}_{ij} = \langle\psi_i|H|\psi_j\rangle \quad .$$

Our procedure will be to begin with a complete set of orthonormal states and thin out this set of states by throwing some away. Thus, we reduce the problem to that of finding a good variational wave function over states generated by this smaller set of independent states. This, however, can be shown to be equivalent to diagonalizing a new Hamiltonian of the same form but having different coefficients. We carry out this procedure of thinning out the remaining set of states and generating a new effective Hamiltonian until our new Hamiltonian takes a form which can be solved. At each stage we have our decision of which states to keep and which to discard on a simple physically intuitive algorithm.

In order to make the ideas more clear let us abandon generalities and dive into our analysis. I will begin with a discussion of a thinning procedure based upon a very simple algorithm and then discuss the results of a slightly more sophisticated analysis.

Let us begin by introducing the notation $j = 2p+r;\ r = 0,1$ and rewriting H as

$$H = \sum_j [\frac{\varepsilon_o}{2}\sigma_z(j) - \Delta_o\sigma_x(j)\sigma_x(j+1)]$$

$$= \sum_p [\frac{\varepsilon_o}{2}(\sigma_z(2p)+\sigma_z(2p+1)) - \Delta_o\sigma_x(2p)\sigma_x(2p+1)]$$

$$- \Delta_o \sum_p \sigma_x(2p+1)\sigma_x(2(p+1)) \quad . \tag{2.13}$$

By this device we divide the lattice into blocks labeled by the integer 'p' containing two sites each, and at the same time we divide it into two terms – the first containing operators referring to a single block and the second containing products of operators in neighboring blocks.

Having done this, we now turn for inspiration, to a study of the Hamiltonian describing any single block 'p'; i.e. we study any one

$$h_p \equiv \frac{\varepsilon_o}{2}(\sigma_z(2p)+\sigma_z(2p+1)) - \Delta_o\sigma_x(2p)\sigma_x(2p+1) \quad . \tag{2.14}$$

If we label the states which correspond to the different possible values $\sigma_z(2p)$ and $\sigma_z(2p+1)$ as $|\downarrow\downarrow\rangle$, $|\downarrow\uparrow\rangle$, $|\uparrow\downarrow\rangle$ and $|\uparrow\uparrow\rangle$ respectively then we see that

$$\begin{aligned} h_p|\downarrow\downarrow\rangle &= -\,\varepsilon_o|\downarrow\downarrow\rangle - \Delta_o|\uparrow\uparrow\rangle \quad, \\ h_p|\downarrow\uparrow\rangle &= -\,\Delta_o|\uparrow\downarrow\rangle \;, \\ h_p|\uparrow\downarrow\rangle &= -\,\Delta_o|\downarrow\uparrow\rangle \;, \\ h_p|\uparrow\uparrow\rangle &= \varepsilon_o|\uparrow\uparrow\rangle - \Delta_o|\downarrow\downarrow\rangle \quad . \end{aligned} \tag{2.15}$$

The four eigenstates of h_p are shown in Table 1, where we have also given their eigenvalues and the difference in energy between the lowest state in each block and the 3 excited states within a block.

Table 1*

State_p	Energy_p	Gap from lowest state
$(\lvert\downarrow\downarrow\rangle + a_o\lvert\uparrow\uparrow\rangle)/\sqrt{1+a_o^2}$	$-\sqrt{\varepsilon_o^2+\Delta_o^2}$	0
$(\lvert\downarrow\uparrow\rangle + \lvert\uparrow\downarrow\rangle)/\sqrt{2}$	$-\Delta_o$	$\sqrt{\varepsilon_o^2+\Delta_o^2} - \Delta_o$
$(\lvert\downarrow\uparrow\rangle - \lvert\uparrow\downarrow\rangle)/\sqrt{2}$	$+\Delta_o$	$\sqrt{\varepsilon_o^2+\Delta_o^2} + \Delta_o$
$(-a_o\lvert\downarrow\downarrow\rangle + \lvert\uparrow\uparrow\rangle)/\sqrt{1+a_o^2}$	$\sqrt{\varepsilon_o^2+\Delta_o^2}$	$2\sqrt{\varepsilon_o^2+\Delta_o^2}$

* $a_o = (\sqrt{\varepsilon_o^2 + \Delta_o^2} - \varepsilon_o)/\Delta_o$

Our thinning procedure will be to define the two states per block $|\downarrow>_p$ and $|\uparrow>_p$ where

$$|\downarrow>_p \equiv (|\downarrow\downarrow> + a_o|\uparrow\uparrow>)/\sqrt{1+a_o^2} \quad ,$$
$$|\uparrow>_p = (|\uparrow\downarrow> + |\downarrow\uparrow>)/\sqrt{2} \tag{2.16}$$

and then observe that the orthonormal set is formed by taking all possible products of these two states over all blocks 'p'. That these states should be able to provide a reasonably good approximation for the ground state of the theory is intuitively obvious, since the states we have thrown away have higher energy.

Having decided upon which two states (out of the possible four states per block) to keep, our next step is to compute the truncated Hamiltonian. This is easily done. Note that

$$H = \sum_p [h_p - \Delta_o\sigma_x(2p+1)\sigma_x(2(p+1))] \tag{2.17}$$

and since the way h_p acts on a given product state is given in Table I, we need only see how the terms $-\Delta_o\sigma_x(2p+1)\sigma_x(2(p+1))$ act.

The way $\sigma_x(2p)$ acts upon a state $|\downarrow>_p$ or $|\uparrow>_p$ is given by

$$\sigma_x(2p)|\downarrow>_p = (|\uparrow\downarrow> + a_o|\downarrow\uparrow>)/\sqrt{1+a_o^2} \tag{2.18}$$

and so

$${}_p<\downarrow|\sigma_x(2p)|\downarrow>_p = {}_p<\uparrow|\sigma_x(2p)|\uparrow>_p = 0 \tag{2.19}$$

and

$${}_p<\uparrow|\sigma_x(2p)|\downarrow>_p = {}_p<\downarrow|\sigma_x(2p)|\uparrow>_p = (1+a_o)/\sqrt{2(1+a_o^2)} \ . \tag{2.20}$$

Similarly, $\sigma_x(2p+1)$ has the matrix elements

$${}_p<\uparrow|\sigma_x(2p+1)|\uparrow>_p = {}_p<\downarrow|\sigma_x(2p+1)|\downarrow>_p = 0$$

and

$${}_p<\uparrow|\sigma_x(2p+1)|\downarrow>_p = {}_p<\downarrow|\sigma_x(2p+1)|\uparrow>_p = (1+a_o)/\sqrt{2(1+a_o^2)}. \tag{2.21}$$

Hence, combining (2.17)-(2.21) we find that we can write H^{TR} in terms of 2×2 matrices referring to each block 'p', i.e.

$$H^{TR}_{(1)} = \sum_p \left[c_1 \mathbb{1}_p + \frac{\varepsilon_1}{2}\sigma_z(p) - \Delta_1 \sigma_x(p)\sigma_x(p+1)\right] \quad (2.22)$$

where

$$c_1 = -\frac{1}{2}\left(\Delta_o + \sqrt{\varepsilon_o^2 + \Delta_o^2}\right)$$

$$\varepsilon_1 = \sqrt{\varepsilon_o^2 + \Delta_o^2} - \Delta_o \quad (2.23)$$

$$\Delta_1 = \frac{\Delta_o(1 + a_o)^2}{2(1 + a_o^2)} \quad .$$

This effective Hamiltonian embodies all the information contained in our choice of a family of trial wave functions and, by construction, it follows that diagonalizing $H^{TR}_{(1)}$ will provide an upper bound upon the true ground state energy. If either $\varepsilon_1 = 0$ or $\Delta_1 = 0$ we could diagonalize $H^{TR}_{(1)}$ exactly. If $\varepsilon_1 << \Delta_1$ or $\Delta_1 << \varepsilon_1$ we could use perturbation theory to study the structure of the theory. However, in general neither case will apply and our only recourse will be to apply the same thinning procedure to the theory defined by $H^{TR}_{(1)}$. In this way we generate a sequence $H^{TR}_{(2)}$, $H^{TR}_{(3)}$, etc. and exactly diagonalizing any one of them will yield an upper bound of the ground state energy. The process is carried out until $H^{TR}_{(n)}$ takes a simple diagonalizable form or until $H^{TR}_{(n+1)} = H^{TR}_{(n)}$ at which point further iteration will avail us little.

To be specific, we could follow the general procedure just outlined and generate from

$$H^{TR}_{(n)} = \sum_p \left[d_n \mathbb{1}_p + \frac{\varepsilon_n}{2}\sigma_z(p) - \Delta_n \sigma_x(p)\sigma_x(p+1)\right] \quad (2.24)$$

a new Hamiltonian

$$H^{TR}_{(n+1)} = \sum_p \left[d_{n+1} \mathbb{1}_p + \frac{\varepsilon_{n+1}}{2}\sigma_z(p) - \Delta_{n+1} \sigma_x(p)\sigma_x(p+1)\right] \quad (2.25)$$

where

$$\varepsilon_{n+1} = (\varepsilon_n(1-a_n^2) - \Delta_n(1-a_n)^2)/(1+a_n^2) \; ,$$

$$\Delta_{n+1} = \frac{\Delta_n}{2}\frac{(1 + a_n)^2}{(1 + a_n^2)} \quad ,$$

$$c_{n+1} = -\left(\frac{\varepsilon_n(1-a_n^2) + \Delta_n(1+a_n)^2}{2(1+a_n^2)}\right) , \tag{2.26}$$

$$d_{n+1} = c_{n+1} + 2d_n ,$$

$$a_n = (\sqrt{\varepsilon_n^2 + \Delta_n^2} - \varepsilon_n)/\Delta_n .$$

Clearly, this recursion relation for the coefficients d_n, c_n, Δ_{n+1} and ε_{n+1} can be easily studied numerically and I will now summarize the results of such a study. Actually, for this simple recursion relation it is very helpful to observe that the ratio $\varepsilon_{n+1}/\Delta_{n+1}$ is given in terms of the ratio ε_n/Δ_n alone. If we let $y_n = \varepsilon_n/\Delta_n$ we can learn almost everything about the way in which (2.26) will iterate if we plot the function

$$R(y_n) = y_{n+1} - y_n$$

$$= \frac{2(1-a_n^2)(1+a_n)}{(1+a_n^2)^2}[y_n(1+a_n) + a_n - 1] - y_n \tag{2.27}$$

shown in Fig. 9, which tells us how the ratio ε_n/Δ_n changes with each iteration. Clearly, for any starting value of 'y' such that $R(y) < 0$ we have, after one iteration, a value of $\varepsilon_{n+1}/\Delta_{n+1}$ which is smaller than it was; similarly, for $R(y) > 0$ we are driven to still a larger value of y. Thus Fig. 9 tells us that for $y < y_c$ successive iterations drive us to a limiting form of the H_n^{TR} for which $\varepsilon_n \to \varepsilon_\infty > 0$ and $\Delta_n \to 0$ as $n \to \infty$. So, the theory for $y < y_c$ is a theory with a degenerate ground state. For $y > y_c$

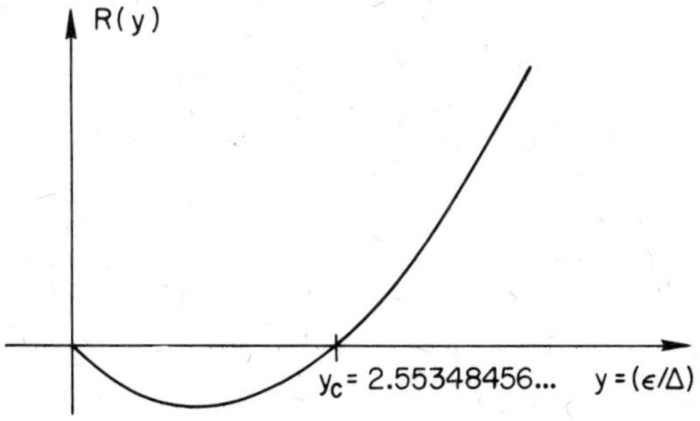

Fig. 9. The graph of $y_{n+1} - y_n$ ($\equiv R(y_n)$) vs. y.

we are driven to $\Delta_n \to \Delta_\infty > 0$ and $\varepsilon_n \to 0$. Hence for $y > y_c$ the theory has a unique ground state. The special point $y = y_c$ is a critical point at which the symmetry properties of the theory change. Since $R(y_c) = 0$ we see that the ratio of ε_n/Δ_n is unchanged with successive iterations and a more complete study of (2.26) shows that $H_{n+1}^{TR} = \rho_c H_n^{TR}$ where $\rho_c < 1$. This tells us that the Hamiltonian reproduces itself up to a scale factor, and thus at the critical point the physics of all length scales is the same — as the folklore would have it.

Location of the value $y_c = 2.55...$ for which $R(y_c) = 0$ is easily accomplished and recalling that the exact value of y_c is 2 we see that this primitive algorithm doesn't do too badly.

2.6. A More Sophisticated Algorithm

Going back to (2.26) we see that the algorithm we have adopted depended upon two distinct choices. First, we assumed that we would keep only two states per box at each iteration. Second we chose $a_n(y_n)$ to be given by the naive algorithm that we should diagonalize the 2-site Hamiltonian at each stage. A more sophisticated algorithm is to let $a_n(y_n)$ be an undetermined function of y and then carry out that iteration for 80 or so steps for a class of functions parametrized by one or more variables and then vary over these parameters so as to minimize the ground state energy density

$$\varepsilon(y_o) = \lim_{n\to\infty} (d_n/2^n) \quad .$$

In this way, except for specifying its general form, the recursion relation itself is undetermined and one varies over a system of possible "renormalization group transformations" to obtain the best possible upper bound on the ground state energy density. The output of such a calculation for the 1-parameter family of functions

$$a_\rho(y) = \tan^{-1}[\tfrac{\pi}{4}\,(1 - \tanh(\rho\, y))] \quad . \qquad (2.28)$$

Figure 10 shows a plot which compares the values of $\varepsilon_o(y_o)$ obtained from the exact solution to the problem (solid line) with the result of our one-parameter variational calculation (dashed line) over the only part of the range of y_o for which the difference is at all significant. Figure 11 plots the exact form of $\langle\sigma_x\rangle$ as a function of y_o (Eq. (2.5)) against that of our approximate calculation. As you can see the value of critical point becomes slightly worse, $y_c^{var} \cong 2.7$ but the dashed curve provides a better than one percent fit to $(1 - (y/y_c)^2)^{.19}$ from $0 \le y \le y_c^{var} - 1 \times 10^{-5}$. This again is not bad for such a crude approximation. Finally, Fig. 12 shows that this crude calculation is capable of revealing a structure

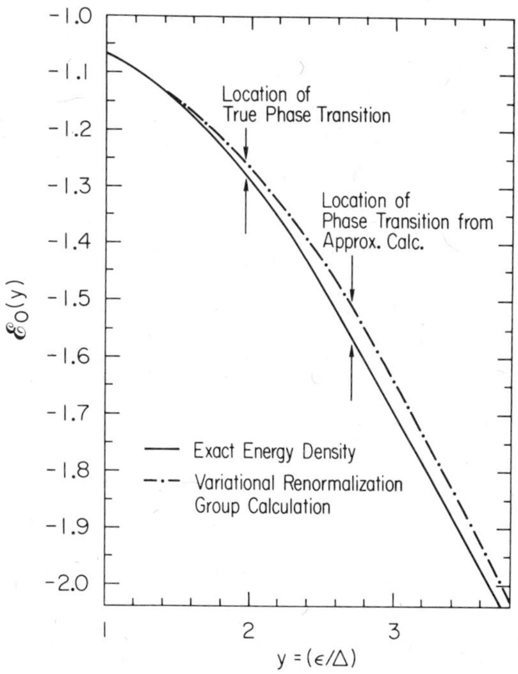

Fig. 10. Exact and variational calculation values of ε_o as a function of y.

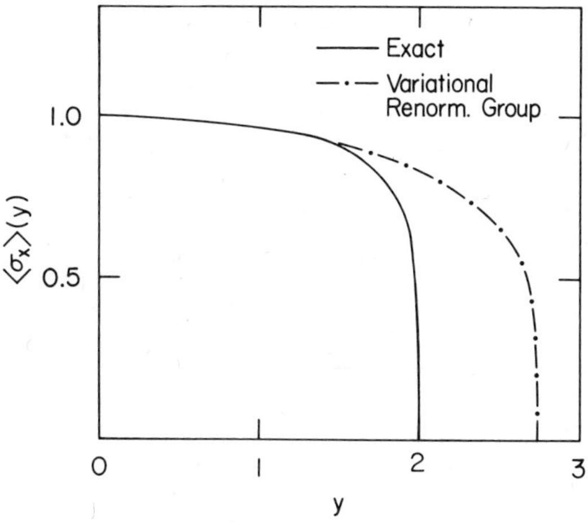

Fig. 11. Exact and variational calculation values of $<\sigma_x>$ as a function of y.

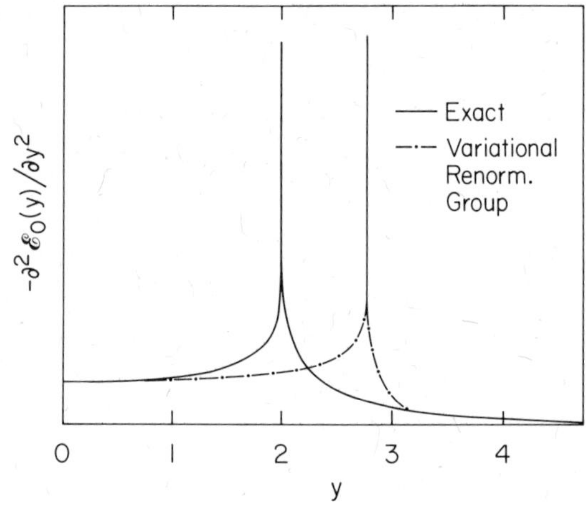

Fig. 12. Exact and variational calculation values of $-\partial^2\varepsilon_0/\partial y^2$ as a function of y.

surprisingly similar to the logarithmic singularity in $-\partial^2\varepsilon_0/\partial y^2$ possessed by the exact solution for $\varepsilon_0(y)$. This is quite a subtle non-perturbative property of the theory and the fact that one is reproducing this phenomena is quite striking.

2.7. Summary

To conclude this lecture let me summarize what we have seen. First we have shown that even a very naive calculation exhibits the fact that a phase transition exists. Second, we have seen that even the slight change of going to a variational calculation allows us to do a remarkably accurate calculation of the ground state energy over the entire range of ε_0 and Δ_0. Third, we see that the general behavior of the order parameter $\langle\sigma_x\rangle(y)$ is done reasonably well even with a crude calculation. Finally, although I have not discussed it, one can show that one also does a similar job in predicting the spectrum of excited states for all values of ε_0/Δ_0.

3. LATTICE THIRRING MODEL

In today's lecture I want to present an analysis of what I shall call a lattice version of the Thirring model. The method of analysis will be exactly the same as that used to analyze the Ising model in a transverse magnetic field except now we will be dealing with a theory of fermions. Before presenting the analysis let us first discuss the reasons I have chosen to discuss this model as a second example of the variational renormalization group method presented in the last lecture.

3.1. Why?

There are several reasons why this model is a very attractive one to study. To begin with it is the first serious model with fermions, and there is something more physical about fermions than about lattice spins. The next good reason for studying this model is that, as we saw in the first lecture, fermionic theories are the first one to require the use of the non-nearest (or long-range) gradient operator. Since there has been great skepticism about whether or not it is possible to carry out calculations with this form of the gradient, it is worth demonstrating that it is easily worked with. Third, this is a model whose continuum version is solvable and exhibits a peculiar behavior for the fermion wave-function renormalization, $Z_2(g)$, in that it vanishes for a finite value of g. We will see that this feature also occurs in the lattice theory and that because we know how to go back and forth between the would-be continuum operators of the theory and lattice fields we know how to establish the fact that it is the same phenomenon. Fourth, in our list of reasons, is that this is an example of a theory with explicit chiral invariance and our analysis will show how the iteration procedure works in such a setting. Finally, the lattice model of the theory exhibits an interesting non-perturbative phenomenon in that for values of g for which $Z_2(g) = 0$ the theory exhibits a kind of confinement.

3.2. The Thirring Model

The Hamiltonian we will study is

$$H = \sum_{j_1, j_2} [i\delta'_\Lambda(j_1 - j_2)\psi^\dagger_{j_1} \alpha \psi_{j_2}] - g_o \sum_j (\psi^\dagger_j \beta \psi_j)^2 \tag{3.1}$$

where

$$\delta'_\Lambda(j) = \begin{cases} 0 & \text{for} \quad j = 0 \\ (-)^j/j & \text{for} \quad j \neq 0 \end{cases} \tag{3.2}$$

$$\alpha = \begin{pmatrix} 1 & 0 \\ 0 & -1 \end{pmatrix} \quad ; \quad \beta = \begin{pmatrix} 0 & 1 \\ 1 & 0 \end{pmatrix} \tag{3.3}$$

and ψ_j is a two component fermi field satisfying anticommutation relations

$$\{\psi^\dagger_{j_1}, \psi_{j_2}\} = \delta_{j_1 j_2} \mathbb{1} . \tag{3.4}$$

Our method of analysis is the same as before. First we will dissect the lattice into blocks of three sites. We will then solve the three site problem exactly and truncate the space of states to the subspace generated by products of the lowest Q = 0, Q = ±1 states per block.

In order to simplify the discussion let us begin by showing how the approach will work in the special case g = 0, i.e. the free massless fermion field in 1+1 dimensions. This special case is interesting because it is exactly soluble by going to momentum space and so one might expect this naive truncation procedure to be at its worst. As we will see it will do surprisingly well.

3.3. Free Field: g = 0 Limit

If we denote the separate components of the two component field by

$$\psi(j) = \begin{pmatrix} b_j \\ d^\dagger_j \end{pmatrix} \tag{3.5}$$

so that (3.4) becomes

$$\{b_{j_1}, b^\dagger_{j_2}\} = \{d_{j_1}, d^\dagger_{j_2}\} = \delta_{j_1 j_2} \tag{3.6}$$

all other commutators being zero. Substituting this in (3.1) we rewrite H as:

$$H = \sum_{j_1, j_2} [i\delta'(j_1-j_2)(b^\dagger_{j_1} b_{j_2} - d^\dagger_{j_1} d_{j_2})] . \tag{3.7}$$

Before breaking the problem up into 3-site blocks let us list some useful symmetries of the Hamiltonian. First, inspection of H shows that one can only absorb a b(d) at one point and create a $b^\dagger(d^\dagger)$ at some other point, hence H never changes the total number of b's or d's in a state. If we define

$$n_b(j) = b^\dagger_j b_j \quad ; \quad n_d(j) = d^\dagger_j d_j \tag{3.8}$$

we can define two conserved operators, the electric charge

$$Q = \sum_j (n_b(j) - n_d(j)) = \sum_j q(j) \tag{3.9}$$

and the chiral or axial charge

$$Q_5 = \sum_j (n_b(j) + n_d(j) - 1) = \sum_j q_5(j) .$$

This notation is chosen so that for a single site 'j' we can introduce the state $|0_j>$ such that

$$b_j|0_j> = d_j|0_j> = 0 \tag{3.10}$$

and then define the other three possible states for a site 'j' to have the quantum numbers given in Table 2.

Useful discrete symmetries are C, P and the anti-linear operator, defined by

$$\begin{aligned} C\, b_j\, C^{-1} &= b_j^\dagger \\ C\, d_j\, C^{-1} &= d_j^\dagger \\ C\, Q\, C^{-1} &= -Q \\ C\, Q_5 C^{-1} &= -Q_5 , \end{aligned} \tag{3.11}$$

Table 2

State	$Q_j = n_b(j) - n_d(j)$	$Q_5(j) = n_b(j) + n_d(j) - 1$
$\|0_j>$	0	-1
$\|+_j> \equiv b_j^\dagger\|0_j>$	+1	0
$\|-_j> \equiv d_j^\dagger\|0_j>$	-1	0
$\|\pm_j> \equiv b_j^\dagger d_j^\dagger\|0_j>$	0	+1

$$P\, b_j\, P^{-1} = d^{\dagger}_{-j}$$
$$P\, d_j\, P^{-1} = b^{\dagger}_{-j} \qquad (3.12)$$
$$P\, Q\, P^{-1} = Q$$
$$P\, Q_5 P^{-1} = -\, Q_5 \quad ,$$

and

$$\Theta\, b_j\, \Theta^{-1} = d_j$$
$$\Theta\, d_j\, \Theta^{-1} = b_j \qquad (3.13)$$
$$\Theta\, Q\, \Theta^{-1} = -\, Q$$
$$\Theta\, Q_5 \Theta^{-1} = Q_5$$

with the phase convention

$$C|0_j> = i|\pm_j>$$
$$P|0_j> = |\pm_j>$$

and

$$\Theta|0_j> = |0_j> \quad . \qquad (3.14)$$

With these preliminaries behind us let us now define $j = 3p+r$, $r = -1,0,+1$ and rewrite H as

$$H = \sum_p [\sum_{r\neq r'} i\delta'(r-r')(b^{\dagger}_{3p+r} b_{3p+r'} - d^{\dagger}_{3p+r} d_{3p+r'})]$$
$$+ \sum_{p\neq p'} [\sum_{r,r'} i\delta'(3(p-p')+r-r')(b^{\dagger}_{3p+r} b_{3p'+r'} - d^{\dagger}_{3p+r} d_{3p'+r'})] \; . \qquad (3.15)$$

The first set of terms, $p=p'$, constitute the single block Hamiltonian and the $p\neq p'$ terms give the block-block recoupling terms. Hence, let us restrict attention to a single block 'p' and diagonalize

$$h_p = \sum_{r\neq r'} i\delta'(r-r')(b^{\dagger}_{3p+r} b_{3p+r'} - d^{\dagger}_{3p+r} d_{3p+r'}) \; . \qquad (3.16)$$

Since there are four states per site, i.e. $|0\rangle$, $b^\dagger|0\rangle$, $d^\dagger|0\rangle$ and $b^\dagger d^\dagger|0\rangle$, we see that there are $4^3 = 64$ states per block, and so to diagonalize h_p we must at first blush diagonalize a 64×64 matrix. It's not so bad though. If we look at sectors of definite Q, Q_5, C or P or CP the problem vastly simplifies. For example, in Tables 3 and 4 we see the states of Q = 0 and Q = +1 divided according to their Q_5 eigenvalues. Since the states of $Q_5 = \pm 3$ are the only Q = 0 states of this quantum number they are eigenstates of h_p. Since C maps a Q = 0, $Q_5 = +1$ state into a state of Q = 0, $Q_5 = -1$ one need only diagonalize the 9×9 submatrix corresponding to the Q = 0, $Q_5 = -1$ sector in order to find the lowest Q = 0 eigenstates. There are, of course, two degenerate states at each energy since there is one of $Q_5 = +1$ obtained by applying C to the $Q_5 = -1$ eigenstates. Next we observe that iCP transforms the nine states of Q = 0, $Q_5 = -1$ among themselves and so one can reduce the problem to studying h_p restricted to states of definite iCP. This simplifies the problem to diagonalizing a 6×6 and 3×3 matrix. Actually, having reduced the problem this far we can now straightforwardly diagonalize the 6×6 and 3×3 problems. The same arguments can be used to simplify the Q = +1 problem. Although it is not really necessary for me to do so, in order to make the point I wish to make, in order to prove that it can be done, let me exhibit for you the exact form of the lowest energy state of Q = 0, $Q_5 = -1$ and Q = +1, $Q_5 = 0$; i.e.

$$|0_p\rangle = \frac{1}{18}(3-4i)|{+}0{-}\rangle + \frac{1}{18}(3+4i)|{-}0{+}\rangle$$

$$- \frac{1}{9}(3+i)(|{+}{-}0\rangle + |0{+}{-}\rangle) - \frac{1}{9}(3-i)(|0{-}{+}\rangle + |{-}{+}0\rangle)$$

$$+ \frac{4}{9}\, i(|0\pm 0\rangle + \frac{5}{8}(|\pm 00\rangle + |00\pm\rangle)) \qquad (3.17)$$

and

$$|+_p\rangle = \frac{1}{18}(4-3i)|\pm{+}0\rangle - \frac{1}{18}(4+3i)|0{+}\pm\rangle$$

$$+ \frac{1}{9}(1-3i)(|{+}0\pm\rangle + |0\pm{+}\rangle) - \frac{1}{9}(1+3i)(|\pm 0{+}\rangle + |{+}\pm 0\rangle)$$

$$+ \frac{4}{9}(|{+}{-}{+}\rangle + \frac{5}{8}(|{+}{+}{-}\rangle + |{-}{+}{+}\rangle)) \quad . \qquad (3.18)$$

Having found explicit forms for $|0_p\rangle$, $|+_p\rangle$ we can define $|-_p\rangle = C|+_p\rangle$ and $|\pm_p\rangle = -iC|0_p\rangle$, and go on to computing the form H takes when truncated to the system of states spanned by products of these four states per box over all boxes. Since the eigenvalues of h_p corresponding to these four states are $E_o = -3$, h_p can be replaced by -3 times the unit operator. Hence, the problem of computing

Table 3. Q= 0 sector

Q_5	State	iCP Transform
-3	$\vert 000\rangle$	$-\vert 000\rangle$
-1	$\vert \pm 00\rangle$	$\vert 00\pm\rangle$
	$\vert 00\pm\rangle$	$\vert \pm 00\rangle$
	$\vert 0\pm 0\rangle$	$\vert 0\pm 0\rangle$
	$\vert +-0\rangle$	$\vert 0+-\rangle$
	$\vert 0+-\rangle$	$\vert +-0\rangle$
	$\vert +0-\rangle$	$\vert +0-\rangle$
	$\vert -+0\rangle$	$\vert 0-+\rangle$
	$\vert 0-+\rangle$	$\vert -+0\rangle$
	$\vert -0+\rangle$	$\vert -0+\rangle$
+1	$\vert 0\pm\pm\rangle$	$-\vert \pm\pm 0\rangle$
	$\vert \pm\pm 0\rangle$	$-\vert 0\pm\pm\rangle$
	$\vert \pm 0\pm\rangle$	$-\vert \pm 0\pm\rangle$
	$\vert -+\pm\rangle$	$-\vert \pm -+\rangle$
	$\vert \pm -+\rangle$	$-\vert -+\pm\rangle$
	$\vert -\pm +\rangle$	$-\vert -\pm +\rangle$
	$\vert +-\pm\rangle$	$-\vert \pm +-\rangle$
	$\vert \pm +-\rangle$	$-\vert +-\pm\rangle$
	$\vert +\pm -\rangle$	$-\vert +\pm -\rangle$
+3	$\vert \pm\pm\pm\rangle$	$\vert \pm\pm\pm\rangle$

Table 4. Q = 1 sector

Q_5	State	P Transform
-2	$\vert +00\rangle$	$-\vert \pm\pm+\rangle$
	$\vert 00+\rangle$	$-\vert +\pm\pm\rangle$
	$\vert 0+0\rangle$	$-\vert \pm+\pm\rangle$
0	$\vert +0\pm\rangle$	$-\vert 0\pm+\rangle$
	$\vert 0\pm+\rangle$	$-\vert +0\pm\rangle$
	$\vert 0+\pm\rangle$	$-\vert 0+\pm\rangle$
	$\vert +\pm0\rangle$	$-\vert \pm0+\rangle$
	$\vert \pm0+\rangle$	$-\vert +\pm0\rangle$
	$\vert \pm+0\rangle$	$-\vert \pm+0\rangle$
	$\vert +-+\rangle$	$-\vert +-+\rangle$
	$\vert ++-\rangle$	$-\vert -++\rangle$
	$\vert -++\rangle$	$-\vert ++-\rangle$
2	$\vert +\pm\pm\rangle$	$-\vert 00+\rangle$
	$\vert \pm\pm+\rangle$	$-\vert +00\rangle$
	$\vert \pm+\pm\rangle$	$-\vert 0+0\rangle$

the truncated form of H reduces to computing the truncated forms of operators like $b^\dagger_{3p+r}$, $d^\dagger_{3p+r}$, etc.

To compute $(b^\dagger_{3p+r})^{TR}$, etc., it helps to observe that when $b^\dagger$ operates on a state it raises the charge of that state by one unit and raises the Q_5 of a state by one unit also. On the other hand $d^\dagger$ raises the Q_5 by a unit but lowers Q by one unit. Hence, the only possible non-zero matrix elements of $b^\dagger_{3p+r}$ and $d^\dagger_{3p+r}$ between the states $|0_p\rangle$, $|+_p\rangle$, $|-_p\rangle$, and $|\pm_p\rangle$ are

$$\langle +_p | b^\dagger_{3p+r} | 0_p \rangle = u_r \quad ,$$
$$\langle \pm_p | b^\dagger_{3p+r} | -_p \rangle = t_r \quad ,$$
$$\langle -_p | d^\dagger_{3p+r} | 0_p \rangle = v_r \quad ,$$
$$\langle \pm_p | d^\dagger_{3p+r} | +_p \rangle = w_r \tag{3.19}$$

and it follows from the symmetries C, P and Θ that

$$u_r = t^*_r \quad ; \quad v_r = w^*_r \quad \text{and} \quad u_r = v_{-r} \quad . \tag{3.20}$$

This information can be all summarized in operator form by introducing anti-commuting operators $B^\dagger_p$, B_p, $D^\dagger_p$ and D_p defined by

$$\langle +_p | B^\dagger_p | 0_p \rangle = 1$$
$$\langle -_p | D^\dagger_p | 0_p \rangle = 1 \text{ , etc.} \tag{3.21}$$

and writing

$$(b^\dagger_{3p+r})^{TR} = B^\dagger_p (u_r Q^2_{5p} + u^*_r Q^2_p) \quad ,$$
$$(d^\dagger_{3p+r})^{TR} = D^\dagger_p (u^*_r Q^2_{5p} + u_r Q^2_p) \quad , \tag{3.22}$$

with

$$Q_{5p} = (B^\dagger_p B_p + D^\dagger_p D_p - 1)$$

and

$$Q_p = (B^\dagger_p B_p - D^\dagger_p D_p) \quad . \tag{3.23}$$

Note that if u_r and v_r are real then since $Q^2_p + Q^2_{5p}$ is the identity operator (3.22) becomes

$$(b^\dagger_{3p+r})^{TR} = u_r B^\dagger_p \quad ,$$
$$(d^\dagger_{3p+r})^{TR} = u_r D^\dagger_p \quad . \tag{3.24}$$

Substituting the general expression (3.21) into (3.15) we arrive at the form of $H^{TR}_{(1)}$, and then this process can be carried out iteratively. Actually, for the free field Hamiltonian it is easy to show that u_r is real and so $H^{TR}_{(1)}$ becomes, using (3.24),

$$H^{TR}_{(1)} = -3\left(\frac{2N+1}{3}\right)\mathbb{1} + \sum_{p\neq q}\left[i\left(\sum_{r,r'=-1}^{+1} \delta'(3(p-q)+r-r')u_r u_{r'}\right) \times (B^\dagger_p B_q - D^\dagger_p D_q)\right] \tag{3.25}$$

which has exactly the same form as the original Hamiltonian if we define a new δ' function to be

$$\delta'_{(1)}(p-q) \equiv \sum_{r,r'=-1}^{+1} \delta'(3(p-q)+r-r')u_r u_{r'} \quad . \tag{3.26}$$

If one carries out this iteration in detail we get an upper bound on the ground state energy of a free massless fermion theory which is (in units of the cutoff Λ^2)

$$E_o^{approx} = -\ 1.217..\ L \tag{3.27}$$

which is to be compared to the exact answer which is

$$E_o^{exact} = -\frac{\pi}{2} L \quad . \tag{3.28}$$

The agreement here is to 20 percent which is not bad considering the naive truncation procedure we have adopted. It is simple to show that one can do considerably better by making only slight changes in the procedure, but we will not bother discussing that here.

This discussion sets up the notation of the general calculation. Let us now turn to a discussion of the $g \neq 0$ case.

3.4. Full Thirring Model

Returning to the full Hamiltonian, (3.1), let us now summarise what happens if one carries out exactly the same procedure except for $g_o \neq 0$. In this case u_r is no longer real and one must use the more complicated form (3.22) to compute $H^{TR}_{(n)}$. After the first iteration the Hamiltonian takes the following general form which is then reproduced in each succeeding iteration, i.e.:

$$H^{TR}_{(n)} = \sum_{p_1,p_2} iX_n(p_1-p_2)(B^\dagger_{p_1}B_{p_2} - D^\dagger_{p_1}D_{p_2})$$
$$- g_n \sum_p (B^\dagger_p B_p + D^\dagger_p D_p - 1)^2 + E_n \mathbb{1}_p \tag{3.29}$$
$$+ i \sum_{p_1,p_2} \Big\{ Z_n(p_1-p_2)(B^\dagger_{p_1} Q^2_{p_1} Q^2_{p_2} B_{p_2} - D^\dagger_{p_1} Q^2_{5p_1} Q^2_{5p_2} D_{p_2})$$
$$+ Z^*_n(p_1-p_2)(B^\dagger_{p_1} Q^2_{5p_1} Q^2_{5p_2} B_{p_2} - D^\dagger_{p_1} Q^2_{p_1} Q^2_{p_2} D_{p_2})\Big\}$$

and Tables 5, 6 and 7 show how the various parameters g_n, $X_n(p)$ and $Z_n(p)$ change with succeeding iterations for different values of g_0.

Before discussing the tables in detail let us note that we should expect something very peculiar to happen for $g_0 >> 1$. This is the case because for $g_0 >> 1$ we expect the single site part of H (3.1) to dominate and so we can first study the individual term h_j, where

$$h_j = - g_0(\psi^\dagger_j \beta \psi_j)^2 = - g_0(n_b(j) + n_d(j) - 1)^2 . \tag{3.30}$$

It is obvious from (3.29) that the neutral states $|0_j\rangle$ and $|\pm_j\rangle$ are degenerate and have energy $-g$ whereas the charged states have energy 0. Hence, the cost of creating two separated charges from a distribution of tightly bound pairs is $2g$ and so we would expect that the excitation spectrum of the theory (in the limit Λ large) not to have separated fermions but to have massless bound state excitations. In an earlier paper Drell, Yankielowicz and I[7] showed that for $g_0 >> 0$ this problem could be converted to almost a Heisenberg anti-ferromagnet and that our conclusions about massless bound states could be proven to be correct if the interactions in this system were of nearest neighbor form. In this case the Heisenberg anti-ferromagnet problem is exactly soluble, the solution having been given by Bethe in 1931, and it is indeed massless. I mention this because, as can be seen from Tables 5-7, for g smaller than $g_c \approx 1$ the iterative calculation converges to a theory with $\lim_{n\to\infty} g_n = 0$ and $X_{(n)}(p)$ going over, up to a scale factor, to a function of 'p' which is quite similar to $\delta'(p)$. Hence, for $g < g_c$ we conclude that we are dealing with a theory whose behavior is similar to a theory with massless fermions and both scalar and fermion excitations of zero mass exist. On the other hand for $g > g_c$ we see that $\lim_{n\to\infty} g_n = g_\infty > 0$ and $X_n(p)$ goes over to a nearest neighbor form. This says that after a finite number of iterations we are in the situation of studying a theory which is equivalent to the one studied in Ref. 7 and so we know that it describes massless bound configurations, and essentially

Table 5. g = 0

Iteration (n)	$X_n(1)$	g_{eff}	$X_{norm}(j=1,..,5)$	$(A1)_n/X_n(1)$	$(A2)_n/X_n(1)$
1	-.53333	0	1.00000 .81897 .81187 .81071 .81040	-.625	-.18519
2	-.31376	0	1.00000 .60875 .60204 .60100 .60072	-.46176	-.13878
3	-.18805	0	1.00000 .49614 .49103 .49024 .49003	-.38047	-.10942
15	-2.9897×10^{-4}	0	1.00000 .31256 .30977 .30934 .30922	-.24386	-.06528
20	-1.9376×10^{-5}	0	1.00000 .30931 .30655 .30612 .30601	-.24138	-.06455

Table 6. g = .1

Iteration (n)	$X_n(1)$	g_{eff}	$X_{norm}(j=1,..,5)$	$(A1)_n/X_n(1)$	$(A2)_n/X_n(1)$
1	-.53265	.076589	1.00000 .81896 .81187 .81072 .81040	-.62498	-.18521
2	-.31313	.058668	1.00000 .60872 .60201 .60098 .60070	-.46173	-.13878
3	-.18760	.046113	1.00000 .49611 .49100 .49021 .49001	-.38044	-.10942
15	-2.9803×10^{-4}	.005397	1.00000 .31256 .30977 .30934 .30922	-.24386	-.06528
20	-1.9315×10^{-5}	.002389	1.00000 .30931 .30655 .30612 .30601	-.24138	-.06455

Table 7. g = 2

Iteraction (n)	$X_n(1)$	g_{eff}	X_{norm} (j=1,..,5)	$(A1)_n/X_n(1)$	$(A2)_n/X_n(1)$
1	-.37452	2.9147	1.00000 .81734 .81111 .81034 .81022	-.61755	-.19282
2	-.12898	5.4885	1.00000 .58901 .58319 .58255 .58244	-.43935	-.14330
3	-.032991	16.712	1.00000 .44220 .43814 .43778 .43778	-.32930	-.10871
10	-1.6136×10^{-9}	-3.0816×10^{8}	1.00000 .00119 .00120 .00123 .00124	-6.2370×10^{-4}	-6.4155×10^{-4}
15	4.1286×10^{-15}	$1.2044 \times 10^{+14}$	1.00000 -2.0661×10^{-5} -2.7470×10^{-5} -3.0416×10^{-5} -3.1892×10^{-5}	1.6849×10^{-5}	1.7855×10^{-5}

non-propagating fermion excitations of mass g_∞ (in units of the inverse lattice spacing Λ).

This basically completes the general description of the way in which the iterative variational calculation precedes. While there is much which can be said about other aspects of the theory I would like to conclude our discussion of this model by asking if the existence of a g_c has any counterpart in the solvable continuum theory, and if we can see how our g_c will correspond to the continuum behavior. I would also like to make a few simple remarks about how one could go beyond what I will describe now and study approximations to continuum two point functions, the Schwinger term, etc.

To address the first point, it is true that the continuum theory does exhibit a strange behavior; namely, depending upon the point splitting procedure used to define the composite operators of the theory one finds that there exists a finite coupling $g_0 = \bar{g}_c$ past which the continuum solution fails to exist. This occurs because for $g_0 > \bar{g}_c$ the continuum Hamiltonian written in terms of currents alone develops a negative coefficient — and since the solution is based upon the assumption that the currents are Bose operators — the theory fails to have a ground state. Clearly, this cannot happen in a fermion theory, since the sign of the Hamiltonian is irrelevant in a fermion theory — changing the sign simply changes the way in which we choose to define the filled sea of negative energy states. Since, on the lattice we are dealing with a fermion theory — the currents are not true bosons in that they remember there is an exclusion principle and so for all values of g_0, the ground state of the theory must exist. Question: Is our lattice g_0 related to the continuum $\bar{g}_c$, and one can understand what it is about the lattice theory which makes the passage to a continuum model as $\Lambda \to \infty$ for $g > \bar{g}_c$ impossible? The answer to this question would seem to be that there is indeed a tight correspondence between the change in behavior of the lattice theory and the strange behavior of the continuum theory. To see this we study the combination of lattice fields which go over to the continuum operator $\psi(x)$ as $\Lambda \to \infty$, i.e. we study

$$\psi_\Lambda(x) = \lim_{N\to\infty} \sum_{j=-N}^{N} \left(\frac{\sin(\pi(\Lambda x - j))}{(2N+1)\sin(\frac{\pi(\Lambda x-j)}{2N+1})} \begin{pmatrix} b_j \\ d_j^\dagger \end{pmatrix} \right)$$

and compute the matrix element

$$\langle 0| \int dx\, \psi_\Lambda(x) |+\rangle \equiv \sqrt{Z_2(g_0)}$$

by means of the obvious iteration procedure defined by (3.22). To normalize this calculation to the free field limit ($g_0 = 0$) calculated the same way we directly compute $\sqrt{Z_2(g_0)/Z_2(0)}$ and the result is shown in Fig. 13. As you can see for $g_0 \lesssim g_c$, $\sqrt{Z_2(g_0)/Z_2(0)}$ is finite whereas for $g_0 \geq g_c$ the ratio is zero. If one goes back to the continuum theory this is exactly what happens for $g = \bar{g}_c$.

At this point, I leave you to draw your own conclusions as to how tight the relation between the behavior of the lattice model and continuum theory really is. To really nail it down we should work out things like $<0|\psi^\dagger_\Lambda(x)\psi_\Lambda(0)|0>$, the Schwinger term, and operator equations of motion. We have not studied this in any detail, although we have computed the Schwinger term and obtained entirely reasonable results. I refer you to Ref. 7 since time does not permit further discussion of this point. I would like to conclude with a brief explanation of what is happening for $g > g_c$ and why one cannot conventionally define the continuum Thirring model beyond this point by taking the limit $\Lambda \to \infty$ and multiplicatively renormalizing H. In a sense I have just made the relevant point, what is breaking down for $g > g_c$ is the usual program of multiplicative renormalization. This is so because the usual multiplicative renormalization scheme requires that one rescale the coupling constant and $\psi(x)$ by a constant in such a way as to render the spectrum finite and at the same time so as to make $<0|\int dx\psi(x)|+>$ be unity. As we have seen, for $g_0 > g_c$ this second condition is impossible even for finite value of Λ. This happens because for $g_0 > g_c$ the fermions suddenly acquire a mass proportional to the cutoff and so leave the set of physical states. While states of zero mass do exist for $g_0 > g_c$, they correspond to tightly bound states. Thus, while a finite relativistic theory exists for $g_0 > g_c$ it cannot be obtained by multiplicatively renormalizing H in the manner prescribed by perturbation theory – although it is true

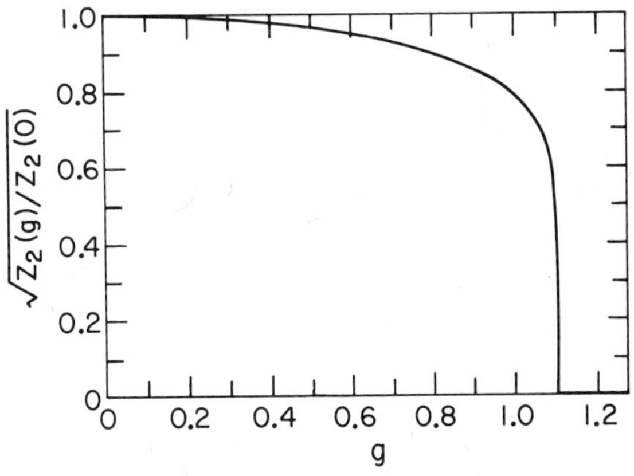

Fig. 13. The quantity $\sqrt{Z_2(g)/Z_2(0)}$ is shown as a function of g.

that such a scheme can be carried out if H is rewritten in terms of composite "magnon" creating operators of the Heisenberg anti-ferromagnet. In some sense this shows that this lattice model exhibits a version of confinement in that, although the theory is simply described by fermionic operators — there are no physical states of the theory created by these operators.

3.5. Summary

To close I would like to highlight the important messages I would like you to get from this analysis. First, computing with our definition of the gradient operation is easy. Second, this variational analysis makes complete sense and does surprisingly well for the free field theory and seems to make sense for the $g_0 \neq 0$ version of the Thirring model. Third, the existence of the critical coupling g_c past which the multiplicative renormalization program breaks down, is an example of the way in which perfectly sensible theories may be derivable from lattice theories even when they cannot be even formulated sensibly within the framework of the usual perturbative approach to renormalization.

4. U(1)-GOLDSTONE MODEL

By this last lecture it has become painfully clear that due to lack of time I will not get to the one lattice gauge theory I had hoped to discuss - i.e. the lattice Higgs model. Rather than try and rush to include a discussion of this model I will conclude this set of lectures with a discussion of the theory which one gauges in order to obtain the Higgs model; namely, the U(1) Goldstone model. However, I will try to keep the discussion of this model as brief as possible — without becoming totally cryptic — so as to leave time for a few remarks about what we do know about the Higgs model and Abelian gauge theories in general.

4.1. Introduction to the U(1)-Model

The model we will discuss is the model of a complex scalar field which was first studied as the simplest example of the spontaneous breaking of a continuous symmetry. The theory is based upon the Lagrangian

$$\mathcal{L} = (\partial_\mu \phi^*)(\partial_\mu \phi) - \lambda(2\phi^*\phi - f^2)^2 \tag{4.1}$$

where

$$\phi = \frac{1}{\sqrt{2}} (\sigma + i\chi)$$

or alternatively, upon the Hamiltonian

$$H = \int dx(\pi_{\phi^*}\pi_\phi + \partial_1\phi^*\partial_1\phi + \lambda(2\phi^*\phi - f^2)^2) \tag{4.2}$$

where we define

$$\pi_{\phi^*} = \partial_o\phi \; ; \qquad \pi_\phi = \partial_o\phi^* \quad .$$

Rather than discussing this continuum theory let us directly transcribe it to a lattice theory and then discuss that. The lattice form we will use will have a nearest neighbor form for the gradient since, as we saw in the first lecture, this makes no important difference for the scalar field and it simplifies the pedagogical part of our discussion. The lattice Hamiltonian we adopt is

$$H = \sum_j\left(\pi_{\phi_j^*}\pi_{\phi_j} + (\phi_{j+1}^* - \phi_j^*)(\phi_{j+1} - \phi_j) + \lambda(2\phi_j^*\phi_j - f^2)^2\right) \tag{4.3}$$

where

$$\phi_j = \frac{1}{\sqrt{2}}(\sigma_j + i\chi_j) \quad .$$

Having transcribed the continuum theory to a lattice theory let us be sure that the important features of the continuum model have not been lost, at least at the classical level, or at the level of perturbation theory. The purpose of the discussion to follow is to reassure those familiar with the continuum Goldstone model that nothing is different, and to introduce those of you who have never studied the problem to the concepts and the level of discussion of these points which one usually encounters.

First, let us discuss the classical limit of this theory. This corresponds to ignoring the $\pi_{\phi_j^*}\pi_{\phi_j}$ term in (4.3) and treating the remaining terms as an expression for the "energy" associated with a function ϕ_j. Since the classical part of (4.3) is a sum of squares, it is a manifestly positive function of ϕ_j; thus, the functions which minimize H are those for which each of these terms separately equal zero. This means that $(\phi_{j+1} - \phi_j) = 0$ or $\phi_j = \phi_o$ independent of 'j', and that

$$2\phi_o^*\phi_o - f^2 = 0 \quad . \tag{4.4}$$

If we write $\phi_o = re^{i\theta}$, then (4.4) implies that $r = f/\sqrt{2}$ and so there is a one parameter family of functions $\phi_o(\theta) = f/\sqrt{2}\ e^{i\theta}$ which have zero energy. In terms of the variables σ_j and χ_j this result is simply that both σ_j and χ_j are independent of j and that $\sigma_0 = f\cos\theta$ and $\chi_0 = f\sin\theta$. This situation, where there is a continuum of degenerate ground state configurations is what is meant by saying that the U(1)-model exhibits spontaneous symmetry breaking, since a U(1) transformation rotates one of the degenerate states into another and so no single "ground state" is rotationally invariant.

Turning from the classical picture let us turn to the usual perturbation discussion of the quantum field theory. For this discussion it is convenient to go to the variables σ_j and χ_j. Substituting $\phi_j = \frac{1}{\sqrt{2}}(\sigma_j + i\chi_j)$ into (4.3)

$$H = \sum_j \left(\frac{1}{2}\pi^2_{\sigma_j} + \frac{1}{2}\pi^2_{\chi_j} + \frac{1}{2}(\sigma_{j+1} - \sigma_j)^2 + \frac{1}{2}(\chi_{j+1} - \chi_j)^2 \right.$$

$$\left. + \lambda(\sigma_j^2 + \chi_j^2 - f^2)^2\right) \quad . \tag{4.5}$$

This Hamiltonian is usually analyzed by observing that in discussing perturbation theory one wants to do an expansion in small vibrations about a stable minimum of the classical potential. In this case, because any one of the field configurations $\sigma_0 = f\cos\theta$ and $\chi_0 = f\sin\theta$ are minima, we have to specify which one of these field configurations one is expanding about; however, the belief – backed up by analyses of models in 3 + 1 dimensions to any finite order in perturbation theory – is that one gets the same theory (in the sense of unitary equivalence) no matter which point we expand about. Assuming that this might be the case, let us choose to expand about $\sigma_0 = f$, $\chi_0 = 0$. In other words let us define a small vibration field σ' by

$$\sigma = \sigma' + f \tag{4.6}$$

and rewrite H as

$$H = \sum_j \left(\frac{1}{2}\pi^2_{\sigma'_j} + \frac{1}{2}\pi^2_{\chi_j} + \frac{1}{2}(\sigma'_{j+1} - \sigma'_j)^2 + \frac{1}{2}(\chi_{j+1} - \chi_j)^2 \right.$$

$$\left. + m^2_{\sigma'}\sigma'^2 + \frac{m^2_{\sigma'}}{2f}\sigma'(\sigma'^2 + \chi^2) + \frac{m^2_{\sigma'}}{8f^2}(\sigma'^2 + \chi^2)^2\right) \tag{4.7}$$

where

$$m^2_{\sigma'} \equiv 8\lambda f^2 \quad .$$

Focusing attention on the quadratic part of (4.7), in order to define the propagators to be used in the perturbation expansion, we see that the σ'-field is a massive free field of mass $m_{\sigma'}^2 = 2\sqrt{2}\ \lambda^{\frac{1}{2}} f$ and χ is a massless field, i.e. it is the infamous Goldstone boson. The usual perturbation analysis is then usually done by holding $m_{\sigma'}^2$ fixed, and expanding in powers of $1/f$. Hence, $f \to \infty$ is the weak coupling region of the theory and is the region where one expects the notion that the ground state of the quantum theory has $\langle\sigma\rangle = f$ and a massless Goldstone boson to be good. In fact, in 2 + 1 and 3 + 1 dimensional theories one believes that the perturbation picture for $f \gg 1$ is essentially correct; however, in the 1 + 1 dimensional theory things are different. This is because in one dimension a massless particle propagator causes infrared divergences so severe as to invalidate the entire perturbation analysis. One might of course conjecture that although the perturbative analysis in 1 + 1 dimensions breaks down, perhaps the general picture of the classical analysis survives. This, however, is known to be false. There is an exact theorem for the continuum theory due to S. Coleman[8], and an earlier version of the same theorem for any lattice theory due to Mermin and Wagner[9], which says that in 1 + 1 dimensions the ground state of the theory cannot have $\langle\phi\rangle \neq 0$. This brings us to the question of why we are so interested in studying this model.

4.2. Why the U(1)-Model?

As I pointed out in the first lecture there is every reason to believe that the ability to compute spontaneous breaking of continuous symmetries is one of the important requirements to put on a purportedly non-perturbative method for analyzing field theories. On the other hand, another important requirement for such a calculational scheme is that it does not fool you and predict spontaneous symmetry breaking when it does not exist. For this reason the 1 + 1 dimensional U(1)-Goldstone model is of great interest to us as – despite the suggestions of the classical and overly naive perturbation theory analysis of the model – there is no Goldstone boson. More precisely, for the 1 + 1 dimensional theory it is impossible for $\langle\phi\rangle$ to be different from zero. Since we have already seen that our iterative calculational scheme does seem to predict phase transitions (or spontaneous symmetry breaking) when we know they do occur, we study this model to be sure that it does not predict them when they are known not to occur.

Another important reason for studying this model is that it provides an example (other than the free scalar field which I will not discuss due to lack of time) of how to handle a boson field theory. This is important, since up until now we have only discussed systems which have a finite number of states associated with each lattice site. For the boson field there are an infinite number of states per site and so there is an interesting new feature of the iteration procedure to be investigated.

Finally, as I alluded to at the outset, gauging the U(1)-symmetry of this model gives the Higgs model which has a rich structure in 1 + 1 dimensions and I will have a few brief remarks to make about this at the end of this talk.

4.3. Non-Perturbative Analysis

At this point of my talk we have discussed why the U(1)-model is interesting and why it is necessary to go beyond perturbation theory in order to properly analyze what is going on. Let us now see how this can be done. To begin let us recast (4.5) as a gigantic Schroedinger problem. This is easily done since

$$[\pi_{\sigma_j}, \sigma_{j'}] = [\pi_{\chi_j}, \chi_{j'}] = -i\delta_{jj'}$$

and so we can think of H as an operator on the space of square integrable functions of the 2(2N + 1) variables (σ_{-N}, χ_{-N}; σ_{-N+1}, χ_{-N+1}; ...; σ_N, χ_N), i.e. we think of the problem of diagonalizing H as equivalent to solving the Schroedinger problem

$$H\Psi(\sigma_{-N}, \chi_{-N}, \ldots, \sigma_N, \chi_N) = E\Psi(\sigma_{-N}, \chi_{-N}, \ldots, \sigma_N, \chi_N) \tag{4.8}$$

where we use for H the form given in (4.5) with the substitutions

$$\pi_{\sigma_j} = \left(\frac{1}{i}\frac{\partial}{\partial\sigma_j}\right) \quad ; \quad \pi_{\chi_j} = \left(\frac{1}{i}\frac{\partial}{\partial\chi_j}\right) \tag{4.9}$$

and we assume

$$\int d\sigma_{-N} \ldots \int d\chi_{+N}\, \Psi^*(\sigma_{-N}, \ldots, \chi_{+N})\Psi(\sigma_{-N}, \ldots, \chi_N) = 1 . \tag{4.10}$$

Having made this observation we will now rewrite H in its Schroedinger form, collect all "single-site" terms and show that if we take the limit $\lambda \to \infty$, f held fixed (i.e. we let $m_\sigma^2 \to \infty$ for fixed f) we greatly simplify the problem to be analyzed without losing any essential features of the model. (To those familiar with the language this limit corresponds to studying the non-linear version of the σ-model instead of the linear version, and at least in 2 + 1 and 3 + 1 dimensions one knows that the Goldstone boson exists in both versions). To be precise let us rewrite (4.5) as

$$H = \sum_j \left(-\frac{1}{2}\frac{\partial^2}{\partial\sigma_j^2} - \frac{1}{2}\frac{\partial^2}{\partial\chi_j^2} + \sigma_j^2 + \chi_j^2 + \lambda(\sigma_j^2 + \chi_j^2 - f^2)^2 \right)$$

$$- \sum_j (\sigma_{j+1}\sigma_j + \chi_{j+1}\chi_j) \quad . \tag{4.11}$$

Note that the terms σ_j^2, χ_j^2, $\sigma_{j+1}\sigma_j$ and $\chi_{j+1}\chi_j$ come from the gradient term. Let us now focus attention on any one of the terms

$$h_j = \left(-\frac{1}{2}\frac{\partial^2}{\partial\sigma_j^2} - \frac{1}{2}\frac{\partial^2}{\partial\chi_j^2} + \sigma_j^2 + \chi_j^2 + \lambda(\sigma_j^2 + \chi_j^2 - f^2)^2 \right) \tag{4.12}$$

and try to solve the single site problem

$$h_j\Psi(\sigma_j,\chi_j) = E\Psi(\sigma_j,\chi_j) \quad . \tag{4.13}$$

Now (4.13) is nothing but a two variable Schroedinger equation and is invariant with respect to rotations in the (σ_j,χ_j) plane. Hence, it is convenient to change variables and define

$$r_j = \sqrt{\sigma_j^2 + \chi_j^2} \quad ; \quad \theta_j = \tan^{-1}(\sigma_j/\chi_j) \tag{4.14}$$

$$-\pi \leq \theta_j \leq \pi$$

and rewrite h_j as

$$h_j = \left(-\frac{1}{2}\left(\frac{1}{r_j}\frac{\partial}{\partial r_j}\left(r_j\frac{\partial}{\partial r_j}\right)\right) - \frac{1}{2r_j^2}\frac{\partial^2}{\partial\theta_j^2} + r_j^2 + \lambda(r_j^2 - f^2)^2 \right) \tag{4.15}$$

and observe that in order to solve the problem one can separate variables and define

$$\Psi(r,\theta) = \sum_m \phi_m(r)\, e^{-im\theta} \quad . \tag{4.16}$$

In this case we see that the problem of finding eigenfunctions of h_j reduces to solving the m-dependent Schroedinger problem

$$\left(-\frac{1}{2r_j}\frac{\partial}{\partial r_j}\left(r_j\frac{\partial}{\partial r_j}\right) + \frac{m^2}{2r_j^2} + r_j^2 + \lambda(r_j^2 - f^2)^2 \right)\phi_m(r_j) = E\phi_m(r_j) \quad . \tag{4.17}$$

Clearly, as $\lambda \to \infty$ the potential

$$V(r_j) = + \frac{m^2}{2r_j^2} + r_j^2 + \lambda(r_j^2 - f^2)^2 \tag{4.18}$$

has its minimum at $r_j = f$ and has a curvature which goes like $(\sqrt{2}\lambda^{\frac{1}{2}}f)^2$. Hence, one can readily convince oneself that a Gaussian of the form $e^{-\gamma(r_j-f)^2/2}$ where $\gamma \approx \lambda^{\frac{1}{2}}f/\sqrt{2}$ provides a good ground state wavefunction for any finite m, and so one can choose this for $\phi_m(r)$ for all m and compute the expectation of H in states of this form. Clearly, since the Gaussians get narrower and narrower as $\lambda \to \infty$ this amounts to ignoring the term $-\frac{1}{2r_j}\frac{\partial}{\partial r_j}(r_j \frac{\partial}{\partial r_j})$ in (4.17) and replacing r_j everywhere by f. We therefore observe that in the limit $\lambda \to \infty$ (4.17) becomes

$$h_j^{(\lambda=\infty)} = (\theta \text{ independent constant}) + \frac{1}{2f^2}\left(\frac{1}{i}\frac{\partial}{\partial\theta_j}\right)^2 \tag{4.19}$$

where $h_j^{(\lambda=\infty)}$ is defined, by construction, on a space of function $f(\theta)$ satisfying periodic boundary conditions; i.e., up to a constant — which we can ignore — h_j is the Hamiltonian of a rotor of moment of inertia $1/f^2$. Going back to (4.11) and making the corresponding substitutions

$$\sigma_j = f\cos(\theta_j) \quad ; \quad \chi_j = f\sin(\theta_j) \tag{4.20}$$

we can rewrite (4.11) in the limit $\lambda \to \infty$ as:

$$H^{(\lambda=\infty)} = \sum_j \left[\frac{1}{2f^2}\left(\frac{1}{i}\frac{\partial}{\partial\theta_j}\right)^2 - f^2\cos(\theta_{j+1} - \theta_j)\right] \tag{4.21}$$

where H is defined as a self-adjoint operator on the space of square integrable functions $\Psi(\theta_{-N},\ldots,\theta_N)$ satisfying periodic boundary conditions in each variable $-\pi \le \theta_j \le \pi$.

The previous argument tells us that in the $\lambda \to \infty$ or $m_\sigma \to \infty$ limit of the U(1)-Goldstone the theory goes over to a system of planar rotors of moment of inertia $1/f^2$ coupled to one another by an amount proportional to the difference between the directions in which they point. This same model is also a beloved model of "statistical mechanics" who cryptically call it the x-y model.

4.4. Have We Lost Anything?

Before discussing this model and the structure of the resulting theory as a function of f let us observe that the theory specified by (4.21) has the same features, at the classical level, that the original theory had. Once, again, the classical approximation is to drop the $(\frac{1}{i}\frac{\partial}{\partial\theta_j})^2$ terms. If we do this it is clear that the classical theory has a 1-parameter family of ground states labeled by a parameter θ_o, namely the configuration $\theta_{j+1} - \theta_j = 0$ or $\theta_j = \theta_o$.

Having argued that the classical level of the theory is unchanged, let us now argue that the "small vibration analysis" of the quantum theory suggests that we are studying, for large f, the theory of a weakly interacting massless field. The easiest way to do this is to let

$$\theta'_j = f\theta_j \quad ; \quad -\pi f \leq \theta_j \leq \pi f \tag{4.22}$$

and rewrite H as

$$H = \sum_j \frac{1}{2}\left(\frac{1}{i}\frac{\partial}{\partial\theta'_j}\right)^2 - f^2 \cos(\theta'_{j+1} - \theta'_j) \quad . \tag{4.23}$$

If we now assume that $\theta'_j = \theta'_o + \delta_j$ and expand H in terms of the small vibration field δ_j we obtain

$$\begin{aligned} H = & - f^2(2N)\,\mathbb{1} + \frac{1}{2}\sum_j\left[\left(\frac{1}{i}\frac{\partial}{\partial\delta_j}\right)^2 + (\delta_{j+1} - \delta_j)^2\right] \\ & + \sum_j\left[-\frac{(\delta_{j+1} - \delta_j)^4}{4!f^2} + \frac{(\delta_{j+1} - \delta_j)^6}{6!f^4} + \ldots\right] \end{aligned} \tag{4.24}$$

which as $f \to \infty$ goes over to the Hamiltonian of a free massless field. This, of course, is what we saw for the original U(1) model, i.e. that perturbation theory corresponded to an expansion in f^{-1} about the theory of a massless scalar field.

Hence, we see that our specialization to the $\lambda \to \infty$, f held fixed limit of the U(1) model (or the x-y model) loses none of the important features we wished to study. Our goal will be to show that our calculation for $f << 1$ agrees with the Mermin-Wagner (or lattice versions of Coleman's theorem) in that it predicts that

$$\langle e^{i\theta_j}\rangle \equiv \frac{1}{f}\langle\phi_j\rangle = 0 \ ,$$

i.e. it is not $e^{i\theta_0}$ as we would expect from the classical argument.

4.5. Discussing The Lattice Model Variationally

Time will not permit me to give all the details of the iterative calculation for this model, nor is it very interesting for you to check my arithmetic. I will spend the remainder of my time trying to give you a feeling for what we did and how a typical iteration looks.

To begin, let us consider some general physics associated with the Hamiltonian (4.4), or its rescaled form (4.23). I would like to point out, that as f decreases towards zero there is a big difference between (4.21) and the theory obtained by replacing $\cos(\theta_{j+1} - \theta_j)$ by $\frac{1}{2}(\theta_{j+1} - \theta_j)^2$. This is true because if we study the theory

$$H = \sum_j \left[\frac{1}{2f^2} \left(\frac{1}{i}\frac{\partial}{\partial\theta_j}\right)^2 + \frac{f^2}{2}(\theta_{j+1} - \theta_j)^2 \right] \tag{4.25}$$

we see that we can always perform the canonical transformation

$$\theta_j = \theta_j'/f \quad ; \quad \left(\frac{1}{i}\frac{\partial}{\partial\theta_j}\right) = f\left(\frac{1}{i}\frac{\partial}{\partial\theta_j'}\right) \tag{4.26}$$

and rewrite H as a massless free field theory, hence the physics of (4.25) is independent of f. However, the physics of (4.21) is quite dependent upon f, as can be readily seen by studying H in the limit f >> 1. In this case if we define

$$H_o = \frac{1}{2f^2} \sum_j \left(\frac{1}{i}\frac{\partial}{\partial\theta_j}\right)^2 \tag{4.27}$$

and treat

$$V = -f^2 \cos(\theta_{j+1} - \theta_j) \tag{4.28}$$

as a perturbation, we see that the eigenstates of H_o are wavefunctions of the form

$$\Psi(\theta_{-N}, \ldots, \theta_N) = \prod_{j=-N}^{N} (e^{-im_j\theta_j}) \tag{4.29}$$

with eigenenergies

$$E(m_j) = \frac{1}{2f^2} \sum_j (m_j)^2 \ . \tag{4.30}$$

Hence, the ground state of H_0 is the unique state

$$\Psi(\theta_{-N},\ldots,\theta_N) = 1 \tag{4.31}$$

and the gap to the lowest excited state is $1/2f^2$, which becomes large as $f \to 0$. Clearly for f sufficiently small the perturbation V has no way of wiping out a gap of order $1/2f^2$ and so the theory is a theory of massive excitations. The question is, how is it that the small f limit of (4.21) is a massive theory, whereas the theory of (4.25) is a massless theory independent of the value of f? The answer is that the arguments, m_j, of (4.29) are integers because the $\psi(\theta_j)$ are defined to be periodic in the variables θ_j — and so for small enough f the gap becomes large. Hence, it is the fact that the $\psi(\theta_j)$ know about boundaries of the defining region which allows the theory to go massive.

This comment is important because the gist of our iterative solution will be to show that for f's greater than some constant f_c the ground state wave function never sees the fact that the Hamiltonian is periodic, that θ_j runs over a finite range. To be specific, let us define $j = 2p+r$, $r = 0,1$ and rewrite H as

$$H = \sum_p \left[-\frac{1}{2f^2}\left(\frac{\partial^2}{\partial\theta_{2p}^2} + \frac{\partial^2}{\partial\theta_{2p+1}^2}\right) - f^2 \cos(\theta_{2p+1} - \theta_{2p}) \right]$$

$$- f^2 \sum_p \cos(\theta_{2(p+1)} - \theta_{2p+1}) \ , \tag{4.32}$$

and let us define

$$h_p = -\frac{1}{2f^2}\left(\frac{\partial^2}{\partial\theta_{2p}^2} + \frac{\partial^2}{\partial\theta_{2p+1}^2}\right) - f^2 \cos(\theta_{2p+1} - \theta_{2p}) \tag{4.33}$$

in analogy to earlier iterations. As before our next step is to analyze this 1-block problem in detail, identify the eigenstates of the system and truncate away those combinations of fields corresponding to "high mass modes". To analyze (4.33) it is suggestive to introduce the variables

$$2\psi_p = (\theta_{2p} + \theta_{2p+1}) \ ,$$

$$2\phi_p = (\theta_{2p} - \theta_{2p+1}) \ , \tag{4.34}$$

and rewrite

$$h_p = -\frac{1}{2f^2}\left(\frac{\partial^2}{\partial\psi_p^2}\right) - \frac{1}{2f^2}\left(\frac{\partial^2}{\partial\phi_p^2}\right) - f^2\cos(2\phi_p) \ . \tag{4.35}$$

Expanding the $\cos(2\phi_p)$ and fixing attention on quadratic terms suggest that the variable ϕ_p behaves like an oscillar of frequency $\omega_\phi = 2$ and mass $m_\phi = f^2$, and the variable ψ_p acts like a rotor in that $-\partial^2/\partial\psi_p^2$ is diagonalized by functions of the form $e^{iK\psi_p}$. Actually, this apparent decoupling of the 2-site Hamiltonian is deceptive since the requirement that h_p acts on wavefunctions $\Psi(\theta_{2p},\theta_{2p+1})$ which are periodic on the square $-\pi \leq \theta_{2p} \leq \pi$, $-\pi \leq \theta_{2p+1} \leq \pi$ requires that variables ψ_p and ϕ_p be coupled; since, for fixed ψ_p we have

$$-\pi \leq \phi_p \leq \pi \quad \text{but} \quad -(\pi-|\phi_p|) \leq \psi_p \leq (\pi-|\phi_p|) \ . \tag{4.36}$$

This recoupling of the variables through the boundary conditions requires that one carefully handle the Schroedinger problem. Having said that one must be careful, let me promise you that we have been. The key point is that for $\omega_{\phi_p} = 2$ we have the ground state wavefunction

$$e^{-m_\phi\omega_\phi(\phi_p^2)/2} = e^{-f^2\phi_p^2}$$

and for large f (f >> 1) it is clear that the system doesn't see the boundary to any great degree. Moreover, it is clear that the main value of ϕ_p is in the range 1/f, hence the ψ_p variable (up to terms on the order of 1/f) can be considered to be a rotor with periodic boundary conditions on the interval $(-\pi,\pi)$. The aim of this brief discussion is to show why a careful study shows that for large f naively treating h_p as the Hamiltonian of an uncoupled rotor and oscillator is O.K. The next step is to couple two such blocks together through the typical coupling

$$-f^2\cos(\psi_{p+1} - \psi_p + (\phi_{p+1} + \phi_p)) \ , \tag{4.37}$$

show that it becomes a system of one rotor and three oscillators, and then truncate away states generated by having the two oscillators of highest mass out of their ground state. This truncation brings us back to an effective Hamiltonian of a system which is one rotor and one oscillator. All that will have changed is the coefficients of the various terms. Let me briefly sketch how this goes by noting that at the nth iteration the truncated Hamiltonian takes the generic form

$$H_n = \sum_p \Big[c_n \mathbb{1}_p + \frac{1}{2}\alpha_n^2\Big(-\frac{\partial^2}{\partial\psi_p^2}\Big) + \frac{1}{2}\Big(-\frac{\partial^2}{\partial\phi_p^2} + \omega_n^2\phi_p^2\Big)$$

$$- \beta_n \cos(\psi_{p+1} - \psi_p + \delta_n(\phi_{p+1} + \phi_p)) \Big] \qquad (4.38)$$

and if we define a superblock of 4-sites by letting $p = 2\ell+s$, $s = 0,1$, then we get an effective superblock Hamiltonian

$$h_\ell = \Big[2c_n \mathbb{1} - \frac{1}{2}\alpha_n^2\Big(\frac{\partial^2}{\partial\psi_{2\ell}^2} + \frac{\partial^2}{\partial\psi_{2\ell+1}^2}\Big) - \frac{1}{2}\Big(\frac{\partial^2}{\partial\phi_{2\ell}^2} + \frac{\partial^2}{\partial\phi_{2\ell+1}^2}\Big)$$

$$+ \frac{\omega_n^2}{2}(\phi_{2\ell}^2 + \phi_{2\ell+1}^2) - \beta_n \cos(\psi_{2\ell+1} - \psi_{2\ell} + \delta_n(\phi_{2\ell+1} + \phi_{2\ell})) \Big] \qquad (4.39)$$

If we have β_n/α_n^2 sufficiently large then we can justify expanding the cosine and keeping only quadratic terms. This reduces the h_ℓ problem to a system of one rotor and three coupled oscillators, diagonalizing this and truncating away the two highest oscillators gives c_{n+1}, α_{n+1}^2, β_{n+1}, ω_{n+1}^2 and δ_{n+1}. The results of one such iteration for two values of $x_0 = 1/f$ are given in Tables 8 and 9. The variable K_n is defined by $\beta_{n+1} = \beta_n e^{-K_n x_o^2}\beta_n$. Although I will not prove it now, it is easy to convince oneself that $\lim_{n\to\infty}(\beta_n^{\frac{1}{2}})$ is the expectation value, in the variationally constructed ground state, of $\frac{1}{f}e^{i\theta_j}$. From the fact that K_n goes to a constant after a few iterations it follows that after a few iterations we can write

$$\beta_n^{\frac{1}{2}} = \bar{\beta}^{\frac{1}{2}} e^{-(.368..)n/2f^2} \quad . \qquad (4.40)$$

Since the volume of the block under consideration goes as $2(2^n)$ we have that

$$n = \frac{\log(V)}{\log(2)} - 1 \qquad (4.41)$$

or

$$\beta_n^{\frac{1}{2}} = \bar{\beta}' e^{-(.368..)\log(V)/f^2}$$

$$= \bar{\beta}' \Big(\frac{1}{V}\Big)^{(.368..)/f^2} \quad . \qquad (4.42)$$

Hence, we find that as the volume of the world goes to infinity we predict that $\langle e^{i\theta(j)}\rangle$ tends to zero as a constant times $(1/V)^{(0.36821/f^2)}$. This is the result predicted from analyses based on a treatment of $\theta(j)$ as a massless free field and is not easily reproduced from a first principles calculation.

Table 8. Iteration for $x_0 = .1$. The notation in this table conforms to that given in Eq. (3.15) except for the definition of K_n. This is defined by the relation $\beta_{n+1} = e^{-K_n x_0^2} \beta_n$. Note in particular that ω_n^2 decreases by a factor of 2^2 for each iteration and hence $\omega_N \propto 1/2^N \propto (\text{volume})^{-1}$.

Iteration (n)	$(\alpha_n)^2$	$(\omega_n)^2$	β_n	δ_n	K_n	c_n
0	.01	2	50	.1		-50
1	.005	.76393	49.906	.097325	.18831	-149.08
2	.0025	.22226	49.77	.077379	.27347	-347.38
3	.00125	.059722	49.612	.05778	.31801	-744.05
8	3.9062×10^{-5}	6.3112×10^{-5}	48.733	.010803	.36635	-25336
9	1.9531×10^{-5}	1.5802×10^{-5}	48.554	.0076468	.36727	-50721
10	9.7656×10^{-6}	3.9535×10^{-6}	48.376	.0054099	.36774	-101490
14	6.1035×10^{-7}	1.5454×10^{-8}	47.669	.0013531	.36819	-1624562
15	3.0518×10^{-7}	3.8637×10^{-9}	47.493	9.5682×10^{-4}	.3682	-324171
16	1.5259×10^{-7}	9.6594×10^{-10}	47.319	6.7658×10^{-4}	.36821	-6498389
17	7.6294×10^{-8}	2.4149×10^{-10}	47.145	4.7841×10^{-4}	.36821	-12996826

Table 9. Iteration for $x_0 = .01$. This table is included to show that for both $x_0 = .1$ and $x_0 = .01$ the iteration is basically the same up to a scale factor. The fact that α_n^2 and ω_n^2 both drop rapidly with respect to β_n tells us that the oscillator approximation is valid at all stages.

Iteration (n)	$(\alpha_n)^2$	$(\omega_n)^2$	β_n	δ_n	K_n	c_n
0	.0001	2	5000	.01		-5000
1	5×10^{-5}	.76393	4999.9	.0097325	.18831	-14999
2	2.5×10^{-5}	.22233	4999.8	.0077385	.27346	-34998
3	1.25×10^{-5}	.059757	4999.6	.0057788	.31801	-74994
9	1.9531×10^{-7}	1.5813×10^{-5}	4998.5	7.6469×10^{-4}	.36717	-5114571
10	9.7656×10^{-8}	3.9562×10^{-6}	4998.3	5.4099×10^{-4}	.36763	-10234139
11	4.8828×10^{-8}	9.8941×10^{-7}	4998.2	3.8264×10^{-4}	.36786	-20473277
14	6.1035×10^{-9}	1.5465×10^{-8}	4997.6	1.3531×10^{-4}	.36807	-1.64×10^{8}
15	3.0518×10^{-9}	3.8663×10^{-9}	4997.4	9.5682×10^{-5}	.36808	-3.2765×10^{8}
16	1.5259×10^{-9}	9.6658×10^{-10}	4997.2	6.7658×10^{-5}	.36809	-0.65530×10^{9}
17	7.6294×10^{-10}	2.4165×10^{-10}	4997.1	4.7841×10^{-5}	.36809	-1.31060×10^{9}

4.6. Summary

What we have shown in this discussion is, that the same iterative procedure we have used to discuss theories describing systems having finite number of states at a site works well in describing boson systems which have an infinite number of states at a site – if we truncate so as to preserve a full field operator per site. Moreover, we have seen that the physical insight obtained from the Hamiltonian picture makes it trivial to see that the theory undergoes a transition to a massive phase for f sufficiently small and that it describes a massless theory for $f \ll 1$. Although we have not discussed the transition region in these lectures the same iteration procedures — with greater care spent upon the recoupling of oscillators in a block through the boundary conditions could be used to calculate for all f and study the nature of the transition. We have not chosen to do so simply because little is known in the way of precise information about the behavior at the critical point and we are using this model to test our method. This brings us to the final point; we have shown that the naive truncation procedure is powerful enough to predict a transition to a Goldstone phase when it occurs, and to predict that $f\langle e^{i\theta_j}\rangle = \langle\phi(j)\rangle = 0$ in one-dimension, where — contrary to classical arguments — the transition to a Goldstone phase is known not to occur.

5. CLOSING REMARKS

Although I will have no time to speak about it, we are now actively studying the application of the same methods to gauge theories. At present, we have focused attention on Abelian theories in all dimensions in order to find out whether or not the prescriptions given by Wilson and Kogut and Susskind correctly describe quantum electrodynamics for any range of couplings. This is an important question for Abelian theories since they can be formulated in a different manner than that given by Wilson and this formulation manifestly reproduces QED; hence, if the Wilson prescription winds up confining electrons for all couplings in 1 + 3 dimensions, one will worry whether or not confinement — as discussed within the framework of lattice theories — has anything to do with properties (such as asymptotic freedom) associated with the continuum theory.

To date, we have found that these methods can be straightforwardly extended to gauge theories once one properly understands the significance of local gauge invariance. In particular, Helen Quinn and I have used this method to analyze the 1 + 1 dimensional Higgs model — obtained by gauging the U(1) Goldstone model just discussed. We have shown that from the Hamiltonian point of view one can trivially, in a strictly physical way, obtain all of the results of analyses based upon instantons.

Up to this point we have seen no reason to believe that the same methods will not extend in a straightforward manner to non-Abelian gauge theories. However, that work is currently in progress and so we have no solid statements we would care to make about it at this time.

REFERENCES

1) A review of the ideas of Kadanoff and Wilson can be found in "Lectures on the Application of Renormalization Group Techniques to Quarks and Strings", Leo P. Kadanoff, PRINT-76-0772 (Brown). First of five lectures at University of Chicago, "Relativistically Invariant Lattice Theories", K.G. Wilson, CLNS-329 (February 1976, Coral Gables Conference); "Quarks and Strings on a Lattice", K.G. Wilson, CLNS-321 (November 1975, Erice School of Physics); K.G. Wilson and J. Kogut, Phys. Rev. 12, 75 (1974); L.P. Kadanoff, "Critical Phenomena", Proc. International School of Physics "Enrico Fermi", Course LI, ed. M. S. Green (Academic, New York, 1972). Thomas L. Bell and Kenneth G. Wilson, "Nonlinear Renormalization Groups", CLNS-268 (May 1974). For a discussion of the ideas of these authors as applied to more physical models, as well as references to earlier works see T. Banks, S. Raby, L. Susskind, J. Kogut, D.R.T. Jones, P.N. Scharbach and D.K. Sinclair, Phys. Rev. D15, 1111 (1977); J. Kogut, D.K. Sinclair and L. Susskind, Nucl. Phys. B114, 199 (1976); L. Susskind, PTENS 76/1 (January 1976); L. Susskind and J. Kogut, Phys. Reports 23C, 331 (1976).

2) See lectures in this volume for a list of references.

3) C.G. Callan, R. Dashen and D. Gross, "Toward a Theory of Strong Interactions", Institute for Advanced Study preprint COO-222-115.

4) See ref. 1 as well as J. Kogut and L. Susskind, Phys. Rev. D11, 395 (1975); L. Susskind, Lectures at Bonn Summer School, 1974 (unpublished); T. Banks, J. Kogut and L. Susskind, Phys. Rev. D13, 1043 (1976).

5) S.D. Drell, M. Weinstein and S. Yankielowicz, Phys. Rev. D14, 487 (1976); D14, 1627 (1976); "Quantum Field Theories on a Lattice: Variational Methods for Arbitrary Coupling Strengths and the Ising Model in a Transverse Magnetic Field", Stanford Linear Accelerator Preprint SLAC-PUB-1942 (1977), to appear in Phys. Rev.

6) Helen R. Quinn and M. Weinstein, "Multiple Vacua in a Lattice Formulation of the Two Dimensional Higgs Model", SLAC-PUB-2034, submitted to Phys. Rev. D.

7) S.D. Drell, B. Svetitsky and M. Weinstein, "Fermion Field Theory on a Lattice", SLAC-PUB-1999, to be published in Phys. Rev. D.

8) S. Coleman, Commun. Math. Phys. 31, 259 (1973).

9) N.D. Mermin and H. Wagner, Phys. Rev. Lett. 17, 1133 (1966).

SOME RECENT ADVANCES IN NEUTRINO PHYSICS*†

A. K. Mann

University of Pennsylvania

Philadelphia, Pennsylvania 19104, U. S. A.

1. SOME PERSPECTIVES

One answer to questions concerning the subject matter of elementary particle physics and its position within the physical sciences is that particle physics is the study of creation. To justify this idea, at least in part, consider the "big bang" model of creation of the universe. In that model the primeval fireball involved very high temperatures, estimated to be greater than 10^{11} degrees Kelvin (^{o}K), which is to be compared with the interior temperature of the sun at about 10^7 degrees Kelvin. We believe the sun to be a nuclear furnace, generating its energy by individual nuclear interactions in the million electron volt (MeV) energy range. In the "big bang" the energies involved in the individual particle interactions would have been more than 10^4 times larger than in the sun, and thus in the tens of giga-electron volts (GeV). This is precisely the energy region of present-day high energy or elementary particle physics laboratory experiments. In this sense, then, laboratory experiments that study, even indirectly, lepton-lepton scattering or quark-quark interactions are revisiting creation.

One consequence attributed to the "big bang" is the presence of a "sea" of black body (3°K) radiation, the cooled-down remnant of the electromagnetic energy in the initial fireball, which has been detected with radio telescopes. Another consequence of the "big bang", predicted by the theory, is a neutrino equivalent (also at about 3°K) of the electromagnetic radiation, giving approximately as many neutrinos as photons (a few hundred per cm^3) throughout the universe. This almost sterile, i.e., almost non-interacting neutrino "sea" has not been observed. It is of interest to recognize, however, that if neutrinos have even a very small rest mass, of the order of a few electron-volts, the total mass represented by neutrinos (since there are so many of them) would be comparable to and perhaps exceed the total stellar mass. According to the dramatic description of Cowsik and McClelland, neutrinos with a mass of a few electrons would "dominate the gravitational dynamics of the universe". The present upper limit on the mass of the electron-type neutrino (ν_e) from laboratory experiments is about 35 eV.

Thus, neutrinos constitute a proper and interesting subject for study. Neutrino physics may lead us, as we have seen, to astrophysics and cosmology. Or we may study them in their own right to determine intrinsic properties such as mass, charge and magnetic moment. Still another alternative, and a particularly fruitful one, is to use neutrinos to probe the nature of matter. Because neutrinos are the only particle that interacts (as far as we know) only weakly, and because that interaction is (as far as know) "pointlike", neutrino interactions with matter allow us to study the elementary constituents of matter - the leptons and quarks - and also the elementary weak currents. We might describe this area of particle physics in which laboratory produced neutrinos are used as the projectiles in scattering experiments as neutrino microscopy.

In these lectures we shall concentrate on the contributions of neutrino microscopy to the recent developments in particle physics. The material we shall cover here divides itself very simply into a study of neutrino-nucleon interactions with (a) no final state muon, i.e., weak neutral current (WNC) processes; (b) one final state muon, i.e., deep inelastic scattering and its implications; and (c) final states with two or more muons, i.e., new particle production and decay. There is, as you might expect, more to neutrino microscopy than can be covered in a few lectures. We shall, however, have to be content to point out other topics of interest as we go along without elaborating on them.

2. SEMILEPTONIC WEAK NEUTRAL CURRENTS

2.1 Historical Introduction

Let me begin with a few minutes of history. Almost exactly five years ago the material shown in Table 1 was presented by Professor D. Perkins in a review[1] of experiments on neutrino interactions. One sees in Table 1 the experimental limits available near the end of 1972 on purely leptonic weak neutral current (WNC) neutrino scattering reactions and also on certain $\Delta S = 0$ WNC semileptonic neutrino scattering processes. These limits, taken in

Table 1. Limits on weak neutral current couplings from D.H. Perkins, Proc. XVI Int'l. Conf. on High Energy Physics, Chicago-Batavia, Vol. 4, p. 189 (1972).

Cross Section	Author	90% CL Upper Limit
$\sigma(\bar{\nu}_e + e^- \rightarrow \bar{\nu}_e + e^-)$	Gurr, Reines, Sobel PRL 28, 1406 (1972)	$<3.0\sigma_{V-A}$
$\sigma(\nu_\mu + e^- \rightarrow \nu_\mu + e^-)$	CERN Gargamelle (paper #785, this Proc.)	$<0.44\sigma_{V-A}$
$\sigma(\bar{\nu}_\mu + e^- \rightarrow \bar{\nu}_\mu + e^-)$	CERN Gargamelle (#785)	$<0.62\sigma_{V-A}$
$\dfrac{\sigma(\nu_\mu n \rightarrow \nu_\mu n\pi^o + \nu_\mu p \rightarrow \nu_\mu p\pi^o)}{2\sigma(\nu_\mu n \rightarrow \mu p\pi^o)}$	W. Lee (#239) CERN Gargamelle (#785)	<0.14* <0.11*
$\dfrac{\sigma(\nu_\mu p \rightarrow \nu_\mu \Delta^+)}{\sigma(\nu_\mu p \rightarrow \mu^- \Delta^{++})}$	Cundy *et al*** PL 31B 478 (1970) Cho *et al* (#473)	<0.46 <0.31
$\dfrac{\sigma(\nu_\mu p \rightarrow \sigma_\mu p)}{\sigma(\nu_\mu n \rightarrow \mu^- p)}$	Cundy *et al* (ibid)	<0.22
$\dfrac{\sigma(\nu_\mu N \rightarrow \nu_\mu + \text{anything})}{\sigma(\nu_\mu N \rightarrow \mu^- + \text{anything})}$	CERN 1.2 m HLBC (unpublished)	$\lesssim 0.2$

* See comments in text.

** The experiment gave $\dfrac{\sigma(\nu_\mu p \rightarrow \nu_\mu n\pi^+)}{\sigma(\nu_\mu p \rightarrow \nu_\mu p\pi^+)} \leq 0.08 \pm 0.04$.

Table 2. Limits on weak neutral current couplings in $|\Delta S| = 1$ transitions. Data are from Rev. Mod. Phys. 45, No. 2, Part II (1973).

Decay Mode	Upper Limit
$\dfrac{\Gamma(K^+ \to \pi^+ + e^+ + e^-)}{\Gamma(K^+ \to \pi^0 + e^+ + \nu_e)}$	0.8×10^{-4}
$\dfrac{\Gamma(K_L^0 \to \mu^+ + \mu^-)}{\Gamma(K^+ \to \mu^+ + \nu_\mu)}$	0.8×10^{-9}
$\dfrac{\Gamma(K^+ \to \pi^+ + \nu + \bar{\nu})}{\Gamma(K^+ \to \pi^0 + e^+ + \nu_e)}$	0.3×10^{-4}

conjunction with the much lower experimental upper limits on the rates of $|\Delta S| = 1$ WNC decays of K mesons shown in Table 2, had led to the widely held conclusion that weak neutral current processes were probably absent in nature.

For many years we have visualized charged lepton scattering reactions, elastic and inelastic, as occurring through an electromagnetic neutral current as in the Feynman diagrams of Figs. 1a and 1b. On the other hand, on the basis of the evidence in Table 1, it was assumed in the phenomenological theory of the weak interaction (prior to present gauge theories) that neutrino scattering reactions occurred only through charged current weak interactions as illustrated in Figs. 1c and 1d, i.e., through the exchange of a (real or virtual) charged vector boson (W^+), and never through neutral vector boson (W^0) exchange as in Figs. 2a and 2b. Consistency with experiment (Table 2) also required that the diagram of Fig. 2c be forbidden.

Failure to observe WNC processes was clearly a stumbling block for theories seeking a unifying connection between the weak interaction and electrodynamics[2]. The absence of a WNC was also a serious problem for the phenomenological weak interaction theory with only weak charged currents because WNC processes should appear in higher order in a charged current theory[3] at a rate relative to charged current processes determined by the magnitude of the mass of the propagator (W^+) in Figs. 1c and 1d; for $M_W \gtrsim G_F^{-\frac{1}{2}}$, the WNC rate should be of the same order of magnitude as the WCC rate.

Less than one year after the summary in Table 1 was presented at Fermilab, two experiments (in one of which the author of Table 1 was a participant) were in fact observing neutrino-induced reactions without a charged lepton in the final state.

One of these experiments was done at CERN in the large volume (11.5 m^3) heavy liquid freon bubble chamber Gargamelle, with neutrinos (ν_μ) and antineutrinos ($\bar{\nu}_\mu$) of average energy about 1 GeV. One of the inelastic reactions observed in Gargamelle is illustrated

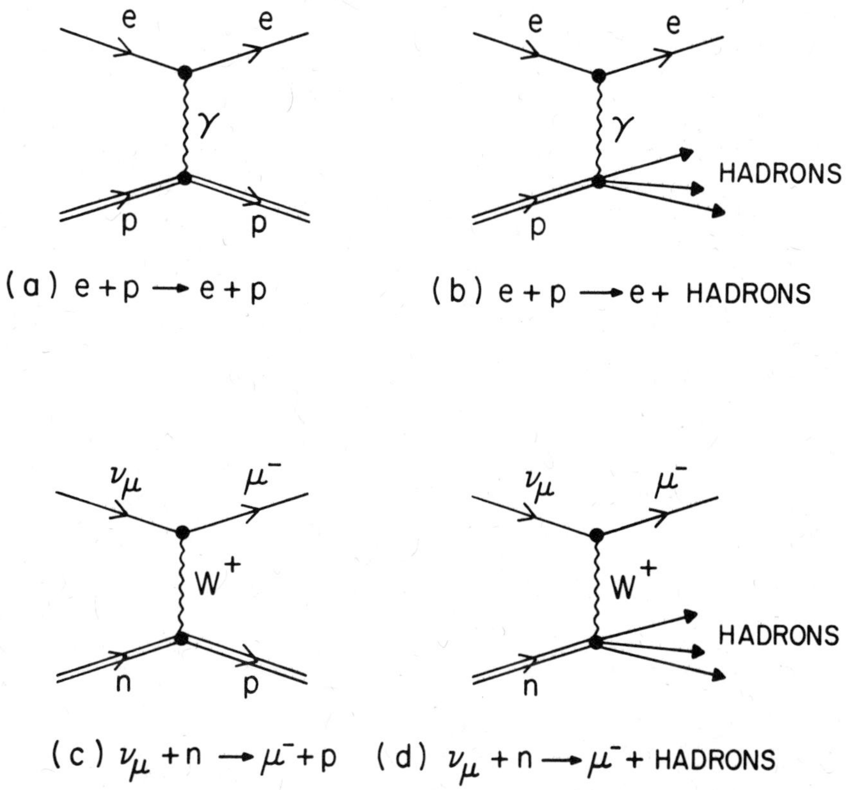

Fig. 1. Feynman diagrams for (a) elastic scattering and (b) inelastic scattering of charged leptons by protons and the corresponding diagrams (c) and (d) for weak charged current (WCC) scattering of neutrinos by neutrons.

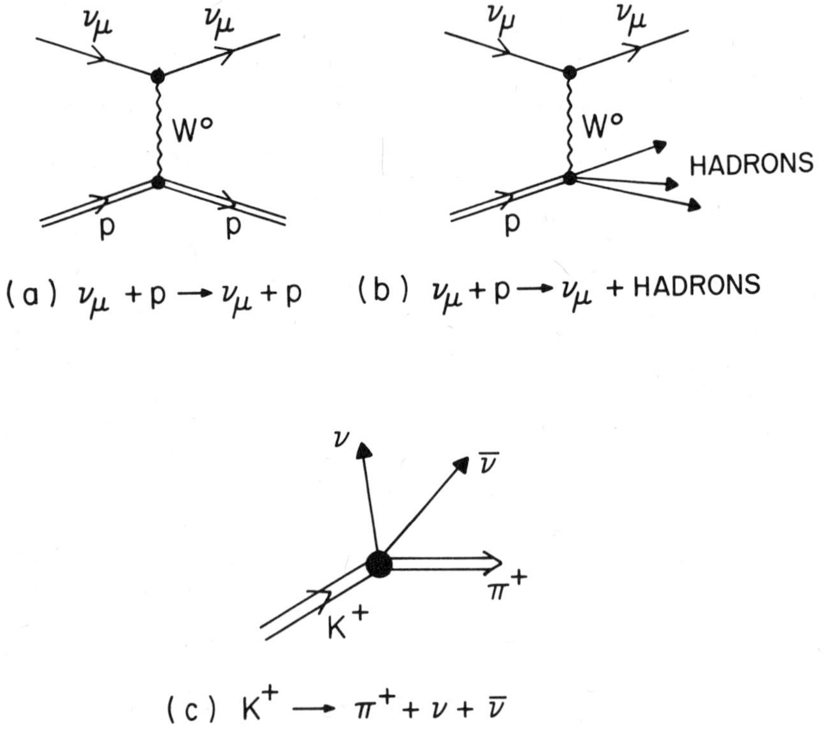

Fig. 2. Diagrams for weak neutral current (WNC) (a) elastic and (b) inelastic neutrino scattering by protons and (c) WNC ($\Delta S = 1$) decay of a charged K meson.

in Fig. 3 which shows the tracing of an actual event observed in the chamber. In that event all of the tracks can be identified as those of strongly interacting particles, from which it was concluded that no charged lepton emerged from the neutrino-nucleon interaction.

The other experiment was carried out at Fermilab by a group of physicists from Harvard, Pennsylvania, Wisconsin and Fermilab (HPWF) using a scintillation counter-spark chamber apparatus and incident neutrinos and antineutrinos with an average energy of about 20 GeV. Here the target-detector of the neutrino interactions was a large ionization calorimeter consisting of optically separated segments, each of which was $(3 \times 3)\text{m}^2$ in area of 0.46 m long and filled with liquid scintillator of density 0.87 gm/cm^3. Each segment was looked at by 12 photomultiplier tubes, the summed output of which measured the ionization energy deposited by the

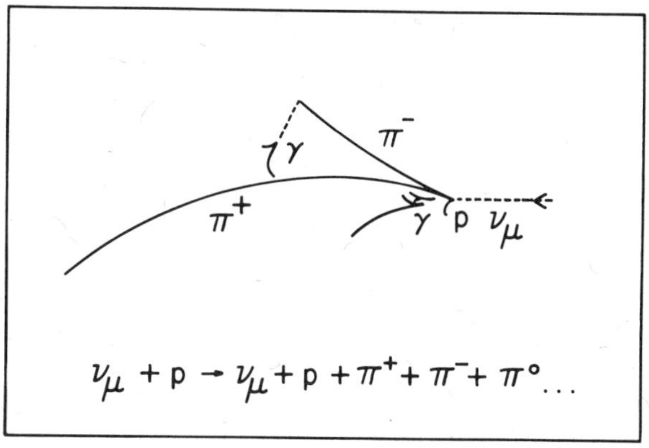

Fig. 3. Tracing of an example of a candidate for an interaction involving a neutral current in the CERN experiment. The neutrino enters from the right and no charged lepton emerges from its interaction with a proton. All the observed tracks can be identified as strongly interacting particles by measuring their curvatures, observing their range in the heavy liquid and their subsequent interactions. CERN Courier 13, No. 10 (October, 1973).

products of a neutrino (antineutrino) interaction. Because of the large volume and mass of the target-detector, most of the energy in the hadronic-electromagnetic cascade from a neutrino interaction was contained within the ionization calorimeter. Interspersed between every four segments of the calorimeter were optical spark chambers which gave a visual display of the secondary particle cascade. Immediately downstream of the ionization calorimeter was a muon detector, consisting of large area scintillation counters and optical spark chambers, to identify muons from WCC reactions and to measure the momentum of a subsample of the muons. The experiment was triggered by the deposition of energy in the target detector greater than a preset minimum of a few GeV. A diagram of a somewhat later version of the HPWF apparatus is shown in Fig. 4 which also includes the reproduction of a WNC event. Another such event - centrally located in the target-detector - is shown in Fig. 5.

These experiments had progressed to the extent that by August, 1973, a review of the subject of weak neutral currents by G. Myatt[4] contained the material shown in Table 3 and later published separately by the two experimental teams[5,6]. It was particularly impressive that experiments using experimental methods so dissimilar and mean neutrino energies so different should independently

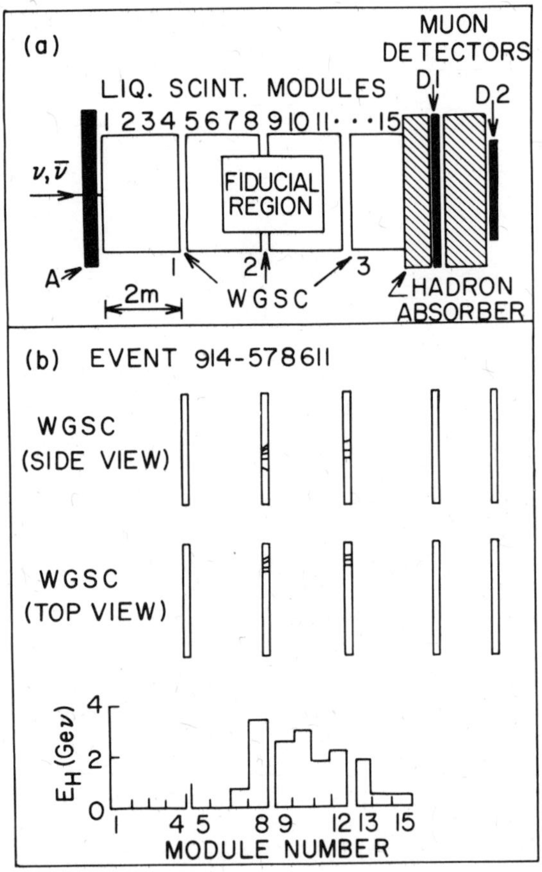

Fig. 4. (a) Schematic view of that portion of the target-detector used to identify neutral current and charged current neutrino interactions in the HPWF experiment. Not shown are the remaining three sections of iron toroidal magnet. (b) Display of the calorimeter and spark chamber information for a neutral current event. The side view of the wide gap spark chambers (WGSC) is from one of the 15° stereo views; the top view of the WGSC is from the 90° stereo view. The pulse heights in the calorimeter modules illustrate the development and absorption of the hadron shower, which had E_H = 16.5 GeV. (The gaps in the pulse height distribution show the location of the WGSC.) No particles appear in D1 or D2 behind the hadron absorber.

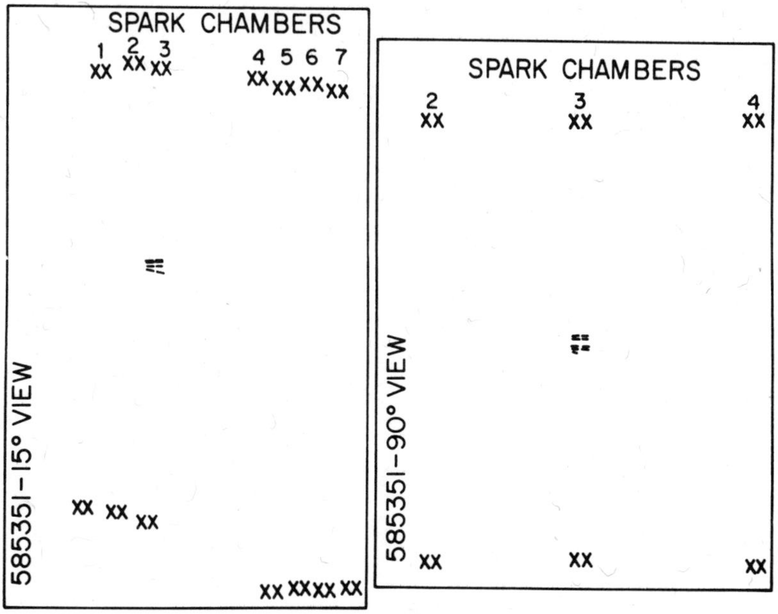

Fig. 5. Stereoscopic views of a WNC event centrally located in the HPWF apparatus. The X-marks are fiducial marks on the wide gap spark chambers. The actual locations of WGSC 1-5 are shown in Fig. 4.

Table 3. Extract of material from a talk entitled "Neutral Currents" by G. Myatt in Proc. Int'l. Symposium on Electron and Photon Interactions at High Energies, Bonn, 1973, p. 389.

2.2 Inclusive Semileptonic Neutrino Reactions

2.2.1 Gargamelle Experiment at CERN

$$(NC/CC)_{\nu} = 0.23 \pm 0.03$$

$$(NC/CC)_{\bar{\nu}} = 0.46 \pm 0.09$$

2.2.2 A New Result from NAL

$$\frac{\text{Events With No Muon}}{\text{Events With Muon}} = 0.26 \pm 0.07$$

for a "natural" mixture of ν and $\bar{\nu}$ from an unfocussed beam.

obtain a positive result of approximately the same magnitude for the semileptonic inclusive weak neutral current reactions. $\nu_\mu(\bar{\nu}_\mu) + \text{nucleon} \rightarrow \nu_\mu(\bar{\nu}_\mu) + \text{hadrons}$. It is fair to say that a new era in the study of weak interactions began then.

What has happened in the experimental sector in the four years since the initial experiments? A partial list of accomplishments is:

(i) Much more data has been acquired on the semileptonic inclusive reactions

$$\nu_\mu(\bar{\nu}_\mu) + N \rightarrow \nu_\mu(\bar{\nu}_\mu) + X \tag{1}$$

where N is a nucleon and X is the sum of hadronic final states.

(ii) The elastic semileptonic neutral current processes

$$\nu_\mu(\bar{\nu}_\mu) + p \rightarrow \nu_\mu(\bar{\nu}_\mu) + p \tag{2}$$

have been observed and measured roughly.

(iii) Measurements have been made of exclusive reactions, particularly single pion production, that bear directly on the isospin properties of the hadronic WNC.

(iv) Preliminary data on purely leptonic WNC reactions have been obtained.

(v) Ambitious programs to search for WNC induced parity non-conservation effects in atoms and nuclei, as well as in charged lepton scattering processes, have been started.

In this talk I shall concentrate on topics (i) and (ii), although there is significant overlap with certain of the other topics. These latter topics are in a somewhat more preliminary state than the first two topics. They will be important subjects at future schools.

2.2 Semileptonic Inclusive Neutral Current Measurements

Deep inelastic neutral current scattering of neutrinos and antineutrinos has been observed in many experiments using different techniques in the last four years. For example, in Fig. 6 is shown the method of distinguishing neutral current events from charged current events first employed in a CalTech-FNAL experiment which used a sampling type hadron calorimeter of alternate layers

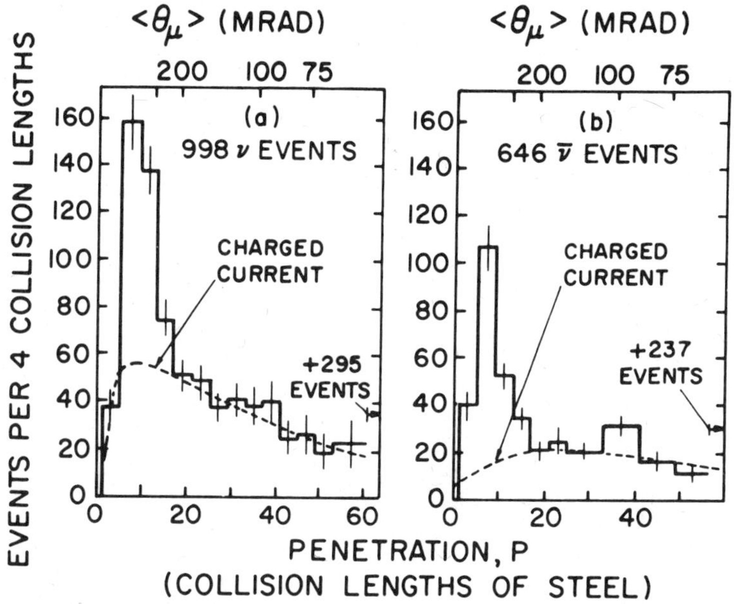

Fig. 6. Penetration distributions of events with no particles existing through the end of the apparatus for (a) ν, and (b) $\bar{\nu}$ from the CalTech-FNAL experiment (reference 7). The smooth curves show the expected shape from CC events alone (wide-band contribution is included). The typical CC muon angle is shown on the top scale.

of relatively thick iron plates and scintillator[7]. The penetration of particles makes possible a clear separation of the two event types. This method has also been adopted by the CERN-Dortmund-Heidelberg-Saclay (CDHS) experiment[8] in which the thick Fe plates of the hadron calorimeter are also magnetized to provide an integrated function detector. The CDHS data are shown in Fig. 7. A similar method using a different variable, the transverse momentum, $p_\perp$, of charged particles relative to the direction of the hadron cascade, yielded the data in Fig. 8 taken in the large (15') bubble chamber at FNAL[9]. Again, there is a clear separation (at low $p_\perp$) of the neutral current from the charged current events.

Refinements of technique by the Gargamelle collaboration have led to better measurement of the neutron background in that experiment and significantly improved data. Corresponding refinements in the HPWF experiment have led to better understanding of the systematic effects in that experiment as shown, for example, in Fig. 9a where the raw results for the ratio

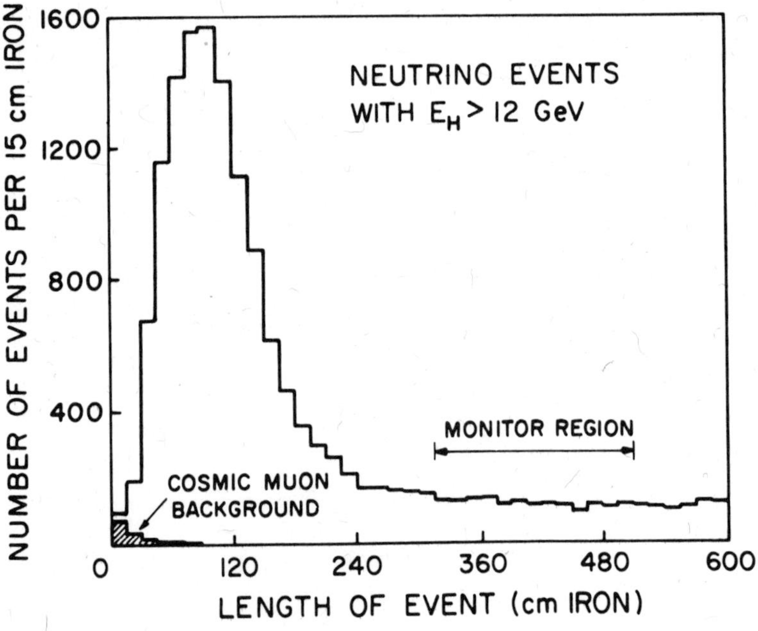

Fig. 7. Distribution in event length from the CDHS experiment (reference 8). Note the similarity to Fig. 6a.

$$R^{\nu}_{INC} = \frac{\sigma(\nu_{\mu} + \text{nucleon} \to \nu_{\mu} + \text{hadrons})}{\sigma(\nu_{\mu} + \text{nucleon} \to \mu^{-} + \text{hadrons})}$$

are given for the two muon detectors in Fig. 4. These detectors have very different properties: detector 1 subtends a larger solid angle for muons than does detector 2 and hence the correction for escaping muons is less for detector 1 than for detector 2; on the other hand, there is a correction of about -10% on average to the charged particles observed in detector 1 due to hadron penetration of the iron shield in front of detector 1, while that correction does not exist for detector 2. The raw results for R^{ν}_{INC} of Fig. 9a are therefore quite different for the two detectors, but the corrected results of Fig. 9b are essentially the same, except for a possible small residual systematic difference, demonstrating that the corrections applied to these data are moderately well understood.

It seems safe to conclude that the semileptonic inclusive reactions of Eq. (1) have in fact been observed using different methods in a number of experiments. A summary of the most recent results on R^{ν}_{INC} and $R^{\bar{\nu}}_{INC}$ is given in Table 4 which is derived in

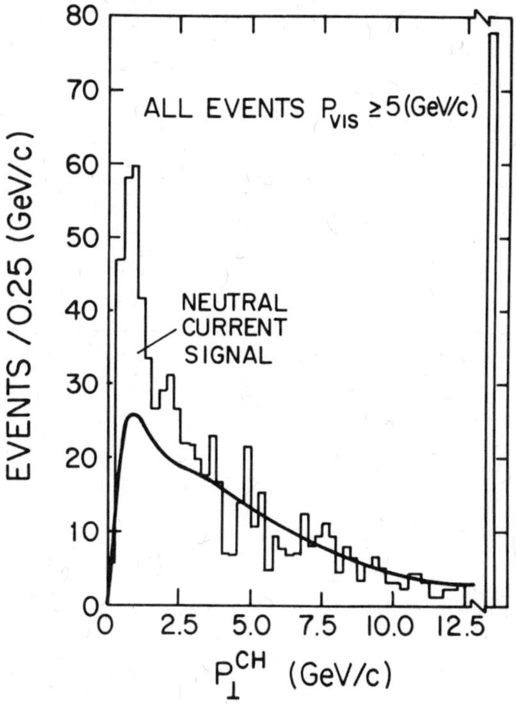

Fig. 8. Experimental distribution in transverse momentum of the charged particles in antineutrino events observed in the 15' FNAL bubble chamber (reference 9). The curve is the expected distribution for charged current events.

part from a recent review of this subject by P. Musset[10]. The agreement among the experiments is remarkable, even though the hadronic energy (E_H) cut-off is not the same for all the experiments. A measure of that effect in the HPWF data is shown by the two entries for $E_H > 0$ GeV and $E_H > 4$ GeV given for that experiment. The effect on $R_{INC}^{\bar{\nu}}$ will be smaller for most of the other high energy experiments which used a harder $\bar{\nu}$ spectrum than did HPWF.

The data in Table 4 bear directly on the space-time structure of the hadronic WNC. They may be used in the first instance to obtain the cross section ratio $\sigma_{NC}^{\bar{\nu}}/\sigma_{NC}^{\nu}$, since

$$\sigma_{NC}^{\bar{\nu}}/\sigma_{NC}^{\nu} = [R^{\bar{\nu}}(E_H > 0)/R^{\nu}(E_H > 0)](\sigma_{CC}^{\bar{\nu}}/\sigma_{CC}^{\nu}) \quad .$$

Table 4. Experimental values of R^{ν}_{INC} and $R^{\bar{\nu}}_{INC}$.

~Isoscalar Targets			
Expt.	R^{ν}_{INC}	$R^{\bar{\nu}}_{INC}$	E_H(GeV)
D_2BC(7'B)	0.25 ± 0.05	-	
GGM (C)	0.26 ± 0.04	0.39 ± 0.06	> 0
HPWF(F)	0.30 ± 0.04	0.33 ± 0.09	> 0
	0.29 ± 0.04	0.39 ± 0.10	> 4
CITFR(F)	0.28 ± 0.03	0.35 ± 0.11	> 12
CDHS (C)	0.29 ± 0.01	0.35 ± 0.03	> 12
BEBC (C)	0.33 ± 0.04	0.35 ± 0.09	> 15 (ν)
			> 30 ($\bar{\nu}$)
Proton Target			
FBHM (15'F)	0.48 ± 0.17		> 0
AC-MP(15'F)		0.49 ± 0.14	

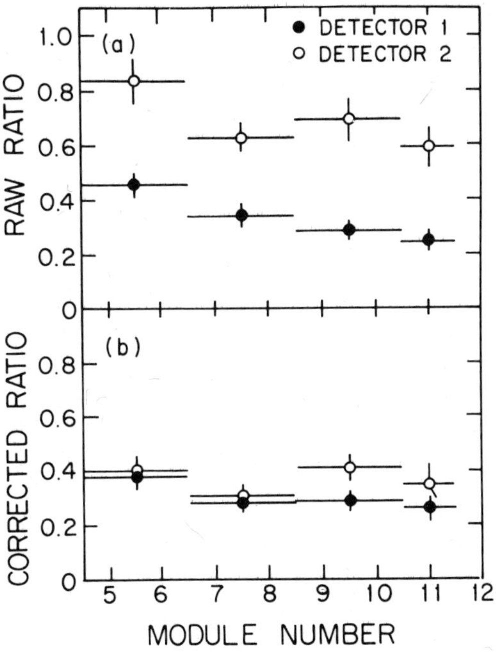

Fig. 9. The ratios of neutral current events to charged current events in the HPWF data (reference 11) plotted against longitudinal position (calorimeter module number). (a) Directly observed data and (b) after corrections for muon detection efficiency and hadron penetration.

To make the correction from $R^{\bar{\nu}}(E_H > E_C)/R^{\nu}(E_H > E_C)$, where E_C is the experimental cutoff in hadronic energy, to $R^{\bar{\nu}}(E_H > 0)/R^{\nu}(E_H > 0)$, we use the simple quark-parton model (QPM) with the measured values of $\sigma^{\bar{\nu}}_{CC}/\sigma^{\nu}_{CC}$, and assume various couplings for the WNC. This procedure[11] leads to corrected experimental values of $\sigma^{\bar{\nu}}_{NC}/\sigma^{\nu}_{NC}$ for each assumed WNC coupling, which may be directly compared with the value of $\sigma^{\bar{\nu}}_{NC}/\sigma^{\nu}_{NC}$ predicted for that type of coupling.

Thus, in the QPM, with $y = E_H/E_\nu$, the cross sections for the reactions in Eq. (1) are

$$\frac{d\sigma^{\nu}}{dy} = \frac{G^2 M E_\nu}{\pi}\left[a + b(1-y)^2 + cy^2\right]$$

$$\frac{d\sigma^{\bar{\nu}}}{dy} = \frac{G^2 M E_{\bar{\nu}}}{\pi}\left[b + a(1-y)^2 + cy^2\right] \qquad (3)$$

where, for example,

(i) any V, A combination implies c = 0

(ii) pure V or pure A coupling implies a = b

(iii) pure S or pure T coupling implies a = b = 0

(iv) pure T coupling implies a = b = c.

The total cross sections for any assumptions concerning a, b and c are obtained by integrating over y.

A comparison of the corrected measured values $\sigma_{NC}^{\bar{\nu}}/\sigma_{NC}^{\nu}$ from the HPWF experiment[11] with the expected (QPM) values is shown in Table 5. Observe that the couplings (V+A) pure S or pure P, or (S,P) or (S,P)T mixtures are appreciably less likely than the other coupling forms in Table 5. Also pure V (or A) and pure T are less probable than (V-A) by about 3 standard deviations.

Similar conclusions are obtained from the distributions in E_H measured in the HPWF experiment[11] when compared with the distributions expected for different couplings. There are shown in Fig. 10

Table 5. Measured and expected values $\sigma_{NC}^{\bar{\nu}}/\sigma_{NC}^{\nu}$ from the HPWF experiment.

Form of WNC	Corrected Experimental Value	Expected Value
V-A	0.61 ± 0.25	0.38
V or A	0.40 ± 0.17	1.00
V+A	0.37 ± 0.16	2.65
S or P*	0.26 ± 0.09	1.00
(S,P)T	0.31 ± 0.11	1.00
T	0.45 ± 0.15	1.00

* or (S,P) mixture.

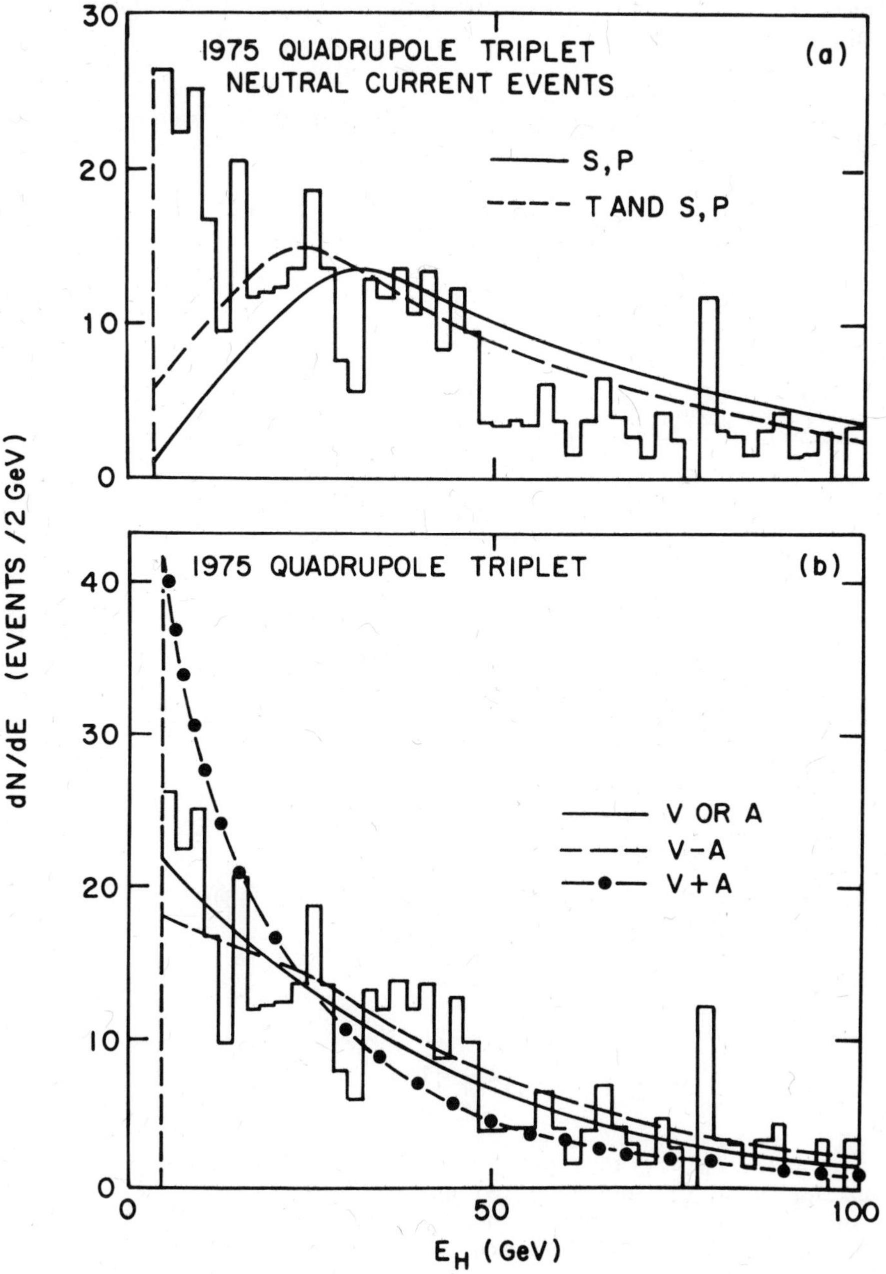

Fig. 10. Plots of the distributions in hadron energy (E_H) of the WNC events observed in the HPWF experiment (reference 11) compared with the distributions expected for different types of coupling.

comparisons which confirm the lower probability of S, P and T combinations and of (V+A).

If, somewhat arbitrarily, we restrict the WNC amplitudes to those involving only V and A and isospins 0 and 1, we can write in place of Eq. (3), providing scale invariance violating effects are neglected,

$$\frac{d\sigma^{\nu}_{NC}}{dy} = \frac{G^2_{NC} M E_{\nu}}{\pi} [g_L + (1-y)^2 g_R]$$

and (4)

$$\frac{d\sigma^{\bar{\nu}}_{NC}}{dy} = \frac{G^2_{NC} M E_{\bar{\nu}}}{\pi} [g_R + (1-y)^2 g_L]$$

with

$$g_L \sim |\varepsilon_L|^2 + \frac{\bar{Q}}{Q} |\varepsilon_R|^2 \qquad (5)$$

$$g_R \sim |\varepsilon_R|^2 + \frac{\bar{Q}}{Q} |\varepsilon_L|^2$$

and

$$|\varepsilon_L|^2 = |\varepsilon_L(u)|^2 + |\varepsilon_L(d)|^2$$

$$= |\tfrac{1}{4}(\alpha + \beta + \gamma + \delta)|^2 + |\tfrac{1}{4}(-\alpha - \beta + \gamma + \delta)|^2 \qquad (6)$$

$$|\varepsilon_R|^2 = |\varepsilon_R(u)|^2 + |\varepsilon_R(d)|^2$$

$$= |\tfrac{1}{4}(\alpha - \beta + \gamma - \delta)|^2 + |\tfrac{1}{4}(-\alpha + \beta + \gamma - \delta)|^2$$

where $\varepsilon_R(u)$, $\varepsilon_R(d)$, etc. are the strengths of the weak neutral coupling to up (u) and down (d) quarks, respectively, $\bar{Q}$ is the antiquark ($\bar{u}$ and $\bar{d}$) content and Q the quark (u and d) content of the nucleon, and α, β, γ, δ are coupling coefficients involving the space-time and isospin structures: α, vector-isovector; β, axial-isovector; γ, vector-isoscalar; δ, axial-isoscalar. Detailed analyses have been carried out in several papers[12] to extract the combination of α, β, γ and δ that best fits the WNC data, including single pion production data and data on the elastic scattering reactions in Eq. (2).

For our purpose here we limit the discussion to a determination of the values of g_L and g_R from which we can draw directly some very general conclusions regarding the WNC coupling. We show in Table 6 the experimental values of g_L and g_R (normalized so that $g_L + g_R = 1$). The weight of this evidence strongly indicates that pure right-handed ($g_R = 1$, $g_L = 0$: V+A) coupling is ruled out, and also that pure V or pure A coupling ($g_L = g_R = 0.5$) is ruled out. Furthermore, these data suggest that pure left-handed coupling ($g_L = 1$, $g_R = 0$: V-A) is disfavored, but the strength of this conclusion is somewhat less in view of the possibility of scale breaking corrections even at low energy.

In Fig. 11 is shown the plot of $R^{\bar{\nu}}_{INC}$ against R^{ν}_{INC} for the data in Table 4 and also the prediction of the SU(2) × U(1) gauge theory model of Weinberg[13] and Salam[14] with $\bar{Q}/Q = 0.1$ and $\bar{Q}/Q = 0.14$. The agreement of the data with this model is seen to be excellent. The value of the single parameter of the model, $\sin^2\theta_W$, is determined from Fig. 10 to be in the interval

$$0.25 \leq \sin^2\theta_W \leq 0.30 \ .$$

In view of small systematic differences among the experiments, it seems better to present limits on $\sin^2\theta_W$ in this way rather than to give in the usual way a value with an assigned error.

Finally, we present in Fig. 12 the ratio R^{ν}_{INC} as a function of average neutrino energy $\langle E_\nu \rangle$ of the beams in the several experiments of Table 4. We see that R^{ν}_{INC} is essentially independent of $\langle E_\nu \rangle$, indicating that the cross section for inelastic neutral current neutrino scattering tracks the approximate linear dependence on laboratory neutrino energy of inelastic charged current neutrino scattering. Without attempting here a detailed fit to the data of Fig. 12, it is clear that a limit on the mass of the weak neutral current propagator (the Z^0 in the SU(2) × U(1) model) can be determined roughly and is greater than about 10 GeV/c^2. It is therefore

Table 6. Experimental values of g_L and g_R.

Expt.	g_L	g_R
GGM	0.77 ± 0.16	0.23 ± 0.07
HPWF	0.85 ± 0.12	0.15 ± 0.12
CITRF	0.64 ± 0.09	0.36 ± 0.09
CDHS	0.85 ± 0.03	0.15 ± 0.03

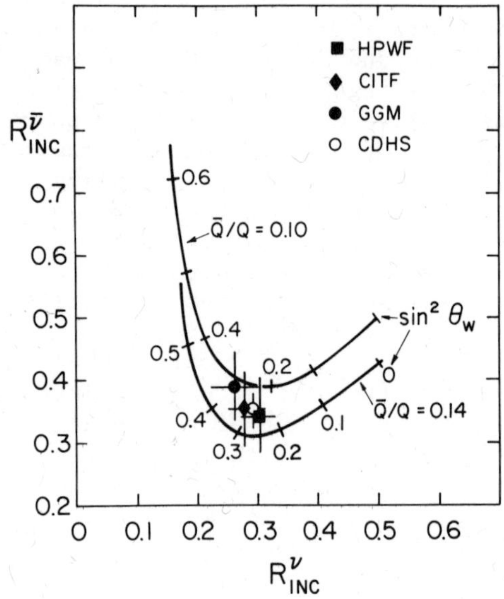

Fig. 11. Plot of $R_{INC}^{\bar{\nu}}$ against R_{INC}^{ν} for much of the data of Table 4. The solid curves are calculated for the SU(2) × U(1) model of Weinberg and Salam for antiquark to quark ratios of 0.1 and 0.14.

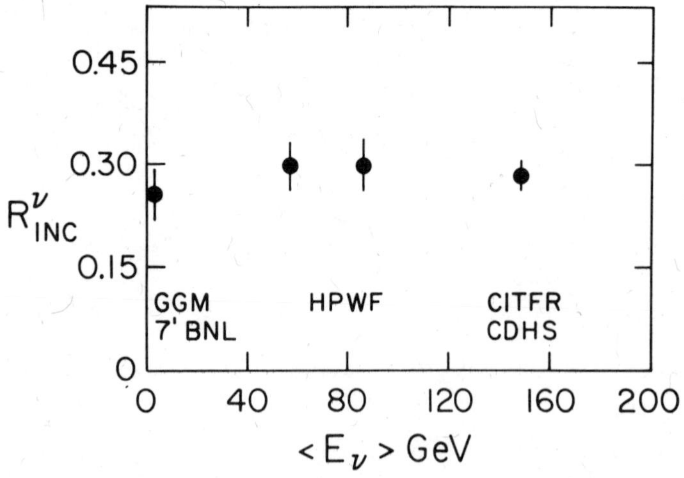

Fig. 12. Plot of R_{INC}^{ν} for different values of $<E_\nu>$.

unlikely that the structure observed at 9.4 GeV/c^2 in the invariant mass distribution of dimuons from proton-nucleus collisions[13] is due to the production of a Z^o.

2.3 Semileptonic Elastic Scattering Measurements

We turn now to measurements of the elastic scattering of neutrinos and antineutrinos by protons, the reactions of Eq. (2). Three experiments, utilizing ν_μ and $\bar{\nu}_\mu$ in the 0.5 to 10 GeV energy region, have yielded data on these reactions: the Columbia-Illinois-Rockefeller (CIR) experiment[14] and the Harvard-Pennsylvania-Wisconsin (HPW) experiment[15], both at BNL, and, very recently, the Aachen-Padova experiment[16] at CERN.

There are several advantages to the study of elastic scattering. Theoretically, the reaction is simple enough to hold the promise that a model independent analysis of elastic scattering data may be made. Even if that promise is not fully realized, the predictions of current models are unambiguous. Experimentally, the differential cross section with respect to the square of the momentum transferred to the proton, $d\sigma/dq^2$, may be obtained directly from measurement of the kinetic energy T of the recoiling proton, since $|q^2| = 2m_pT$. Furthermore, comparison of the ν and $\bar{\nu}$ total cross sections leads to the nature of the hadronic weak neutral current coupling; for example, equality of the cross sections implies pure V (or A) coupling, while any difference of the cross sections means that polar vector-axial vector interference is present.

It is interesting to compare the apparatus of the CIR experiment, shown in Fig. 13, with that of the HPW experiment, outlined in Fig. 14.

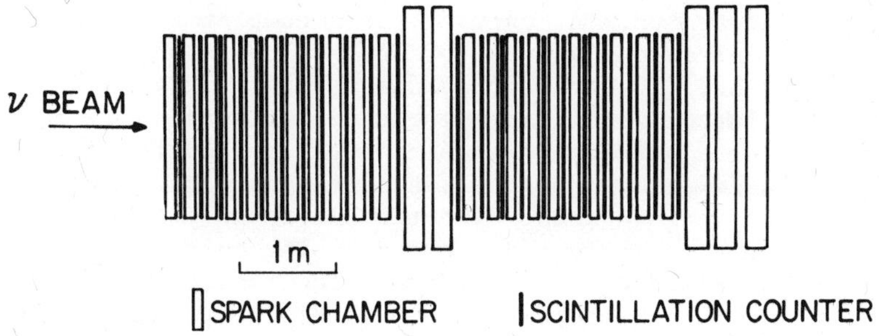

Fig. 13. Schematic representation of the Columbia-Illinois-Rockefeller detector.

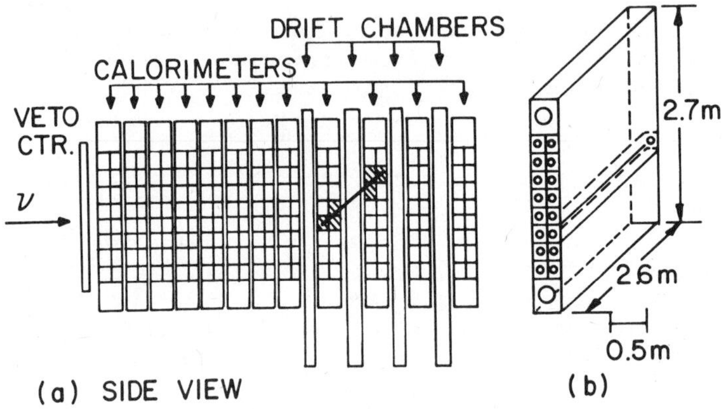

Fig. 14. (a) Side view of the Harvard-Pennsylvania-Wisconsin detector showing a typical recoiling proton. (b) Details of a single calorimeter module.

The former arrangement utilizes thick plate optical spark chambers and liquid scintillator counters as the neutrino target. The spark chambers provide good visual identification of an event and good angular resolution. The liquid scintillation counters allow precision timing of an event relative to the radiofrequency structure of the extracted proton beam from the BNL synchrotron, and also measurement of dE/dx, but no measurement of the total energy of a final state charged particle is available. The CIR apparatus is unshielded, i.e., no massive shielding is nearby, and relies on time-of-flight to discriminate against neutron-induced background. The apparatus is approximately 1.8 m × 1.8 m in transverse dimensions and 28 tons in total mass, with 8 tons of fiducial volume. The Aachen-Padova apparatus is similar to that of CIR; it is described in detail in Professor Faissner's talk at the Ben Lee Memorial Conference[16].

The HPW apparatus, in contrast, is a completely active liquid scintillator ionization calorimeter of total mass 33 tons and fiducial mass 10 tons. The transverse dimensions are 2.6 m × 2.7 m. It is divided into 192 optically isolated cells in which dE/dx is measured along the particle path, the total energy is obtained from summation of dE/dx along the path, and the angle of the trajectory relative to the incident beam direction is roughly determined. To verify the method of angle measurements using only the cell structure of the calorimeter, the downstream section of the calorimeter includes four drift chambers as shown in Fig. 14, which are utilized as illustrated in Fig. 15. Also illustrated in Fig. 15 is the method of particle identification by means of the dE/dx measurement applied to an actual event. The basic idea behind this ionization

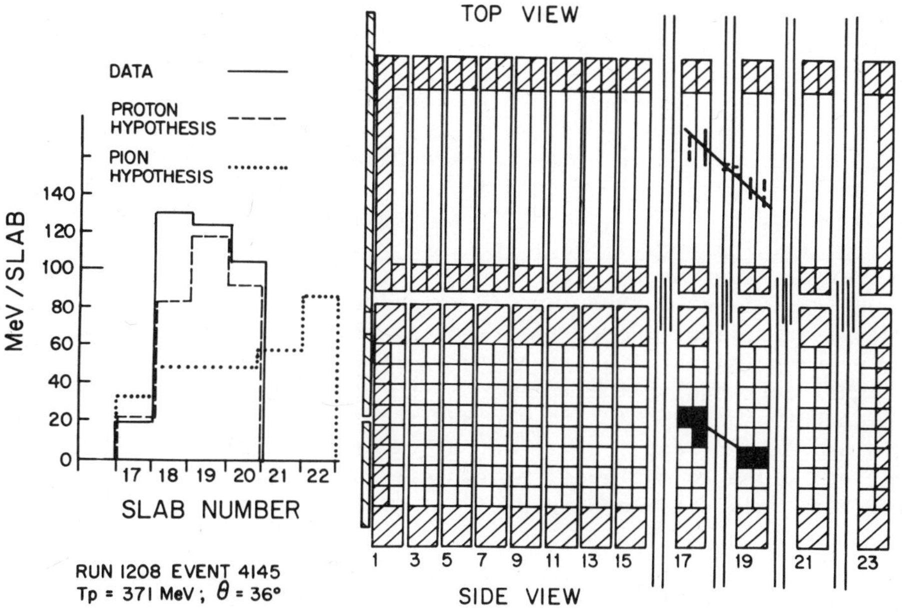

Fig. 15. Illustration of the method of measurement of proton angle and energy in the HPW experiment. Shaded areas are the active shield of the fiducial volume. Trajectory location in the top view is done by both pulse height and time comparison. Also shown is the method of particle identification for an actual event.

calorimeter is that the fiducial (inner) volume is actively shielded by the exterior liquid scintillator elements as shown in Fig. 15. Because of its sensitivity to neutrons and low energy charged particles, this apparatus is protected by a massive, passive shield, and relies on time-of-flight as well. Furthermore, much of the out-of-time neutron background and a part of the in-time neutron background (arising from neutrino interactions in the passive shield) is eliminated by observing a second interaction of the neutron as it traverses the large volume of the detector and the active shield; an energy deposition in the detector or shield larger than 2 MeV serves to veto (off-line) an event if it is spatially separated from the single charged particle trajectory that constitutes a useful event.

The time-of-flight spectrum of events identified as $\nu_\mu + p \to \nu_\mu + p$ in the HPW experiment is shown in Fig. 16a. This is to be compared with the corresponding spectrum for the charged current elastic events, i.e., from $\nu_\mu + n \to \mu^- + p$, shown in Fig. 16b. The spatial distributions (in x, y and z) of the WNC events are shown in Fig. 17; the uniformity of these distributions is evidence that most events are neutrino-induced rather than neutron-induced.

Backgrounds in the CIR, Aachen-Padova and HPW experiments due to other neutral current reactions, e.g., $\nu_\mu + n \to \nu_\mu + n$ with $n + p \to n + p$, $\nu_\mu + p \to \nu_\mu + p + \pi^0$, etc., are substantial [(≈25 to 35)% of the total observed rate], but are different in detail for the different experiments. For example, there is a net correction of roughly -10% in the HPW and CIR experiments and -40% in the Aachen-Padova experiment due to the reaction chains $\nu_\mu + n \to \nu_\mu + n$ with $n + p \to n + p$ (signal increase) and $\nu_\mu + p \to \nu_\mu + p$ with $p + n \to p + n$ (signal decrease). The $\nu_\mu + p \to \nu_\mu + p + \pi^0$ channel, however, contributes between 5 and 10% to the HPW total signal but less to the CIR and Aachen-Padova total signal because photons from π^0 that convert close to the event origin are difficult to recognize in the coarser grained HPW apparatus.

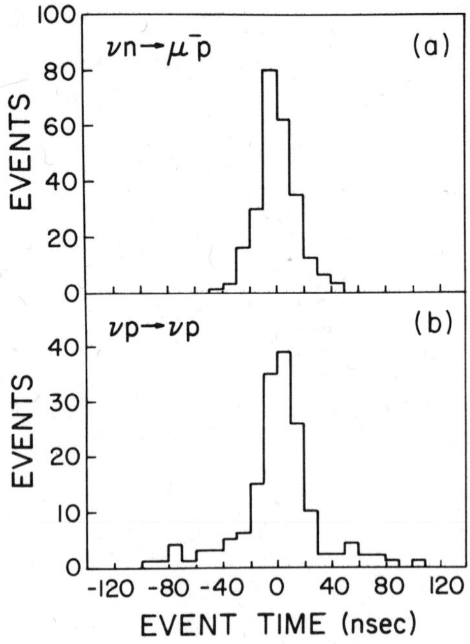

Fig. 16. Distribution of event times for (a) neutral current and (b) charged current elastic events in the HPW experiment. Event times are measured with respect to the 12 radio-frequency bunches of the AGS beam; each bunch is 50 ns fwhm, and bunch spacing is 220 ns.

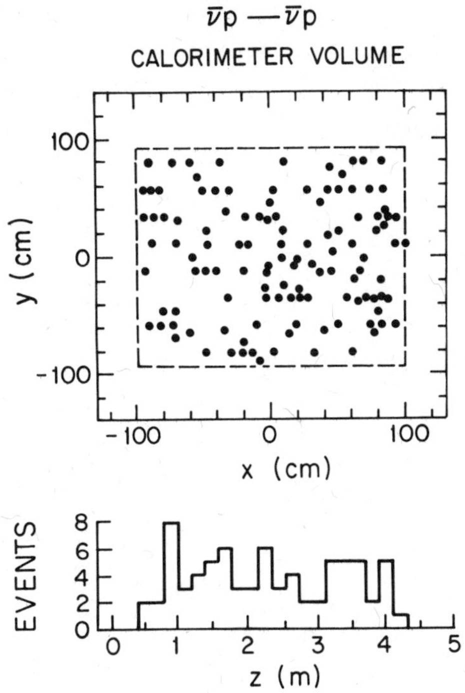

Fig. 17. Distributions of $\nu p \to \nu p$ event origins in the HPW data. Coordinate Z is along the incident beam direction. The dashed line indicates the fiducial volume.

Despite differences in apparatus, signal identification, and backgrounds, the CIR and HPW experiments have obtained values of $R^{\nu}_{el} = \sigma(\nu_{\mu} + p \to \nu_{\mu} + p)/\sigma(\nu_{\mu} + n \to \mu^- + p)$ that are in good agreement within experimental errors, as shown in Table 7. A preliminary value of R^{ν}_{EL} from the Aachen-Padova experiment, presented at this conference, is included in Table 7. An independent experiment by HPW has also determined the value of $R^{\bar{\nu}}_{EL}$ given in Table 7. It should be emphasized that corrections to the data of Table 7, particularly background subtractions, are as yet very crude.

Plots of $d\sigma/dq^2$ for $\nu_{\mu}(\bar{\nu}_{\mu}) + p \to \nu_{\mu}(\bar{\nu}_{\mu}) + p$ obtained by HPW are shown in Figs. 18a and 18b, which also give the q^2 dependence for the charged elastic processes $\nu_{\mu} + n \to \mu^- + p$ and $\bar{\nu}_{\mu} + p \to \mu^+ + n$. Although the data in Figs. 18a and 18b are rough, they do suggest that larger data samples acquired by this experimental method might provide direct, independent determinations of $\sin^2\theta_W$ from the shapes of the q^2 distributions, and perhaps make possible experimental

Table 7. Measured values R_{EL}^{ν} and $R_{EL}^{\bar{\nu}}$ for $0.3 \leq |q^2| \leq 0.9$ $(GeV/c)^2$.

Experiment	Quantity	Value	Reference
CIR	R_{EL}^{ν}	0.23 ± 0.09	14
HPW	R_{EL}^{ν}	0.17 ± 0.05	15
Aa-Pa	R_{EL}^{ν}	0.17 ± 0.04	16
HPW	$R_{EL}^{\bar{\nu}}$	0.2 ± 0.1	15

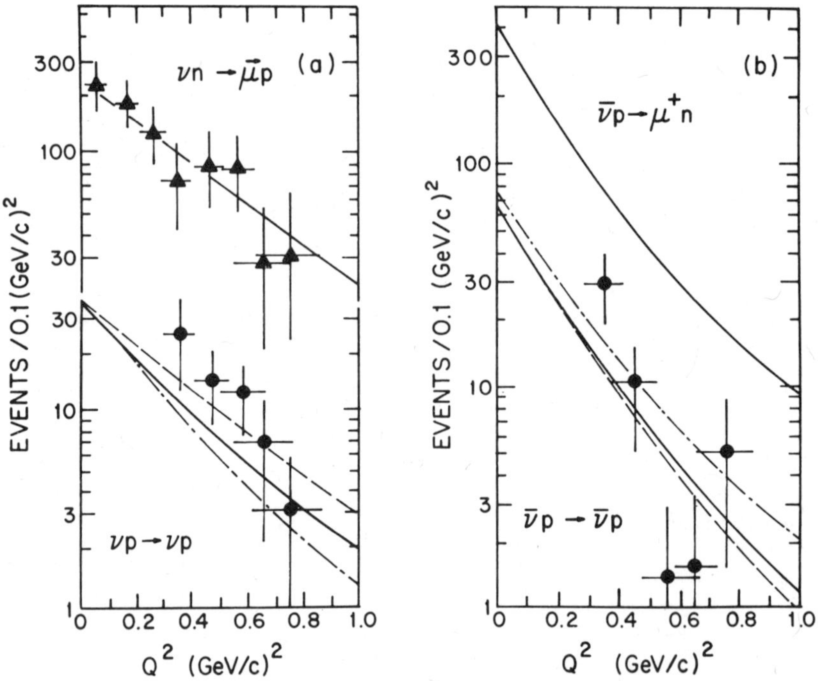

Fig. 18. (a) Observed q^2-distributions for charged and neutral current elastic scattering of neutrinos from the HPW experiment (reference 15). The data have been corrected for geometric acceptance and the $\nu p \to \nu p$ data have had a uniform background of 23% subtracted. The solid line for $\nu n \to \mu^- p$ is calculated for $m_A = 0.9$ GeV/c^2 and $m_V = 0.84$ GeV/c^2. Predictions of the Weinberg-Salam model for $\nu p \to \nu p$ correspond to $\sin^2\theta_W = 0.2$ (dashed line), 0.3 (solid line), and 0.4 (dash-dot line). (b) Observed distributions for antineutrino scattering. A background of 35%, uniform in q^2, has been subtracted. Weinberg-Salam model predictions are for $\sin^2\theta_W = 0.2$ (dashed line), 0.3 (solid line), and 0.4 (dash-dot) line).

confirmation that $d\sigma/dq^2$ is equal for ν and $\bar{\nu}$ at $q^2 = 0$, which must hold in any model if charge symmetry invariance is valid.

A useful comparison of experiment and theory[17] is given in Fig. 19 where the ratio of WNC cross sections $\sigma(\bar{\nu}_\mu + p \to \bar{\nu}_\mu + p)/\sigma(\nu_\mu + p \to \nu_\mu + p)$ is plotted against $\sin^2\theta_W$. For $0.25 \le \sin^2\theta_W \le 0.30$, as suggested by the R^{ν}_{INC} and $R^{\bar{\nu}}_{INC}$ data, the elastic data are consistent with the prediction of the Weinberg-Salam SU(2) × U(1) model (allowing for reasonable uncertainties in nuclear form factors) and approximately 3 standard deviations from pure V coupling.

2.4 Summary and Prospects

We can summarize the conclusions to be drawn from the material presented here as follows:

(i) The space-time structure of the hadronic weak neutral current appears to be an admixture of polar vector (V) and axial vector (A), as is the structure of the hadronic weak charged current. Although pure scalar (S), pseudoscalar (P) and tensor (T) couplings are

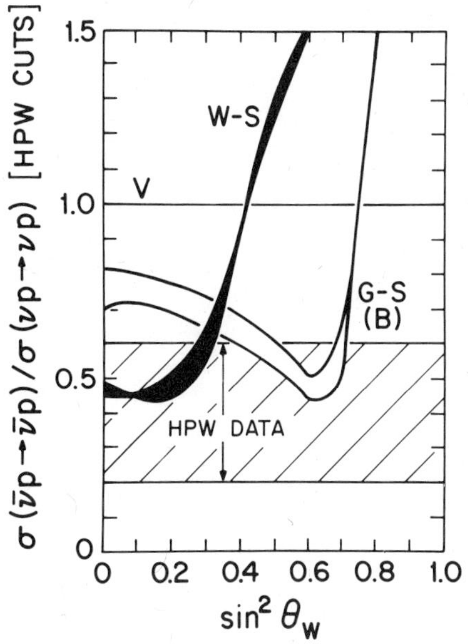

Fig. 19. Comparison of the ratio $\sigma(\bar{\nu}p \to \bar{\nu}p)/\sigma(\nu p \to \nu p)$ measured in the HPW experiment with the ratio predicted in the Weinberg-Salam model and the Gursey-Sikivie model. Graph is from reference 17.

eliminated for both types of weak currents, as are mixtures of only S and P, it is not possible to rule out unequivocally a mixture of S, P and T that would reproduce a given (V,A) mixture. Nor is it possible to rule out small contributions of S, P or T to a dominant (V,A) mixture.

(ii) Assuming that only V and A are present, the data indicate that
- (a) pure (V+A) is ruled out with very high probability
- (b) pure V, pure A are ruled out with high probability
- (c) pure (V-A) is unlikely.

(iii) In any theory of weak interactions with no more than one real neutral vector boson, the presence of both V and A couplings directly implies parity nonconservation in the hadronic weak neutral current.

(iv) The (V,A) admixture determined by experiment is consistent with that of the SU(2)× U(1) model with

$$0.25 \leq \sin^2\theta_W \leq 0.30,$$

specified largely by the inelastic data.

(v) The dependence on energy of the cross section for the inclusive reactions

$$\nu_\mu + N \rightarrow \nu_\mu + X$$

is roughly linear up to about 150 GeV.

It may also be useful to mention the immediate prospects for improvements in the data we have discussed.

(i) We can expect soon to have more precise semileptonic elastic and inelastic differential and total cross section data that will lead directly to
- (a) a unique determination of the left and right-handed weak couplings to u and d quarks
- (b) precise values of $\sin^2\theta_W$ from both inelastic and elastic neutrino scattering data
- (c) detailed and precise tests of scale invariance of weak neutral current phenomena.

(ii) We may hope also to have more intensive searches for flavor changing in weak neutral current processes.

REFERENCES

*) Supported in part by the U.S. Department of Energy.

†) Each section of Professor Mann's lectures has a self contained bibliography, tables and figures.

1) D.H. Perkins, Proc. XVI Int'l. Conf. on High Energy Physics, Chicago-Batavia, J.D. Jackson and A. Roberts, eds., Vol. 4, p. 189, 1972.

2) S. Weinberg, Phys. Rev. Lett. 19, 1264 (1967) and 27, 1688 (1972); Phys. Rev. D5, 1421 (1972); A. Salam and J.C. Ward, Phys. Lett. 13, 168 (1964).

3) See, for example, the talk by J.D. Bjorken at the Ben Lee Memorial Int'l. Conference on Parity Nonconservation, Weak Neutral Currents and Gauge Theories, FNAL, October, 1977.

4) G. Myatt, Proc. Sixth Int'l. Symposium on Electron and Photon Interactions at High Energies, Bonn, 1973, H. Rollnik and W. Pfell, eds. (North-Holland, Amsterdam, 1974), p. 389.

5) F.J. Hasert et al., Phys. Lett. 46B, 138 (1973).

6) A. Benvenuti et al., Phys. Rev. Lett. 32, 800 (1974).

7) B.C. Barish et al., Phys. Rev. Lett. 34, 538 (1975).

8) M. Holder et al., CERN preprint, August (1977).

9) Data from the ANL-Carnegie Mellon-Purdue collaboration presented by A.F. Garfinkel at the 1977 Int'l. Symposium on Lepton and Photon Interactions at High Energies, Hamburg (1977).

10) P. Musset, Review talk given at the 1977 Int'l. Symposium on Lepton and Photon Interactions at High Energies, Hamburg (1977).

11) A. Benvenuti et al., Phys. Rev. Lett. 37, 1039 (1976); P. Wanderer et al., HPWF preprint 77/1 (1977), submitted to Phys. Rev. D.

12) See, for example, J.D. Bjorken, Proc. of Summer Institute on Particle Physics, November, 1976, C. Zipf, ed., SLAC Report No. 198; L.M. Sehgal, Aachen preprint (1977); P.Q. Hung and J.J. Sakurai, preprint UCLA/77/TEP/17 (1977).

13) W.R. Innes et al., Phys. Rev. Lett. 39, 1240 (1977).

14) W. Lee et al., Phys. Rev. Lett. 37, 186 (1976).

15) D. Cline et al., Phys. Rev. Lett. 37, 252, 648 (1976).

16) Data presented at the Ben Lee Memorial Int'l. Conference on Parity Nonconservation, Weak Neutral Currents and Gauge Theories, FNAL, October, 1977.

17) C.H. Albright, C. Quigg, R.E. Schrock and J. Smith, Phys. Rev. D14, 1780 (1976).

3. INELASTIC NEUTRINO-NUCLEON SCATTERING
(A partial summary through criticism of recent data)

A series of experiments on the inelastic scattering of high energy neutrinos (ν) and antineutrinos ($\bar{\nu}$) by nucleons was carried out at Fermilab during the past few years by physicists from Harvard, Pennsylvania, Wisconsin and Fermilab (HPWF collaboration). The relevant processes were

$$\nu_\mu(\bar{\nu}_\mu) + N \to \mu^-(\mu^+) + X \qquad (1)$$

where N is an isoscalar target and X is any hadron state. One result of the experiments was the observation of an apparent discrepancy when the $\bar{\nu}$ data were compared to theoretical predictions based on the body of weak interaction data in general and on lower energy ν and $\bar{\nu}$ scattering experiments in particular[1,2,3]. Within experimental error, the ν data showed no such discrepancy, which was taken as evidence for the correctness of the experimental method and data. Subsequently, because of the interest in this $\bar{\nu}$ discrepancy, it came to be called the "high-y anomaly" by some theoreticians, after the Bjorken scaling variable $y = (E_{\bar{\nu}} - E_\mu)/E_{\bar{\nu}}$ which is a measure of the inelasticity of an interaction.

Each of the new, more recent experiments on high energy ν and $\bar{\nu}$ interactions at Fermilab, and now at CERN, has sought to study the y-anomaly. For various reasons there have been insufficient data from these experiments to confirm or repudiate the effect. This situation is changing, and experiments using higher intensity proton beams as neutrino sources and a variety of experimental methods are now yielding data bearing directly on the y-anomaly. The results of one of these investigations are in circulation as a preprint entitled "Is There a High-Y Anomaly in Antineutrino Interactions?" by M. Holder et al. (CDHSB collaboration)[4]. The authors of this paper conclude that there is no y-anomaly, and implicitly interpret their data as confirming scale invariance, charge symmetry invariance and the simple quark-parton model (QPM).

We have studied the paper of Holder et al., and it is our opinion that serious internal inconsistencies in the CDHSB data call into question the validity of any conclusions obtained from those data. These inconsistencies suggest - in view of the small statistical uncertainties of the CDHSB data - that there exist systematic errors in their experiment which they have failed to take into account. We proceed here to note and discuss these inconsistencies.

Briefly, the simple theoretical framework with which the early data were compared is as follows. In terms of the variables $x = Q^2/2M_N E_H$ and $y = E_H/E_\nu$, the scale invariant differential cross section for process (1) can be written as

$$\frac{d\sigma^{\nu}}{dy} = K_{\nu}\left\{1 - (1-B_{\nu})y + (1-B_{\nu})\frac{y^2}{2}\right\} \quad ; \tag{2}$$

$$\frac{d\sigma^{\bar{\nu}}}{dy} = K_{\bar{\nu}}\left\{1 - (1+B_{\bar{\nu}})y + (1+B_{\bar{\nu}})\frac{y^2}{2}\right\} \quad , \tag{3}$$

where the Callan-Gross relation, $2xF_1(x) = F_2(x)$, was assumed. The parameters $K_{\nu,\bar{\nu}}$ and $B_{\nu,\bar{\nu}}$ are related to the structure functions F_2 and F_3, which are functions of x only, if Bjorken scaling is assumed, by the following definitions:

$$K_{\nu,\bar{\nu}} = \frac{G^2 M_N E_{\nu,\bar{\nu}}}{\pi} \int F_2^{\nu,\bar{\nu}}(x)\,dx \quad ; \tag{4}$$

$$B_{\nu,\bar{\nu}} = -\frac{\int xF_3^{\nu,\bar{\nu}}(x)\,dx}{\int F_2^{\nu,\bar{\nu}}(x)\,dx} \quad . \tag{5}$$

Charge Symmetry Invariance (CSI) means that when scattering on an isoscalar ($I=0$) target, $F_i^{\nu}(x) = F_i^{\bar{\nu}}(x)$; $i = 1,2,3$. Therefore, CSI implies that i) $K_{\nu} = K_{\bar{\nu}}$, ii) $B_{\nu} = B_{\bar{\nu}}$. Relation i) can be checked by comparing $d\sigma/dy$ at $y=0$ between ν_{μ} and $\bar{\nu}_{\mu}$ data while relation ii) can be checked by studying the shape of y-distributions independent of normalization. In the limiting case $B_{\nu} = B_{\bar{\nu}} = 1$, equations (2) and (3) reduce to the well known forms

$$\frac{d\sigma^{\nu}}{dy} = K_{\nu} \quad ; \quad \frac{d\sigma^{\bar{\nu}}}{dy} = K_{\bar{\nu}}(1-y)^2 \quad . \tag{6}$$

In terms of the Quark Parton Model (QPM) this corresponds to the limit in which there is no antiparton ($\bar{Q}$) in the nucleon, as can be seen by the relation

$$\frac{\bar{Q}}{Q+\bar{Q}} = \frac{1}{2}(1 - B) \quad . \tag{7}$$

The y-anomaly can be summarized empirically by the following statements: i) a decrease with increasing energy of the values of $B_{\bar{\nu}}$ for the $\bar{\nu}_{\mu}N$ data[2], manifesting the breakdown of scale invariance; and ii) inequality of B_{ν} and $B_{\bar{\nu}}$ at high energy, and in particular at low x, equivalent to an effective violation of charge symmetry invariance[1]. It was observed[1,2] that the shape of the y-distribution for $\bar{\nu}_{\mu}N$ scattering was flatter at high energy than at low energy, as shown in Figs. 1a and 1b.

The following statements are directly from the CDHSB paper (reference 4):

A) The geometric acceptance in y is near <u>unity</u> for all but the upper end of the y-distribution.

B) Charge symmetry invariance is valid within 10%.

C) Even for $x < 0.1$, the y-distributions are compatible with the picture that valence quarks dominate, but the sea quarks concentrate at small x.

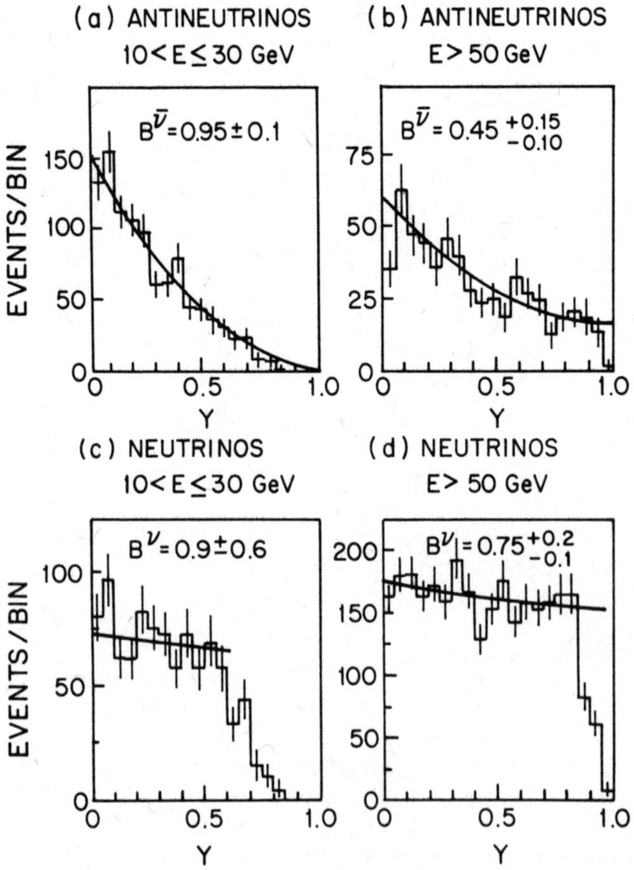

Fig. 1. y-distributions for $x < 0.6$ published by the HPWF collaboration.

D) There is no energy dependence of $<y>^{\bar{\nu}}$ or of $B_{\bar{\nu}}$.

Based on the statements (B), (C) and (D) the authors conclude that there is no y-anomaly.

The CDHSB data of reference 4 are shown in Fig. 2. Figs. 2a and 2b show the normalized y-distributions for ν_μ and $\bar{\nu}_\mu$ scattering with no cut on x. Since the acceptance in y is flat as mentioned in statement (A), it is a simple procedure to estimate the $B_{\nu,\bar{\nu}}$ values given by the data, by using equations (2) and (3) with the constraint $K_\nu = K_{\bar{\nu}}$. The values so obtained are $B_\nu \approx 0.3$ and $B_{\bar{\nu}} \approx 0.8$

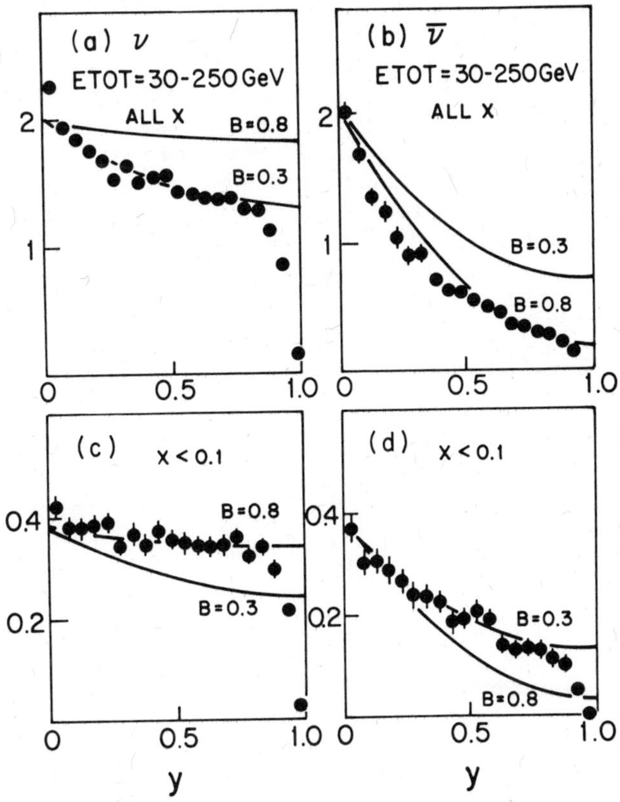

Fig. 2. y-distributions from a recent experiment at CERN by the CDHSB collaboration. Curves for different B values are drawn by the HPWF collaboration.

for all x. Given the statistical precision of the points shown in Fig. 2, these data therefore clearly indicate that $B_\nu \neq B_{\bar{\nu}}$. This contradicts statement (B) of reference 4, since C.S.I directly implies $B_\nu = B_{\bar{\nu}}$.

The y-distributions at $x < 0.1$, from reference 4 are shown in Figs. 2c and 2d for ν_μ and $\bar{\nu}_\mu$, respectively. The estimated B values, contrary to those obtained from the data for all x, gives $B_\nu \approx 0.8$ and $B_{\bar{\nu}} \approx 0.3$. Table 1 is a tabulation of the B values from the CDHSB ν_μ and $\bar{\nu}_\mu$ data for different x regions. Using equation (7), we find that the ν_μ data of reference 4 imply that a) the total antiparton content is $\frac{1}{2}(1 - 0.3) = 35\%$, and b) that almost all of the antipartons appear in the large x region! This clearly invalidates statement (C) which was made in reference 4. Furthermore, observe that $\bar{\nu}_\mu$ data from the same experiment show the exact opposite effect.

We turn now to the plot of $\langle y \rangle$ as a function of ν_μ ($\bar{\nu}_\mu$) energy taken from reference 4 and shown in Fig. 3. The ordinate of this plot should read "$\langle y \rangle$ for $x < 0.6$". It is immediately clear that the curve for $B_\nu = 0.8$, calculated by the CDHSB group, does not fit the ν_μ data. The ν_μ data taken as a whole is many (>20) standard deviations away from the curve, if the CDHSB error bars are taken seriously. Again, a straightforward calculation shows that these data are best fitted by a value of B_ν close to 0.6. Hence it appears that at all energies the CDHSB ν_μ data are consistent with a low value of B_ν ($\simeq 0.3$ for all x; $\simeq 0.6$ for $x < 0.6$).

Table 1. Values of B_ν and $B_{\bar{\nu}}$ obtained from the CDHSB data.

Region	B_ν	$B_{\bar{\nu}}$
All x	0.3	0.8
$x < 0.1$	0.8	0.3

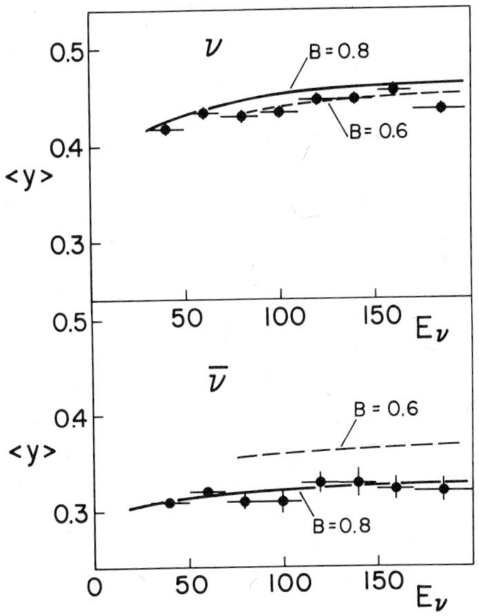

Fig. 3. Energy dependence of <y> (for $x < 0.6$) measured by the CDHSB collaboration. The dashed curves marked B = 0.6 were not included in the original figure.

This data from the same experiment seem to prefer, however, a higher value of $B_{\bar{\nu}}$ (≈0.8) at all energies.

Now there are two possible logical conclusions: (i) If all the CDHSB data of Figs. 2 and 3 are taken as correct, then the principal conclusion of their paper - that there is no anomaly in the y-distributions does not follow from their data. Indeed, the opposite conclusion is better reached from their results. (ii) The other possibility is that there exists a substantial error - presumably systematic - in the CDHSB data. It is important to realize that any systematic error common to both the ν_μ and $\bar{\nu}_\mu$ data could distort the shape of the y-distributions for both ν_μ and $\bar{\nu}_\mu$ in the same direction. This in turn could result in a value of <y> too low for both the ν_μ and $\bar{\nu}_\mu$ data. The decrease in <y> for ν_μ resulting from such an error would be, however, much smaller than the corresponding decrease in <y> for $\bar{\nu}_\mu$. This is illustrated by the calculated curves for $B_\nu = B_{\bar{\nu}} = 0.6$ shown by us as the dashed curves in Fig. 3. Another important point to be noted is that an error in y is correlated to an error in the measurement of E_ν since both variables are calculated from the same measured values of E_μ and E_H. It therefore follows from the above discussion that the content of statement (D) made by Holder et al. in reference 4 is subject to serious doubt.

In summary, we believe that none of the statements (B), (C) and (D) made in the paper by the CDHSB group is justified for the reasons discussed here. We have used the simple scaling theory to illustrate the inconsistencies in the paper of Holder et al. since the data presented in that paper were analyzed in the context of this naive framework. Indeed, a more general analysis of the data, relaxing the assumptions of Bjorken scaling and the Callan-Gross relation, should now be carried out. However, even if the distortions in the y-distributions of reference 4 were due to scale breaking effects of the type predicted, for example, by asymptotically free field theories, there would still remain the problem that the CDHSB data exhibit no energy dependence of the values of $\langle y \rangle$ and $\sigma_{\bar{\nu}}/\sigma_\nu$.[3]

REFERENCES

1) B. Aubert et al., Phys. Rev. Lett. 33, 984 (1974).

2) A. Benvenuti et al., Phys. Rev. Lett. 36, 1476 (1976).

3) A.K. Mann, in New Pathways in High Energy Physics II, A. Perlmutter, ed. (Plenum Press, 1976).

4) M. Holder et al., CERN preprint [since published in Phys. Rev. Lett. 39, 433 (1977)].

4. MULTIMUON PRODUCTION BY NEUTRINOS

4.1 Introduction

The subject of multilepton production by neutrinos and antineutrinos began in the summer of 1974, a few months before the discovery of the J/ψ. The reactions observed then and during the following two years were

$$\nu_\mu + N \to \mu^- + \mu^+ + X \quad ; \quad \nu_\mu + N \to \mu^- + e^+ + X$$

$$\bar{\nu}_\mu + N \to \mu^- + \mu^+ + X \quad ; \quad \nu_\mu + N \to \mu^- + \mu^- + X \qquad (1)$$

$$\bar{\nu}_\mu + N \to \mu^+ + \mu^+ + X$$

where N is a nucleon and X is any hadron state. This early history[1] is summarized in the data[2-5] of Table 1 where it is shown that dilepton final states were observed in small numbers over a wide range of energies in counter-spark chamber experiments and also in heavy liquid bubble chambers.

Table 1. Early dilepton history

Experiment	Reference	Events	Type	<Energy>
HPWF ('74, '75)	2	~50	$\mu^-\mu^+$	$\gtrsim$ 75 GeV
CITF ('76)	3	~10	$\mu^-\mu^+$	$\gtrsim$ 50
BCHW ('76)	4	~15	μ^-e^+	$\gtrsim$ 30
GGM ('76)	5	~ 5	μ^-e^+	$3 \lesssim E < 15$

It became clear after a brief time that a significant fraction of the dilepton events, particularly that with two energetic leptons, was not due to the decay in flight of a π or K meson in the hadron cascade of an ordinary charged current neutrino-nucleon interaction. Possible explanations then centered on only four alternatives[6]: (i) the four-fermion process in which a neutrino goes directly into a pair of oppositely charged leptons and another neutrino, with one of the charged leptons radiating to a nucleus; (ii) charged weak vector boson (W^+) production and decay; (iii) neutral heavy lepton production and decay; and (iv) production of a new type of massive hadron and its subsequent weak decay.

For several reasons, the first two alternatives were found untenable. The third alternative was unlikely as the sole source of the observed dimuons[7] but it could not be ruled out as a contributing process. Hence it was soon concluded that the second lepton and the unseen neutrino (necessary to conserve lepton number) in the dilepton events were probably due to the production and weak decay of members of a hadron family with a new degree of freedom, i.e., a new hadronic quantum number, as illustrated in Fig. 1a. It is the weak decay of the quark "c" in Fig. 1a, leading to the second muon and outgoing neutrino, that signals the presence of a new hadronic quantum number, similar to strangeness in that it is

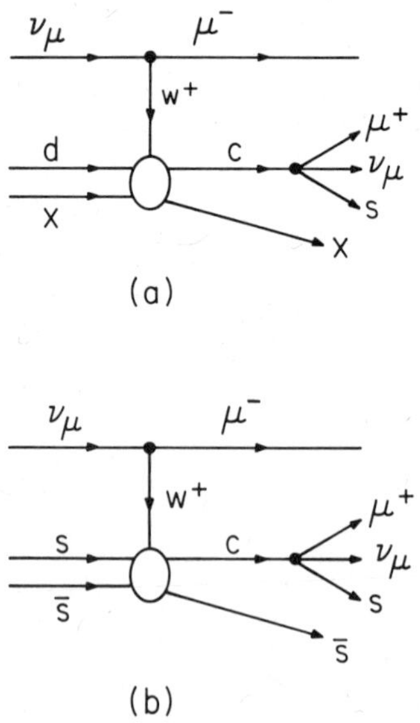

Fig. 1. (a) Neutrino-quark interaction leading to a dimuon final state through the production and decay of a charmed (c) quark. (b) Alternative neutrino-quark interaction leading to a dimuon final state with two strange (s) quarks.

rigorously conserved in strong and electromagnetic interactions but not in weak processes.

Thus neutrino-induced dileptons were taken by many as the first substantial evidence for the existence of a fourth, charmed (c) quark in addition to the ordinary, u and d, quarks and the strange (s) quark. This evidence for bare or explicit single charmed particle production and decay, and the evidence for a bound charm-anticharm state, the J/ψ, were therefore complementary. Furthermore, this interpretation was consistent with observations of the μ-e dileptons. These were expected on the basis of muon-electron universality through the decay mode $c \to s + e^+ + \nu_e$ at a rate approximately equal to the rate for $c \to s + \mu^+ + \nu_\mu$. In addition, the μ-e events exhibited at least one strange particle per μ-e event, presumably due to the s-quark in the c-decays.

In the quark-parton model (QPM) it is expected that charmed quark production would also take place through a neutrino interaction with an s-quark in the $s\bar{s}$, $u\bar{u}$, $d\bar{d}$ "sea" of the nucleon. This reaction is shown in Fig. 1b, and should occur at a rate determined by the magnitude of the $s\bar{s}$ component of the sea, possibly comparable to the rate of c-production via the diagram of Fig. 1a. Observe that two strange quarks are present in the final state of the diagram in Fig. 1b since one is released in the production process and one in the decay. Hence a precise measurement of the number of strange particles associated with neutrino-induced dileptons would be particularly useful in determining the magnitude of the strange particle component of the sea. The number of strange particles per μ-e is still an unsettled question.

Since the data of Table 1 were obtained, there have been acquired significantly larger samples of dileptons in detectors with better geometric acceptance for the leptons, and with different neutrino (ν_μ) and antineutrino ($\bar{\nu}_\mu$) beams. These samples are of interest because, as we shall see, they confirm both the initial conclusions with respect to charm reached from the early dilepton data and the very precise charmed pair production data from e^+e^- interactions. Another interesting aspect of the larger dilepton samples is the opportunity they provide to search for new degrees of hadronic freedom beyond charm and new leptons.

Indeed, this capability is a general feature of neutrino interactions which have more than one lepton in the final state. There have recently been observed at FNAL a few neutrino-induced events with three final state muons, which appear at first glance to possess unusual properties[8,9]. Empirically, the reaction is

$$\nu_\mu + N \to \mu^- + \mu^- + \mu^+ + X \quad . \qquad (2)$$

The trileptons, as well as the dileptons, constitute a relatively sharp probe of new matter because they originate from an inverse beta decay process which is known to violate most symmetries, discrete and group, and they are produced at high energies in "pointlike" (relatively hard) collisions.

In this lecture I will, for simplicity, discuss multilepton data primarily from the new HPWF experiment (E-310). The plan is, first, to spend a few minutes on the experimental method which is somewhat different than in the earlier HPWF experiment (E-1A); then to discuss dimuons, mainly of the type $\mu^-\mu^+$ but also of the type $\mu^-\mu^-$; finally, to describe the properties and speculate a little on the possible origins of the observed trimuons.

4.2 Experimental Method

All of the new HPWF data have been taken with three types of ν_μ ($\bar{\nu}_\mu$) beams. In one type, also used earlier in E-1A, the secondary hadrons produced in the initial collision between the extracted proton beam and the neutrino area target were focussed by a quadrupole triplet and left to decay without charge selection (QT beam)[10]. The other types employed no focussing elements but did charge selection of the secondary hadrons by means of a "dog-leg" arrangement of bending magnets[11]. Hence the energy spectra of the ν_μ and $\bar{\nu}_\mu$ beams from these configurations are derived from bare target spectra and are significantly purer due to the sign selection than is the spectrum of the QT beam. These will be referred to as BTSSν and BTSS$\bar{\nu}$ beams. The calculated spectra of ν_μ and $\bar{\nu}_\mu$ from the QT and BTSS$\bar{\nu}$ beams are shown in Fig. 2. Since no multilepton data taken with the BTSSν beam will be discussed here, the spectra from that beam are not shown. It should be emphasized that the ratio of neutrino to antineutrino interactions in the QT beam is $N(\mu^-)/N(\mu^+) \approx 5/1$, while the same ratio in the BTSS$\bar{\nu}$ beam is 0.10. For our purpose here, i.e., the discussion of multilepton final states, it is therefore a reasonable approximation to consider the QT beam to be dominantly a ν_μ beam and the BTSS$\bar{\nu}$ beam to be primarily a $\bar{\nu}_\mu$ beam, albeit the opposite helicity contaminations are not negligible.

These wide band beams have the advantage of being inexpensive and easy to prepare and to exchange; also they reach to the highest energies and intensities available from an incident proton flux of a given energy. Their primary disadvantage is lack of purity relative to a carefully constructed (and more expensive) dichromatic beam[12].

It is worth noting in this connection that significant improvements in the circulating and extracted proton beam energy and

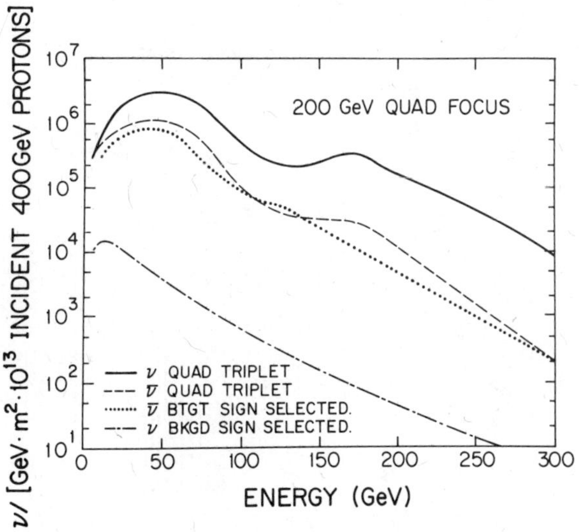

Fig. 2. Calculated neutrino and antineutrino spectra for two of the beam types used in acquiring the data of E-310. Approximately 12% of all neutrino interactions in the QT beam are above 200 GeV.

intensity at FNAL have been made in the past three years. The total number of protons that impinged on the neutrino area target during the 2 and 1/2 year history of E-1A was approximately 10^{18}; this same number of protons can now be put on the neutrino area target for E-310 in about 2 and 1/2 weeks. The improved intensity makes the wideband beams of Fig. 2 particularly valuable as intense sources of neutrinos with energies above 200 GeV.

The detector of E-310 is shown in Fig. 3. It is an enlarged version of the earlier detector of E-1A with important modifications. (i) Of significance in the study of multileptons is a target-detector of three parts, an iron target (FeT), a liquid scintillator calorimeter (LiqC), and a thick plate iron calorimeter (FeC), each part of different density. This makes possible empirical discrimination against the background of muons from π- and K-meson decays in flight which cause ordinary charged current neutrino interactions to simulate multilepton events. (ii) The geometric acceptance of E-310 covers a muon angle relative to the incident ν_μ ($\bar{\nu}_\mu$) beam of about 500 mrad compared with the limiting angle of 225 mrad in E-1A. This leads to a lower cutoff in muon momentum in E-310 (≈2 GeV/c) rather than the ≈4 GeV/c cutoff in E-1A. (iii) The presence of many wide gap spark chambers in the

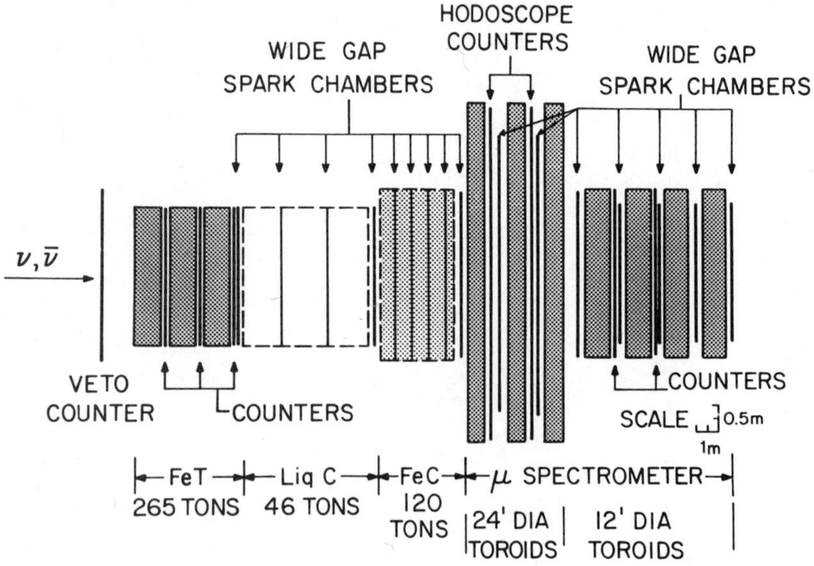

Fig. 3. Schematic representation of the target-detector and magnetic spectrometer of E-310.

ionization calorimeters of E-310 gives excellent, detailed visual information about each multimuon event. (iv) The three targets give more electronic counter information about a multimuon, e.g., the number of minimizing particles, and (v) the presence of FeC at the downstream end of the target-detector assembly gives better longitudinal containment of showers and clearer discrimination against hadron punch-through simulation of a muon. Finally, (vi) there is a wider variety of multilepton event triggers in E-310 due primarily to the hodoscope counters in the 24 ft. diameter magnetic spectrometer.

4.3 Opposite Sign Dimuons

There is shown in Fig. 4 a dimuon event observed in the E-310 detector. The two muons are in time coincidence within 150 nsec, the calorimeter modules through which the muons pass show two minimum ionizing particles, and the hodoscope counters in the 24 foot diameter magnet as well as the additional trigger counters in

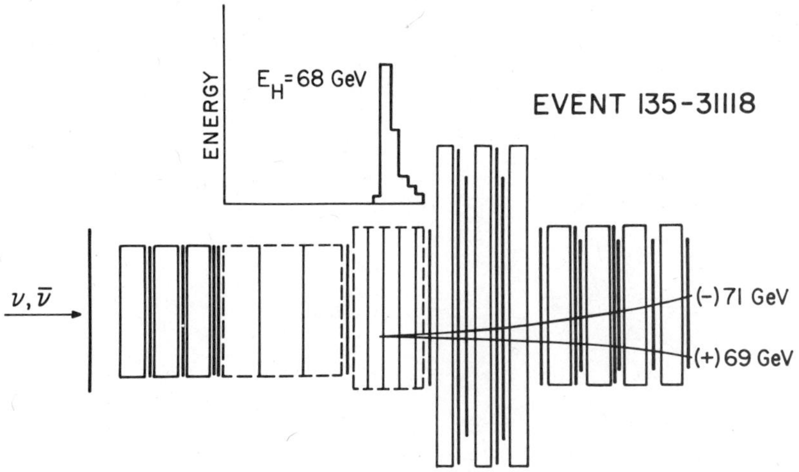

Fig. 4. Tracing of an opposite sign dimuon event observed in E-310 superimposed on the apparatus.

the 12 foot diameter magnet show two hits corresponding to the track position information from the wide gap spark chambers.

We show in Fig. 5 a plot of the momentum of the negatively charged muon p_- against the momentum of the positively charged muon p_+ for a partial sample of the QT data from E-310. For comparison we include also in Fig. 5 the QT data from E-1A and part of the E-310 data from the BT$\bar{\nu}$ beam. The E-310 and E-1A QT data show the same features: the distributions are heavily weighted toward quite low μ^+ momenta while the μ^- momenta cover a much larger interval. The BT$\bar{\nu}$ data show the opposite tendency, i.e., low μ^- momenta but μ^+ momenta covering the momentum region determined by the spectrum of the incident $\bar{\nu}_\mu$ beam. The BT$\bar{\nu}$ data also suggest that some of the QT beam-induced dimuons in the region of low p_- ($\lesssim$25 GeV/c) may result from the $\bar{\nu}$ component of the QT beam. The projections of Fig. 5 onto the p_- and p_+ axes are shown separately in Fig. 6 for the QT and BT$\bar{\nu}$ data. This permits a direct, if approximate, comparison of ν-induced with $\bar{\nu}$-induced dimuons. The label of the abscissas in Fig. 6 is chosen to be "leading" or "opposite" sign to coincide with expectations based on the incident ν_μ ($\bar{\nu}_\mu$) beam.

In Fig. 7 are shown event distributions in the variable $\alpha \equiv (p_L - p_O)/(p_L + p_O)$, again for the QT and BT$\bar{\nu}$ data separately, where the subscripts L and O refer to leading and opposite sign dimuons, respectively. Despite the limited $\bar{\nu}$ statistics, the ν- and $\bar{\nu}$-distributions in α are essentially similar and indicate that on

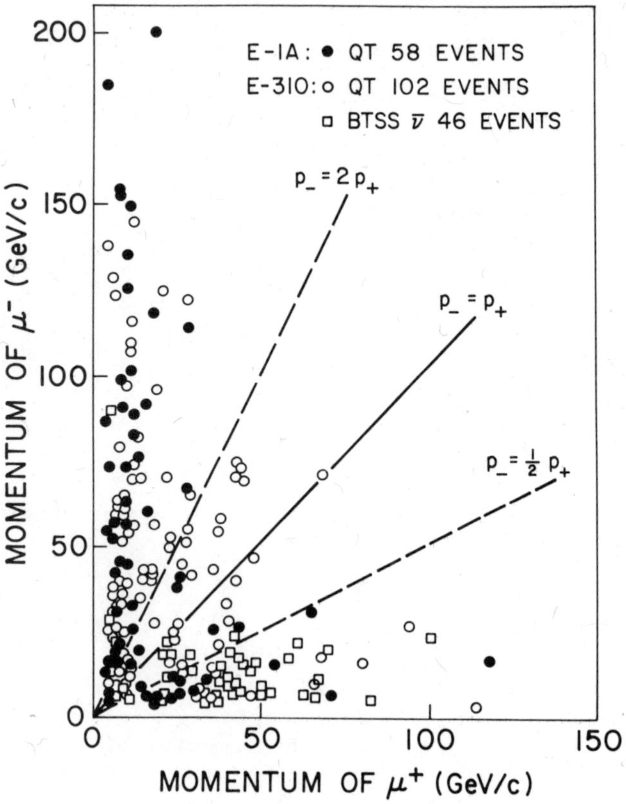

Fig. 5. Scatter plot of the momentum of the negatively charged muon (p_-) against the momentum of the positively charged muon (p_+) for different dimuon samples.

average the opposite sign muon carries away a small fraction of the total muon momentum.

Fig. 8 shows the distributions in $\Delta\phi$ the included angle between the components of p_- and p_+ transverse to the incident ν_μ ($\bar{\nu}_\mu$) beam direction (p_{TL} and p_{TO}). We see that the $\Delta\phi$-distributions are similar for ν_μ- and $\bar{\nu}_\mu$-induced dimuons and that, as the insert in Fig. 8a indicates, the transverse components P_{TL} and p_{TO} tend on average to be almost oppositely directed. In contrast, the distributions in $\Delta\gamma$, the included angle between the components of P_L and p_O transverse to the direction of the exchanged vector boson (W) indicated in Fig. 9, are approximately uniform.

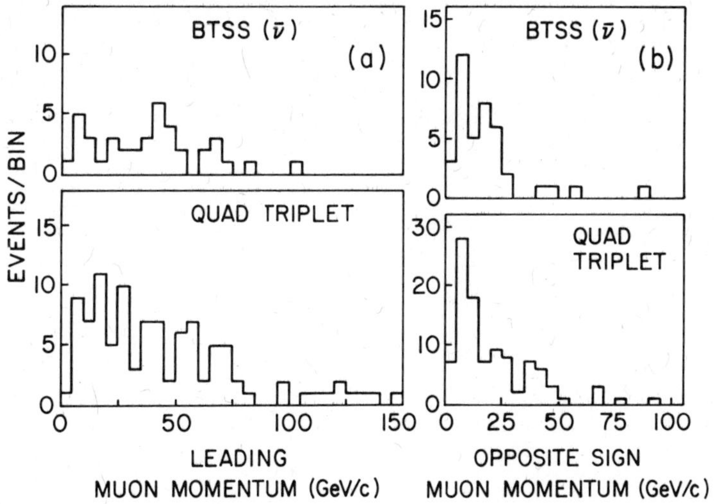

Fig. 6. Projections of the scatter plot of Fig. 5 onto the two axes for the QT and BT$\bar{\nu}$ data.

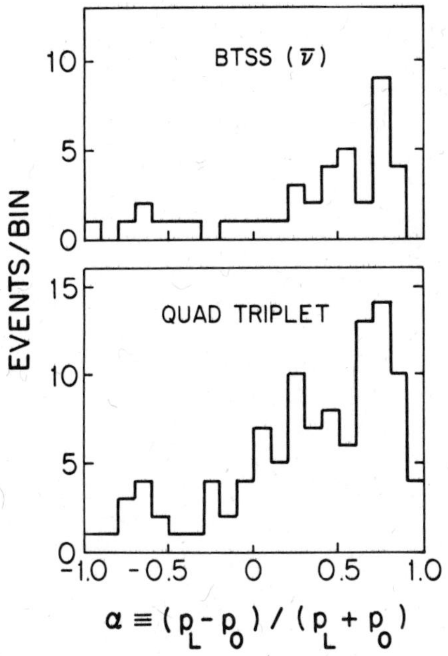

Fig. 7. Event distributions in the variable $\alpha \equiv (p_L - p_O)/(p_L + p_O)$ for the QT and BT$\bar{\nu}$ data.

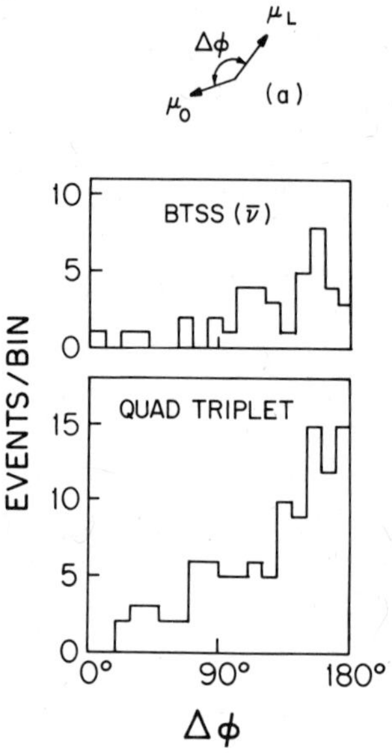

Fig. 8. Distributions in the angle $\Delta\phi$ between the transverse components (relative to the incident ν_μ ($\bar{\nu}_\mu$) direction) of the leading and opposite sign muon momenta.

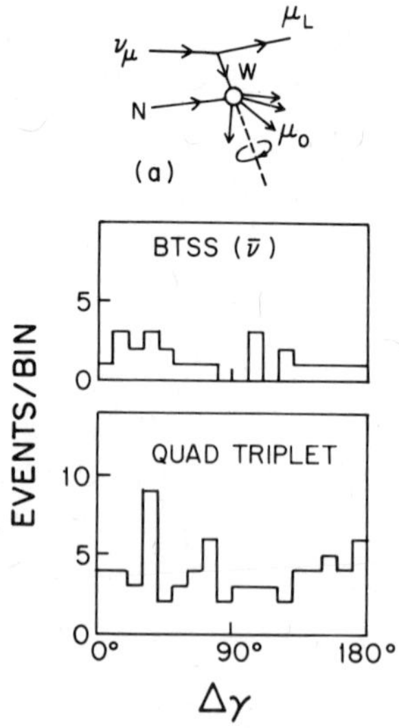

Fig. 9. Distributions in the angle $\Delta\gamma$ between the transverse components of the leading and opposite sign muon momenta. Here the transverse components of p_L and p_0 are transverse to the direction of the exchanged vector boson (W).

The distributions in transverse momentum of the opposite sign muon, p_{TO}, are given in Fig. 10 with respect to the W-axis and in Fig. 11 with respect to the production plane. Note the similarity of the values of $\langle p_{TO} \rangle$ in Fig. 10 and Fig. 11 for the BT($\bar{\nu}$) and QT beams. Incidentally, it was the distribution in p_{TO} (relative to the production plane) that provided the first rough measurement of the mass of the charmed hadron[13], as shown in Fig. 12.

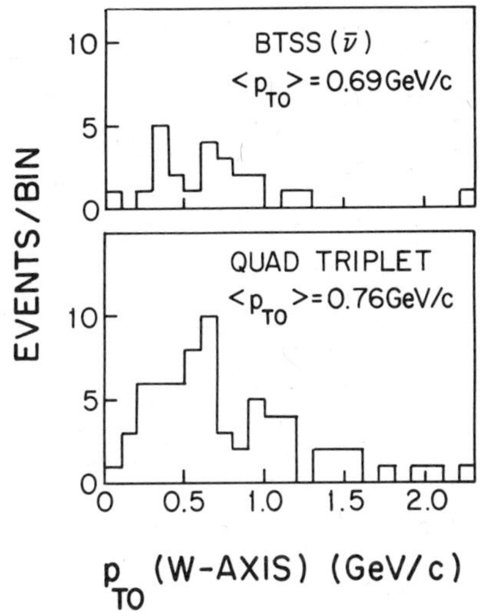

Fig. 10. Distributions in transverse momentum relative to the W-direction of the opposite sign muons.

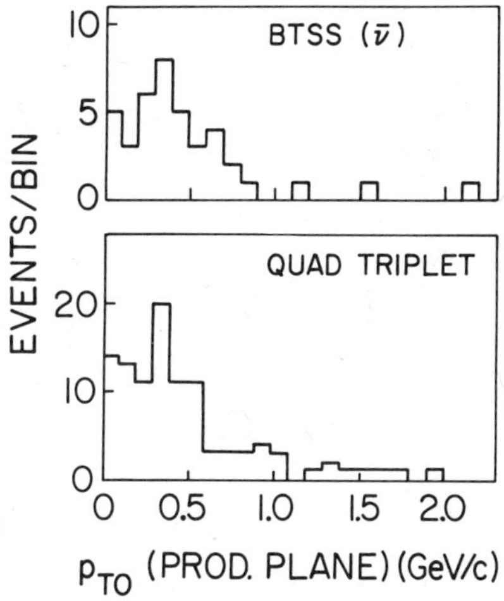

Fig. 11. Distributions in transverse momentum relative to the production plane of the opposite sign muons.

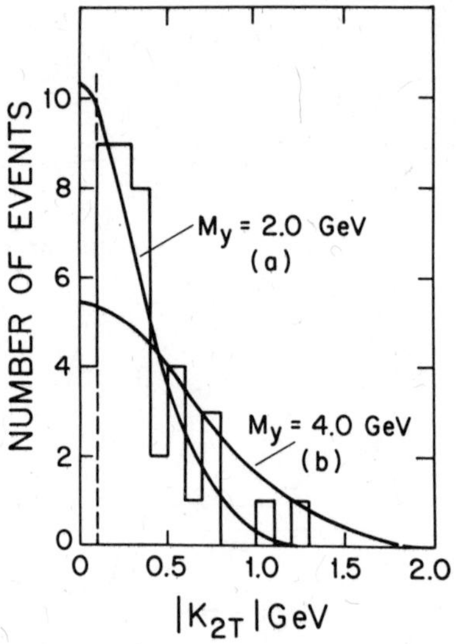

Fig. 12. Illustration of the mass dependence of early data similar to that in Fig. 11. The curves were calculated in the QPM by Sehgal and Zerwas (reference 13).

All the kinematic distributions for the leading and opposite sign muons discussed so far: momenta, fractional asymmetry α in muon momentum, angles between various momentum components of the muons of a given event, and transverse components of muon momenta, are essentially the same within statistical uncertainties for the QT and BT$\bar{\nu}$ data. Although, as we stated, there are present in those beams non-trivial amounts of opposite helicity neutrinos, it nevertheless seems safe to conclude that the kinematic distributions above are in the main the same for ν_μ- and $\bar{\nu}_\mu$-induced dimuons. This is also true for the distributions in visible event energy (E_{VIS}) shown in Fig. 13, when the difference between the energy spectra of the ν_μ and $\bar{\nu}_\mu$ components of the incident beams is considered.

The distributions in the dimensionless scaling variables $x = Q^2/2m_pE_H$ and $y = E_H/E_\nu$ are presented in Figs. 14 and 15. Here $Q^2 = 4E_\nu E_{\mu L}\sin^2(\theta_{\mu L}/2)$ with $\theta_{\mu L}$ the angle between the directions of the leading muon and the incident ν_μ ($\bar{\nu}_\mu$) beam, E_H is the hadronic energy in an event, and m_p is the proton mass. It is important to indicate in Figs. 14 and 15 which events have $p_O > p_L$ since some of those events may have been produced by neutrinos of helicity opposite to that of the dominant component of the incident beam. Taking this into account, we see that there is no significant difference within the limited statistical samples between the x-

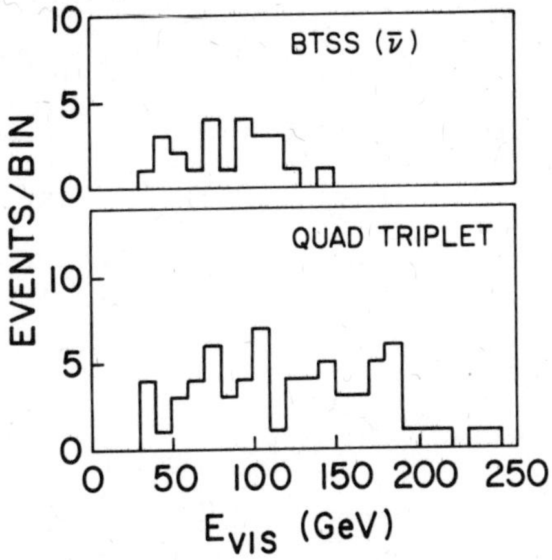

Fig. 13. Distributions in visible energy of QT and BT$\bar{\nu}$ events.

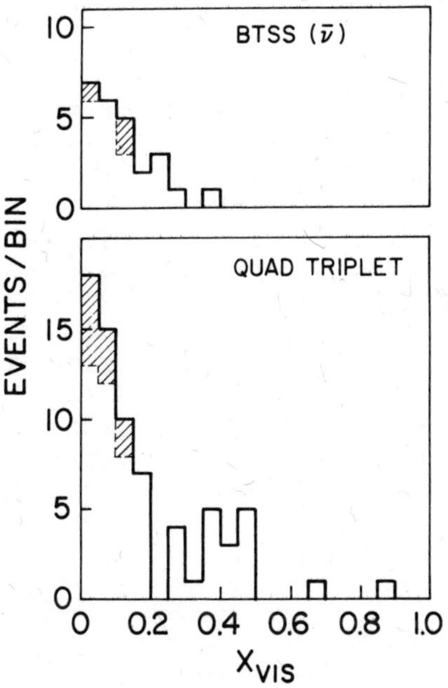

Fig. 14. Distributions in the scaling variable $x = Q^2/2m_pE_H$. The cross hatching here and in Fig. 15 indicates events with $p_0 > p_L$ which may have been produced by opposite helicity neutrinos present in the incident beam.

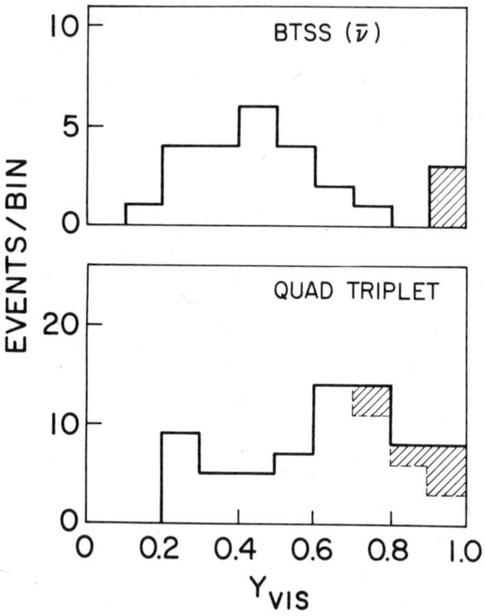

Fig. 15. Distributions in the scaling variable $y = E_\mu/E_\nu$.

distributions for ν_μ- and $\bar{\nu}_\mu$-induced dimuons, or between the corresponding y-distributions.

There are a number of QPM calculations of charmed hadron production by ν and $\bar{\nu}$ and subsequent semileptonic decay[14]. These are in general consistent with the data presented above, which suggests that the majority of opposite sign dimuons (and also μ-e events) produced by both ν and $\bar{\nu}$ are indeed a manifestation of charmed hadrons. It is tempting to go further and argue, primarily on the basis of the overall similarity of the ν and $\bar{\nu}$ dimuon distributions, that no sources of dimuons other than charm are present in the data. But this is clearly too sweeping a conclusion to be reached from the data at this time. In fact, it may be difficult to substantiate that conclusion even with appreciably improved data because heavier quarks, if they exist, may give rise to dimuons in smaller quantities than do charmed quarks, and the kinematic properties of the second muon may not be much different in the two cases if the heavier quark cascades down through a charmed quark or a τ-lepton[15]. For example, a heavy (~5 GeV/c^2) quark of charge $-1/3$ produced in a $\bar{\nu}_\mu$ interaction of unknown strength with a u-quark will decay mainly nonleptonically into a u-quark or possibly a c-quark. Its semileptonic decay modes would involve τ (if it also exists) and e as well as μ. Hence only a small, possibly very small, fraction of all decays of the massive, charge -1/3 quark might lead to muons that are kinematically different from those resulting from charmed quark decays. We must, therefore, be satisfied for the time being with the weaker conclusion that there is no evidence in the present dimuon data in favor of the production and weak decay of quarks beyond charm.

Finally, we summarize in Table 2 the observed rates for dilepton production[16-18] from which we see that, within rather large errors, the values of $R(\ell^+\mu^-)/R(\mu)$ for ν from different experiments are in agreement. The situation is less clear for the $\bar{\nu}$ data where more and better data are necessary. No conclusion concerning the magnitude of the strange quark sea can as yet be drawn from Table 2.

4.4 Same Sign Dimuons

Dimuons with both muons having the same charge have been observed in small numbers. A scatter plot of the muon momenta from a small sample of same sign dimuons is shown in Fig. 16. In general, these events are distributed much like the opposite sign dimuons, i.e. the second muon is in most events significantly

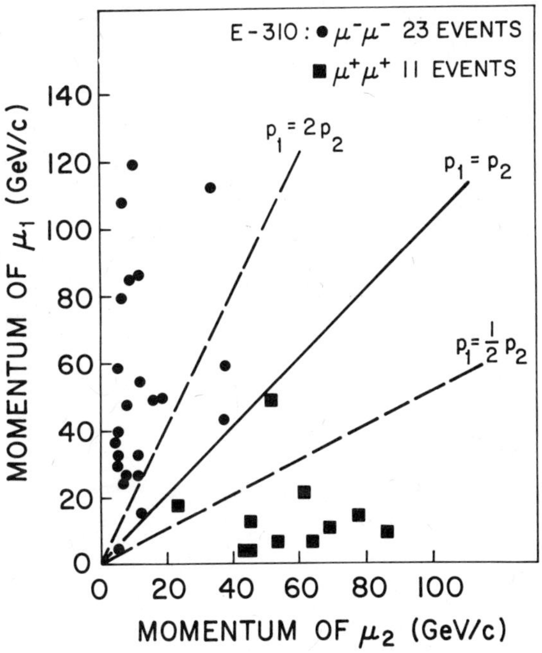

Fig. 16. Scatter plot of the muon momenta from a sample of same sign dimuons.

Table 2. Measured dilepton rates

Experiment	Reference	Lower Limit $E_{\ell 2}$(GeV)	Type	Corr.	$R(\ell^+\mu^-)$ / $R(\mu)$ ν (%)	$\bar{\nu}$ (%)	E_{VIS} † (GeV)
HPWF (1975)	2	4	$\mu^+\mu^-$	GEOM	0.8±0.3	2±1	>40
CTFR (1977)	3	2.4	$\mu^+\mu^-$	GEOM	0.8±0.5	1.4±0.7	≳150
BCHW (1976)	4	0.8	$e^+\mu^-$	GEOM	0.8±0.3	-	>10
COL-BNL 15'BC (1977)	16	4	$e^+\mu^-$	GEOM	0.5±0.2	-	>40
CDHSB (1977)	17	4.5	$\mu^+\mu^-$	NONE	>0.5±0.1	0.5±0.1	>50
BHW (1977)	18	0.8	eμ	GEOM	$0.34^{+0.23}_{-0.13}$	$0.10^{+0.13}_{-0.07}$	$\langle E_\nu \rangle=47$ $\langle E_{\bar{\nu}} \rangle=30$

† The Gargamelle result (ref. 5) is not included since the average incident energy is so much less than that for the experiments listed here.

softer than the leading muon. Nevertheless, in 14 events out of 34 the second muon has an energy greater than 10 GeV, which suggests that such muons at least probably do not arise from the decay in flight of a pion or kaon in the hadronic cascade attending an ordinary charged current neutrino interaction.

To test this hypothesis further we look at Fig. 17 which shows the longitudinal distributions (in target segment number) for both opposite sign and same sign (-/-) dimuons. To a good approximation the ratio of the number of $\mu^-\mu^-$ events to the number of $\mu^-\mu^+$ events, $N(\mu^-\mu^-)/N(\mu^-\mu^+)$, taken for each of the three parts of the target-detector separately should be independent of detection efficiency and other biasses. This ratio is plotted for the three target-detector parts against the reciprocal of the density ($1/\rho$) of those parts in Fig. 18. This large variation in density of the three parts of the target-detector is a particular strength of the E-310 since it permits the plot of Fig. 18 to be made with a relatively large separation along the $1/\rho$ axis and with small extrapolation to $1/\rho = 0$. The data of Fig. 18, despite the large statistical errors, tend to support the idea that many of the same sign dimuons are from a much shorter-lived source than π- or K-mesons.

This tentative conclusion is further supported by the properties of the (-/-) dimuon event shown in Fig. 19. The high energy of both muons in this event make it quite unlikely that it is due to π- or K-decay.

In summary, the preliminary E-310 data on same sign dimuons suggest that at least some of the (-/-) dimuons are directly

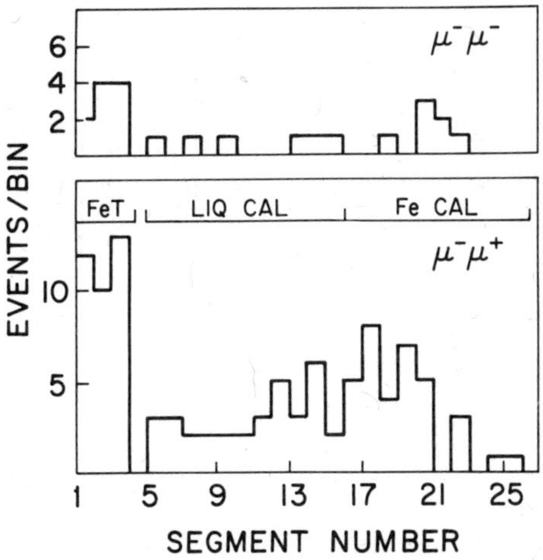

Fig. 17. Longitudinal distributions of opposite sign and same sign (-/-) dimuons.

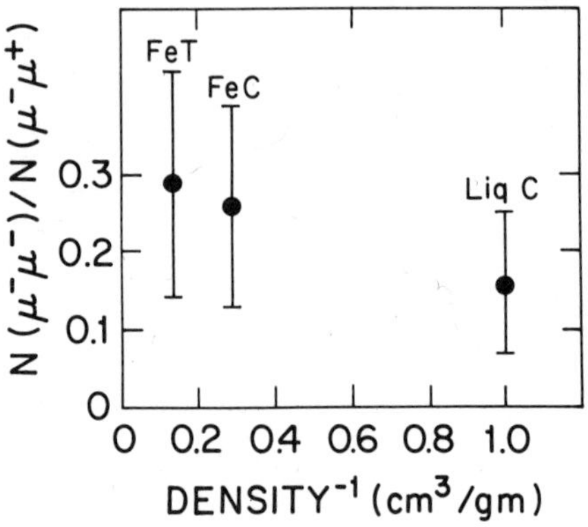

Fig. 18. Plot of the ratio of (-/-) events to (+/-) events for the three parts of the E-310 target-detector against the reciprocal of the density for those parts.

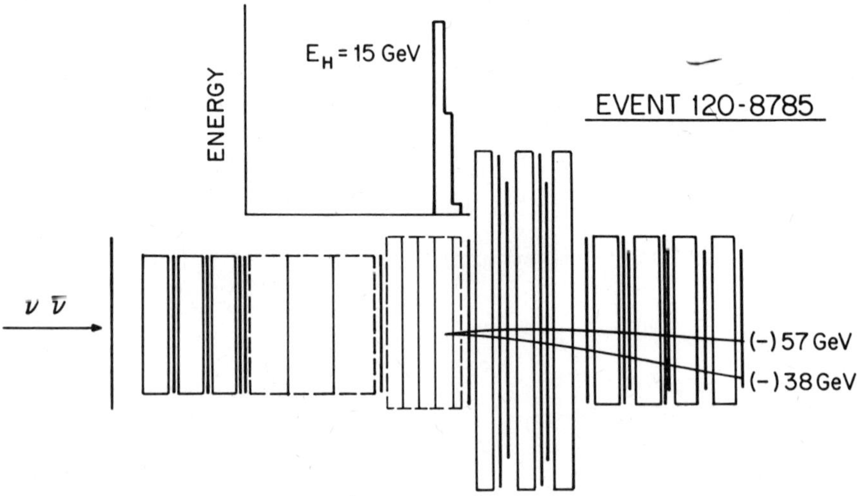

Fig. 19. Tracing of a (-/-) dimuon superimposed on the E-310 apparatus; note particularly the high momenta of the two muons.

produced. Since the statistics of the sample is at present too small to make detailed study profitable, we content ourselves here with posing the question: what is the origin or origins of prompt (-/-) dimuons? Two possibilities that come directly to mind are (i) associated production of charm and (ii) a heavy lepton cascade. It will require more same sign dimuon data to decide which, if either, of these possibilities is the actual source.

4.5 Trimuons

We show in Fig. 20 a trimuon event overlaid on the apparatus of E-310. In this event the origin is in Liq C, the leading muon is negative and the softest muon is also negative with $p_\mu < 10$ GeV/c. The angles between the muon pairs are relatively large (>100 mrad in each case), and the energy of the hadron shower E_H is large (61 GeV). The total visible energy E_{VIS} of the event is also large (132 GeV). In Fig. 21 is shown another trimuon event which differs from the first in that the leading muon is positive and the angle between each muon pair is small (<100 mrad). Other features of the two events are similar, including large E_H and large E_{VIS}.

In Figs. 22 and 23 are shown trimuons with somewhat different properties from the trimuons in Figs. 20 and 21. Events 119 and 281 are distinguished by very large total muonic energy (236 GeV

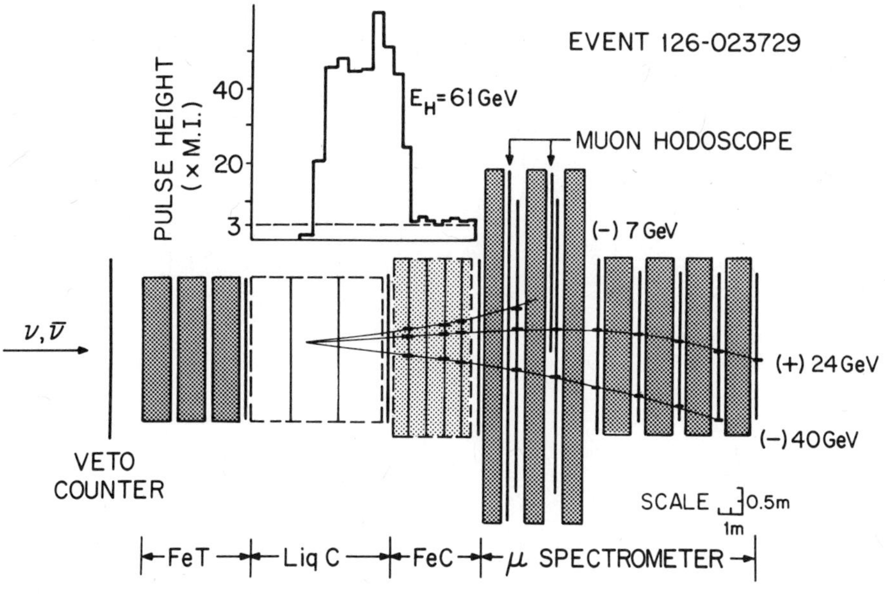

Fig. 20. A trimuon event traced onto the E-310 apparatus.

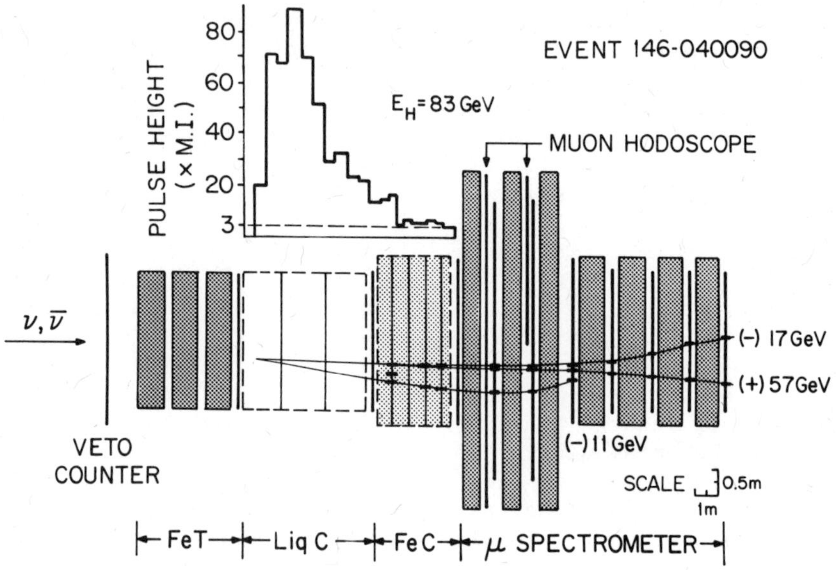

Fig. 21. Another (-/-/+) trimuon event in which the leading muon is positive.

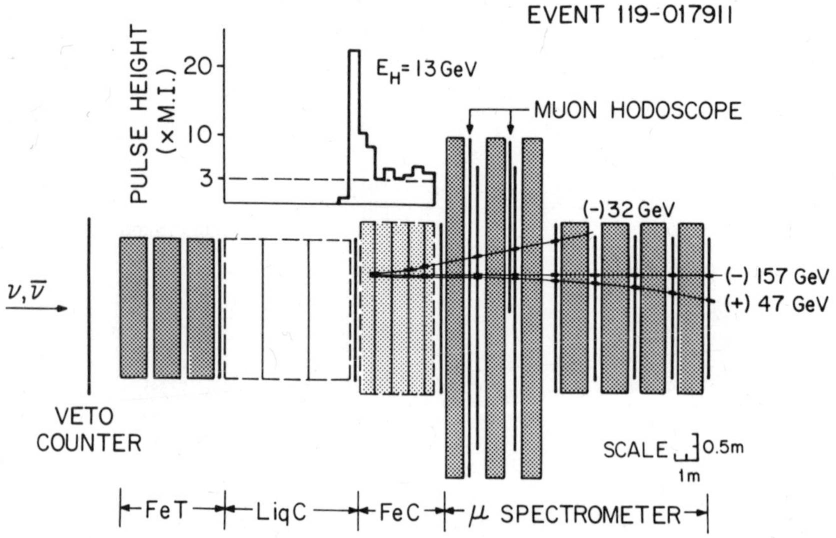

Fig. 22. A trimuon with large total muon momenta and small hadron energy.

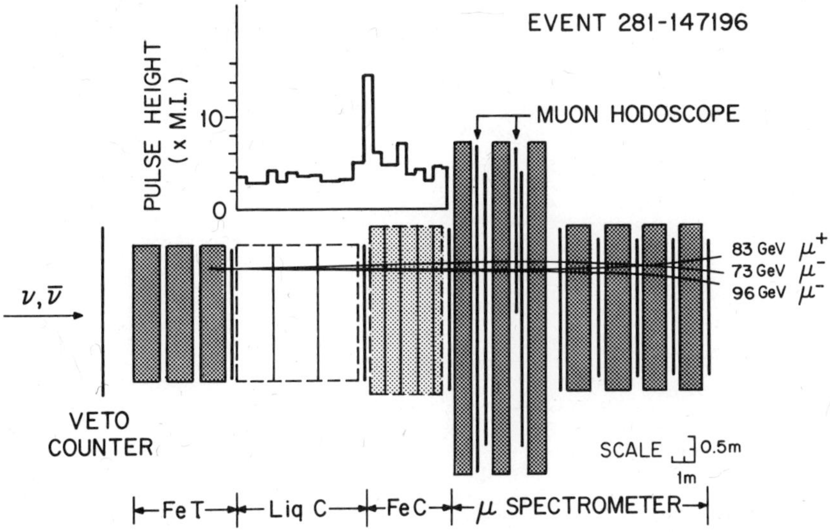

Fig. 23. A trimuon with large total muon momenta and small angles between muon pairs.

and 240 GeV, respectively), and by relatively small values of E_H (13 GeV and $\lesssim$30 GeV, respectively). In both events the angles between muon pairs are also small (<100 mrad for 119, and <20 mrad for 281). Event 119 occurs at the interface between Liq C and FeC, while event 281 originates in the third section of FeT.

Background estimates for events 119 and 281 are given in Table 3. The calculation of random space-time coincidence probability is straightforward. Calculation of the decay in flight probability of a π- or K-meson in the hadronic shower attending dimuon production involves the ratio of the absorption length of the material in which the initial interaction occurred to the mean decay length of a π or K meson of a given (the lowest negative) momentum, multiplied by the probability $f(E_{\pi,K}/E_{CASCADE})$ of finding the meson with that momentum in the momentum spectra of π and K mesons measured in neutrino interactions, multiplied by the probability of finding a dimuon with the properties of the remaining two muons in event 119 and event 281 in the total spectrum of observed dimuons, multiplied by the total probability of dimuon production. We take 50 cm for the relevant absorption length, $f(E_{\pi,K}/E_{CASCADE}) = 1$, a relative dimuon rate of 10^{-2} and our sample of 300 dimuons to calculate these probabilities.

Table 3. Background estimates for two unusual trimuon events.

Type	Event	
	119	281
Random space-time coin	$<10^{-3}$	$<10^{-3}$ event
i. 3 single muons (ν-events)		
ii. 1 single muon, 1 dimuon (ν-events)		
iii. 1 dimuon event, 1 stray muon		
iv. 3 stray muons		
Decay of π/K from dimuon	$<10^{-1}$	$<10^{-1}$ event
Electromagnetic processes		
i. trident from single muon (ν-event)	$<10^{-1}$	$<10^{-1}$ event
ii. dimuon from hadronic vertex		

The contents of Table 3 lead to the conclusion that dimuon production in random coincidence with another neutrino interaction, and dimuon production with attendant π or K decay in flight are unlikely sources of events 119 and 281.

The relatively small angles between the muon pairs in these events, particularly event 281, suggest the possibility of a combined weak-electromagnetic origin, for example, electromagnetic production of a muon pair in conjunction with a neutrino interaction. Our rough estimate of the probability of this process is too small to account for events with the properties of 119 and 281. There are, however, several detailed calculations of this effect in progress[19].

The detailed properties of all the trimuon events observed thus far by the HPWF collaboration are given in the papers of reference 8. These correspond to a trimuon rate relative to all neutrino interactions with $E_\nu > 100$ GeV of between 1×10^{-4} and 5×10^{-4} with no correction for detection efficiency.

One interesting characteristic of events 119 and 281 is evident in Fig. 24 where we plot the momentum components of the three muons transverse to the incident neutrino direction for each event. All the muons lie on the same side of the neutrino beam, and no pair of opposite sign muons is able to balance the transverse momentum of the third muon, as would be expected if an opposite sign pair were produced at the hadron vertex of the event.

Further consideration of possible origins of the trimuon events divides among hadronic and leptonic origins or some mixed

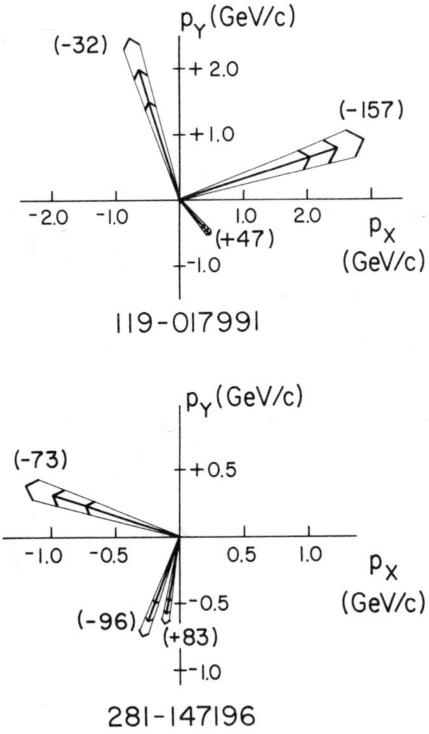

Fig. 24. Plot of the transverse momentum components of trimuon events (119-017991) and (281-147196). The z-axis is along the incident neutrino direction. The dark arrow heads and central lines indicate the measured values of p_T; the shaded areas indicate the measurement errors. The numbers in parentheses are the muon charge and total momentum in GeV/c. Note factor of two difference in scales.

leptonic-hadronic origin. A dimuon might originate at the hadronic vertex of a charged current neutrino interaction through (i) direct muon pair production similar to that in the reaction $\pi + N \rightarrow \mu^+ + \mu^- + \ldots$, or (ii) associated charm production, or (iii) sequential new hadronic production and decay. So far the only mechanism for trimuon production at the leptonic vertex of a neutrino interaction that has been seriously considered is a cascade of new leptons. The mixed origin opens several possibilities.

To compare trimuon data with direct muon pair production data, we show in Fig. 25 the muon pair invariant mass distribution from Anderson et al.[20] and the corresponding distributions for $\mu_F^- \mu^+$ and $\mu_S^- \mu^+$ from the trimuons. Note that in Fig. 25 the vertical scale is linear for the trimuon data but logarithmic for the direct muon production plot, which indicates that there are far too few low

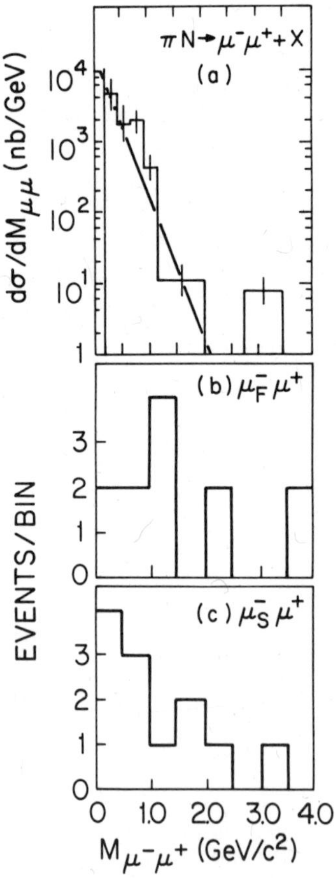

Fig. 25. Muon pair invariant mass distribution from (a) data on direct muon pair production by hadrons (reference 20) compared with (b) and (c) trimuon data.

invariant mass pairs in the trimuon data compared with the distribution for direct muon production. In Fig. 26 are presented the distributions in transverse momentum $p_\perp$ of the muon pair from Anderson et al.[20] (for four different invariant mass intervals) which are to be compared with the corresponding distributions in p_T for the trimuon data, shown in Fig. 27. The average values of $\langle p_T(\text{pair})\rangle$ are approximately 0.4 GeV/c for direct muon production and 1.7 GeV/c for the trimuons. It therefore seems unlikely that the bulk of the trimuons is due to direct lepton pair production at the hadron vertex.

If the trimuons are due primarily to associated charm production in charged current neutrino interactions, we expect that the muons from the semileptonic decays of c and $\bar{c}$ will have properties similar to the muons resulting from single charmed particle production, i.e., the oppositely charged muons in dimuon events. This probably will be true well above the associated production threshold if the production of charm and charmed pairs is deeply inelastic. Hence we compare the p_T distribution for the odd sign muons from dimuons in Fig. 28a with the corresponding distributions for trimuons in Figs. 28b and 28c. Observe that $\langle p_T\rangle$ for the dimuons is different from that for the trimuons. This conclusion is also indicated by the plots of p vs p_T in Fig. 29, again comparing dimuon and trimuon data. Furthermore, we note that an upper limit to the rate for associated charm production is available from the same sign dimuons (-/-), if we assume they are all due to that process with

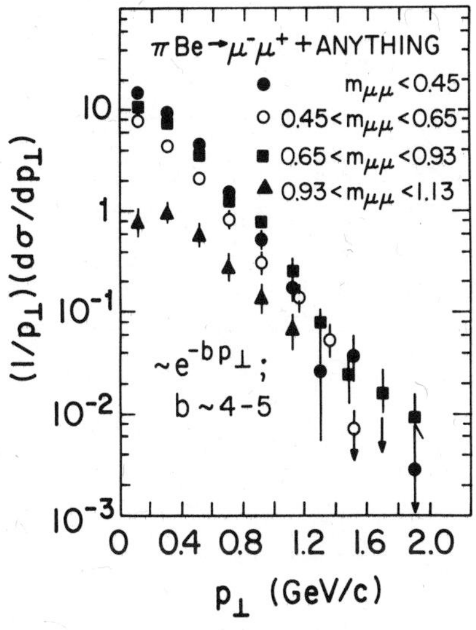

Fig. 26. Distribution in transverse momentum of the muon pairs from the data of reference 20. For these data $\langle p_\perp\rangle \approx 0.4$ GeV/c.

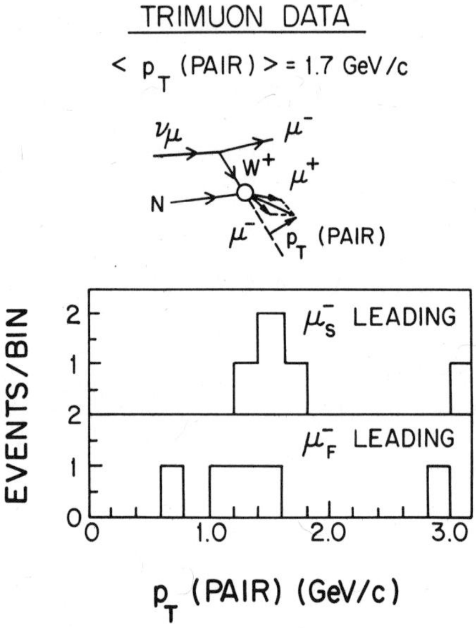

Fig. 27. Distribution in transverse momentum of the muon pairs in the trimuon data. Note that $\langle p_T(\text{pair})\rangle$ for these data is about 1.7 GeV/c.

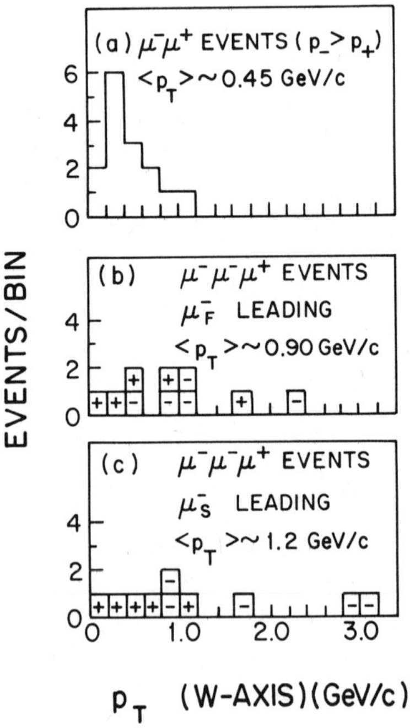

Fig. 28. Comparison of distributions in transverse momentum for (a) the odd sign muons from dimuons and (b) and (c) the muons from trimuons.

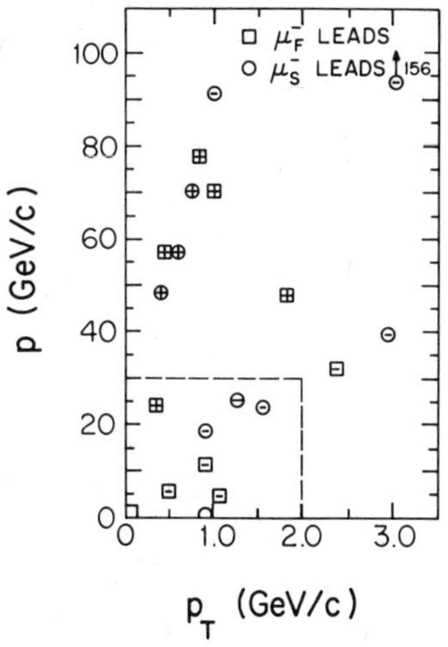

Fig. 29. Scatter plot of momentum vs transverse momentum for dimuon data (limits shown by the dashed line box) and trimuon data. Absence of charge indication means charge not determined.

one charmed hadron (c) decaying nonleptonically and the other ($\bar{c}$) decaying semileptonically. Taking from experiment R(−/−)/R(+/−) ≈ 0.1, and $B[(c \to \mu^+ + \nu_\mu + s)/(c \to \text{all})] \approx 0.1$, we find that the upper limit for trimuon production by associated charm production is $\lesssim 10^{-4}$ of all neutrino interactions, without including any kinematic factors such as those due to differences in momentum and transverse momentum of the muons in the dimuons and trimuons. These factors taken together would be expected to lower that limit by more than an order of magnitude. Thus, it seems improbable that associated charm production is the main source of the trimuons. Similar reasoning also applies to sequential new particle production and decay and leads to an equivalent conclusion.

The implications of a heavy lepton cascade as the mechanism of trimuon production are sketched in the diagrams of Fig. 30, where it is seen that such cascades would contribute to the opposite and same sign dimuon rates as well. Assuming the leptons in Fig. 30 have only the usual weak interaction, it is possible to calculate directly the properties of trimuons produced in this way[21]. The results of the calculations depend primarily on the masses of the new leptons, and are not especially sensitive to the type of coupling of the new leptons. These results with the masses of the heavy charged and neutral leptons taken as 8 GeV/c^2 and 4 GeV/c^2 respectively, are compared with the trimuon data in Figs. 31 to 34. In the main the model is consistent with the data,

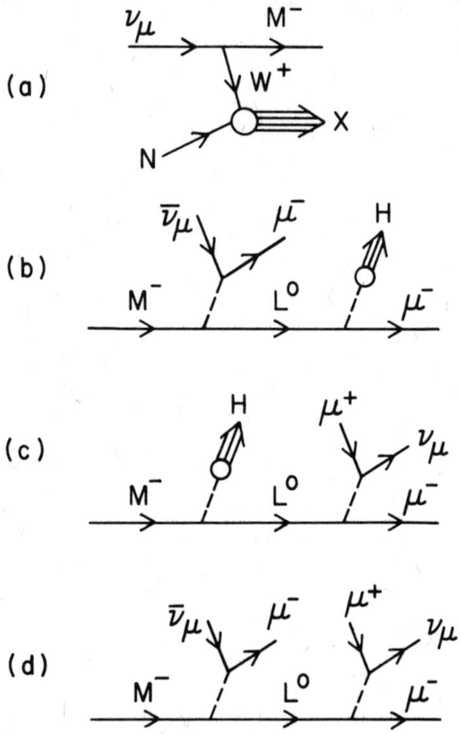

Fig. 30. Feynman diagrams illustrating the implications of a heavy lepton cascade. (a) M^- production, (b) cascade to $\mu^-\mu^-$, (c) cascade to $\mu^-\mu^+$, and (d) cascade to $\mu^-\mu^-\mu^+$. H and X stand for hadrons.

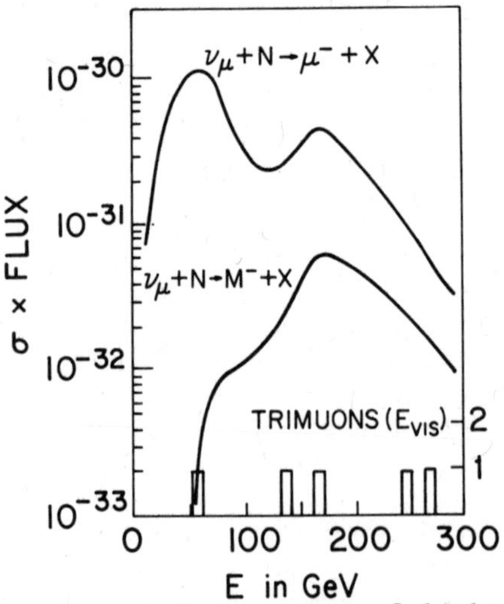

Fig. 31. Energy dependence of the rates of light and heavy charged lepton production by neutrinos. The curves are from reference 21; the data from E-310. The heavy lepton mass was chosen to be 8 GeV/c^2.

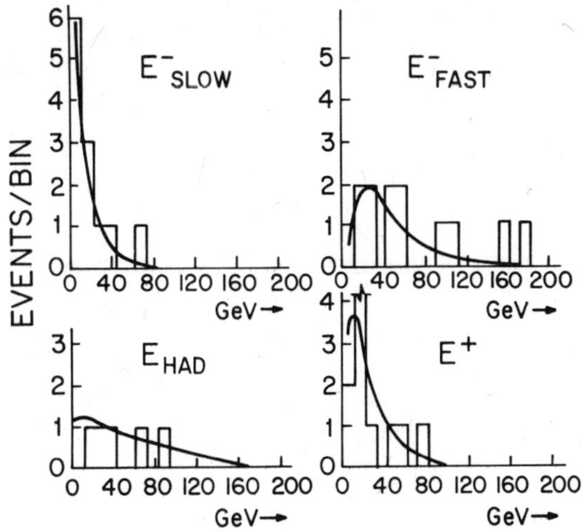

Fig. 32. Muon energies predicted by the lepton cascade model with $M^- = 8$ GeV/c^2 and $L^0 = 4$ GeV/c^2 (reference 21) compared with the E-310 trimuon data.

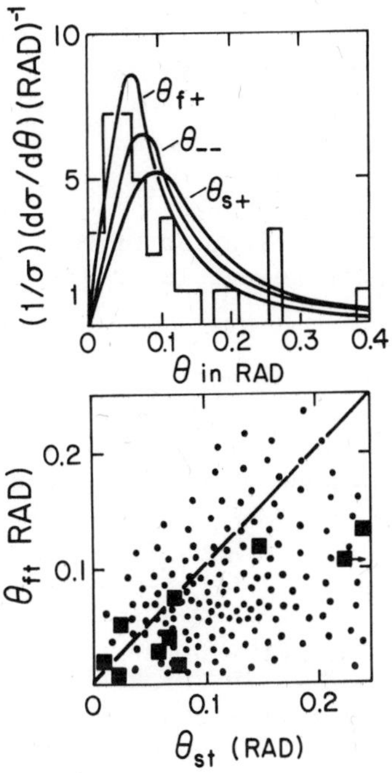

Fig. 33. Angular distributions from the lepton cascade model (reference 21) and from the E-310 data. The small points in Fig. 33b represent the calculated density of events.

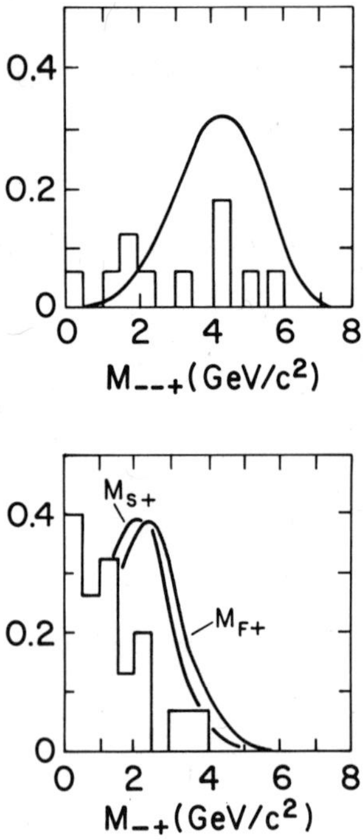

Fig. 34. Invariant mass distributions from reference 21 compared with E-310 data.

except possibly in the distribution of M_{-+} (Fig. 33a) where the data are perhaps concentrated at lower invariant masses than the calculation suggests.

We should emphasize that no part of the discussion above relating to trimuons is conclusive since no clear evidence concerning the origins of trimuons is available at this time. For example, we cannot rule out contributions to trimuons from each of the mechanisms expected to occur at the hadronic vertex, or from the combined electromagnetic-weak process at the leptonic vertex mentioned earlier. Similarly, a cascade of new heavy leptons or a combination of new lepton and new quark other than charm might also contribute to trimuon production. A larger sample of trimuons, particularly one obtained at the highest neutrino energies, is necessary for further clarification. Other vital evidence would be the observation of neutrino-induced trilepton events of the type $\mu^+\mu^-e^-$ and $\mu^-\mu^-e^+$, and of events with more than three muons in the final state. In this connection, one event has been observed in E-310 which suggests the presence of at least three soft muons (<4 GeV/c) in company with a more energetic negative muon that traverses the magnetic spectrometer. This is the only candidate for an event with more than three muons thus far in E-310; hence such events occur at a rate less than about one-fifth that of trimuons.

Summarizing the trimuon data, we find that

(1) the direct production by ν_μ of $\mu^-\mu^-\mu^+X$ has been demonstrated at a rough rate given by $\sigma \approx 10^{-40}$ cm^2 at $E_\nu > 100$ GeV; (2) there appears to be more than one type of trimuon: (a) a type with large angles between muon pairs and moderate total muon energy, (b) a type with small angles between muon pairs and large total muon energy and small hadron energy, and (c) a type with $E_{\mu^+} > \Sigma E_{\mu^-}$ (presumably ν_μ-induced) and total muon energy roughly the same as the total hadron energy; (3) it is unlikely that events of type (b) are produced at the hadron vertex of the neutrino interaction; (4) it is not yet clear that energetic trimuons are produced by $\bar{\nu}_\mu$; and (5) a rough limit on the rate of $4\mu + 5\mu + \ldots$ events is about 1/5 the observed trimuon rate.

To conclude as we began, the subject of multileptons has still a substantial future. Neutrino-induced multimuons, for example, are a probe of new particle production and decay with a relatively clean, sharp signature and moderately well understood backgrounds. The observation of neutrino interactions with three and four final state muons, if not completely incisive, nevertheless has considerable heuristic value in pointing out certain regions for further study in neutrino and other reactions. We can expect more such hints as multimuon data continue to increase during the next few years.

Experiment E-310 is being carried out by a collaboration of physicists from Fermilab, Harvard, Ohio State, Pennsylvania, Rutgers, and Wisconsin. Individual members of the collaboration are: A Benvenuti†, F. Bobisut††, D. Cline, P.S. Cooper, M.G.D. Gilchriese, M. Heagy, R. Imlay, M.E. Johnson, T.Y. Ling, R. Lundy, A.K. Mann, P. McIntyre, S. Mori, D.D. Reeder, J. Rich, R. Stefanski and D. Winn.

I am grateful for the kindness and hospitality shown me by the organizers of the Banff Summer Institute. I wish particularly to express my appreciation to D.H. Boal and A.N. Kamal for their thoughtfulness and patience.

REFERENCES

†) Now at Istituto di Fisica, Universita di Bologna, Bologna, Italy.

††) Visitor at the University of Pennsylvania, on leave from Istituto di Fisica dell' Universita, Padova, Italy.

1) For an account of early experimental evidence from neutrino and e^+e^- interactions see, for example, G. Feldman and A.K. Mann, McGraw-Hill Yearbook of Science and Technology, 1977.

2) B. Aubert et al., in Neutrinos - 1974 (AIP Conf. Proc. No. 22, 1974), C. Baltay, ed.; B. Aubert et al., in Proc. XVII Int'l. Conf. on High Energy Physics, London, 1974, J.R. Smith, ed.; A. Benvenuti et al., Phys. Rev. Lett. 34, 419 (1975); 35, 1199 and 1249 (1975).

3) B.C. Barish et al., Phys. Rev. Lett. 36, 939 (1976); 39, 981 (1977).

4) J. von Krogh et al., Phys. Rev. Lett. 36, 710 (1976).

5) J. Blietschau et al., Phys. Lett. 60B, 207 (1976).

6) A. Benvenuti et al., Phys. Rev. Lett. 35, 1203 (1975).

7) L.N. Chang, E. Derman and J.N. Ng, Phys. Rev. Lett. 35, 6 (1975); A. Pais and S.B. Treiman, Phys. Rev. Lett. 35, 1206 (1975).

8) A. Benvenuti et al., Phys. Rev. Lett. 38, 1110 and 1183 (1977). Also preprint entitled "Further Observation of Trimuon Production by Neutrinos and Antineutrinos", submitted to Phys. Rev. Lett.

9) B.C. Barish et al., Phys. Rev. Lett. 38, 377 (1977).

10) A. Skuja, R. Stefanski, and A. Windelbon, FNAL Technical Note TM469 (1974).

11) R. Stefanski and H.B. White, FNAL Technical Note TM626A (1976).

12) D.A. Edwards and F.J. Sciulli, FNAL Technical Note TM660 (1976).

13) See, for example, L.M. Sehgal and P.M. Zerwas, Phys. Rev. Lett. 36, 399 (1976).

14) L.M. Sehgal and P.M. Zerwas, Nucl. Phys. B108, 483 (1976); E. Derman, Nucl. Phys. B110, 40 (1976); V. Barger et al., Phys. Rev. D16, 746 (1977); Phys. Lett. 70B, 51 (1977).

15) M.L. Perl et al., SLAC-PUB 1997, LBL 6731, Aug. 1977, submitted to Phys. Lett.

16) C. Baltay et al., Phys. Rev. Lett. 39, 62 (1977).

17) M. Holder et al., Phys. Lett. 69B, 377 (1977).

18) H.H. Bingham in Conf. Proc. Neutrino-77 Conference, Elbrus, USSR, (June, 1977).

19) See, for example, J. Smith and J.A.M. Vermaseren, Preprint ITP-SB-77-66 (Nov., 1977); V. Barger, T. Gottschalk and R.J.N. Phillips, Preprint COO-881-9 (Nov., 1977).

20) K.J. Anderson et al., Phys. Rev. Lett. 37, 799 (1976).

21) We use results from C.H. Albright, J. Smith and J.A.M. Vermaseren, Phys. Rev. Lett. 38, 1187 (1977). An extensive list of references on possible trimuon origins is given in a talk by V. Barger, at the Ben Lee Memorial Conference, FNAL (Oct., 1977), Preprint COO-881-7.

PARTICIPANTS

Abarbanel, H.	Fermi National Accelerator Laboratory, Batavia, Illinois 60510.
Adkins, G. S.	2555 La Mesa Drive, Santa Monica, California 90402.
Allen III, E.	307 Nat. Sciences II, Univ. of California, Santa Cruz, California 95064.
Andrei, N.	Physics Department, Princeton University, Princeton, New Jersey 08540.
Ansourian, M. M.	P.O.B. 131, M.I.T. Branch, Cambridge, Massachusetts 02139.
Appelquist, T.	Physics Department, Yale University, New Haven, Connecticut 06520.
Aubrecht II, G. J.	Physics Department, Ohio State University, 1465 Mt. Vernon Ave., Marion, Ohio 43302.
Bailey, D. C.	Eaton Laboratory, McGill University, 750 McGregor, Montreal, Quebec H3A 1A4.
Barish, S. J.	Physics Department, Carnegie-Mellon Univ., Pittsburgh, Pa. 15213.
Basham, C. L.	Physics Department, University of Virginia, Charlottesville, Virginia 22901.
Berg, B.	Inst. f. Theor. Physik, 1000 Berlin 33, Arnimallee 3, West Germany.
Boal, D. H.	Dept. of Physics, University of Alberta, Edmonton, Alberta T6G 2J1.
Brodsky, S.	S.L.A.C., P.O. Box 4349, Stanford, California 94305.

Brown, C. N. — Fermi National Accelerator Laboratory, Batavia, Illinois 60510.

Buchanan, C. D. — Physics Department, U.C.L.A., Los Angeles, California 90025.

Caianiello, E. — Faculty of Science, University of Salerno, via Vernieri 42, 84100 Salerno, Italy.

Campbell, B. A. — Physics Department, MacDonald Chemistry Bldg., McGill University, Montreal, Quebec H3A 2K6.

Capri, A. Z. — Dept. of Physics, University of Alberta, Edmonton, Alberta T6G 2J1, Canada.

Collins, J. C. — Physics Department, Princeton University, Princeton, N.J. 08540.

Crutchfield, W. Y. — Physics Department, Princeton University, Princeton, N.J. 08540.

Cuthiell, D. — Inst. for Theor. Physics, University of Alberta, Edmonton, Alberta T6G 2J1.

Das, K. P. — 155 S, Institute of Theoretical Science, Sci. II, University of Oregon, Eugene, Oregon 97403.

De Kam, J. — Physics Department, University of Alberta, Edmonton, Alberta, T6G 2J1.

De Lillo, S. — Physics Department, University of Alberta, Edmonton, Alberta T6G 2J1.

Drell, S. D. — S.L.A.C., P.O. Box 4349, Stanford, California 94305.

Edwards, B. J. — Department of Physics, University of Alberta, Edmonton, Alberta T6G 2J1.

Eylon, Y. — Lawrence Berkley Lab., Univ. of California, Berkeley, California 94720.

Fang, J. — Physics Department, U.C.L.A., Los Angeles, California 90024.

Feldman, G. J. — S.L.A.C., P.O. Box 4349, Stanford, California 94305.

Fujii, Y.	College of Gen. Education, University of Tokyo, Komaba 3-8-1, Tokyo 153, Japan.
Gildener, E.	Physics Department, Univ. of California, San Diego; La Jolla, California 92093.
Gilman, F. J.	S.L.A.C., P.O. Box 4349, Stanford, California 94305.
Gottlieb, S.	Physics Department, Princeton University, Princeton, New Jersey 08540.
Ha, Y. K.	Physics Department, Yale University, New Haven, Connecticut 06520.
Haacke, E. M.	7 Walmer Rd., #1802, Toronto, Ontario M5R 2W8.
Haber, H. E.	Physics Department, University of Michigan, Ann Arbor, Michigan 48104.
Handy, C.	Dept. of Physics, Box 40, Columbia University, New York, N.Y. 10027.
Hardy, J. E.	Physics Department, Carleton University, Ottawa, Ontario K1S 5B6.
Henty, D. L.	Dept. of Physics, University of Alberta, Edmonton, Alberta T6G 2J1.
Hogan, M. E.	255 Temple Ave., Apt. 7, Long Beach, California 90803.
't Hooft, G.	Institute of Theor. Physics, University of Utrecht, Utrecht, The Netherlands.
Isgur, N.	Physics Department, University of Toronto, Toronto, Ontario M5S 1A7.
Jackiw, R.	Physics Department, M.I.T., Cambridge, Massachusetts 02139.
Jayaprakash,	Dept. of Physics, University of Illinois, Urbana, Illinois 61801.
Johnson, K.	Physics Department, M.I.T., Cambridge, Massachusetts 02139.

Jongeward, G.	Dept. of Physics, University of Illinois, Urbana, Illinois 61801.
Kamal, A. N.	Dept. of Physics, University of Alberta, Edmonton, Alberta T6G 2J1.
Katz, H. J.	Dept. of Physics, University of Illinois, Urbana, Illinois 61801.
Lam, C. S.	Physics Department, McGill University, P.O. Box 6070, Station A, Montreal, Quebec H3C 3G1.
Laue, H.	Physics Department, University of Calgary, Calgary, Alberta T2N 1N4.
Laughton, D.	Jadwin Hall, P.O. Box 708, Princeton, N.J. 08540.
Libby, S. B.	Physics Department, Princeton University, Princeton, N.J. 08540.
Litwin, K.	Niels Bohr Institute, Blegdamsvej 17, 2100 Copenhagen Ø, Denmark.
Lu, A.	Physics Department, University of California, Santa Barbara, California.
Mann, A. K.	Physics Department, University of Pennsylvania, Philadelphia, Pennsylvania 19104.
Marinaro, M.	Matematiche Fisiche, Universita di Salerno, 84100 Salerno, Italy.
Matsumoto, H.	Department of Physics, University of Alberta, Edmonton, Alberta T6G 2J1.
Mizrachi, L.	Physics Department, Tel-Aviv University, Tel-Aviv, Israel.
Moen, I.O.	Physics Department, Trent University, Peterborough, Ontario K9J 7B8.
Ng, J.	Theoretical Physics Inst., University of Alberta, Edmonton, Alberta T6G 2J1.
Ong, C.-L.	Physics Department, University of Toronto, Toronto, Ontario M5S 1A7.

Ore Jr., F. R.	M.I.T., Cambridge, Massachusetts 02139.
Pallua, S.	Ruder Boskovic Institute, 41001 Zagreb, Croatia, Yugoslavia.
Papastamatiou, N.	Physics Department, University of Wisconsin, Milwaukee, Wisconsin 53201.
Parke, S. J.	Lyman Laboratory, Harvard University, Cambridge, Massachusetts 02138.
Perrottet, M.	Lyman Laboratory, Harvard University, Cambridge Massachusetts 02138.
Phillips, R. J. N.	Theory Division, Rutherford Laboratory, Chilton, Didcot, Oxfordshire, OX11 OQX, UK.
Phippen, J. W.	Department of Physics, Weber State College, Ogden, Utah 84408.
Pumplin, J.	Physics Department, Michigan State University, East Lansing, Michigan 48824.
Resnick, L.	Physics Department, Carleton University, Ottawa, Ontario K1S 5B6.
Rno, J. S.	Physics Department, University of Maryland, College Park, Maryland 20742.
Robinson, J. L.	Applied Math Dept., Univ. of Western Ontario, London, Ontario N6A 3K7.
Rogers, E. O.	Physics Department, Univ. of Washington, Seattle, Washington 98195.
Savaria, P.	Physics Department, University of Toronto, Toronto, Ontario M5S 1A7.
Schiff, H.	Physics Department, University of Alberta, Edmonton, Alberta T6G 2J1.
Scott, D.M.	Department of Physics, Ohio State University, Mansfield, Ohio 44906.
Semenoff, G. W.	Physics Department, University of Alberta, Edmonton, Alberta T6G 2J1.

Shafi, Q.	Fakultat fur Physik, University of Freiburg, Hermann-Herderstr. 3, West Germany.
Shamaly, A.	Physics Department, University of Alberta, Edmonton, Alberta T6G 2J1.
Sherry, T.	Ctr. for Particle Theory, University of Texas, Austin, Texas 78712.
Shintani, M.	Physics Department, University of Alberta, Edmonton, Alberta T6G 2J1.
Shirafuji, T.	Physics Department, Saitama University, Urawa, Saitama-ken 338, Japan.
Sidles, J.	Enrico Fermi Inst., Univ. of Chicago, Chicago, Illinois 60637.
Sodano, P.	Physics Department, University of Alberta, Edmonton, Alberta T6G 2J1.
Sundaresan, M. K.	Physics Department, Carleton University, Ottawa, Ontario K1S 5B6.
Svetitsky, B.	S.L.A.C., Stanford, California 94305.
Takahashi, Y.	Department of Physics, University of Alberta, Edmonton, Alberta T6G 2J1.
Tassie, L. J.	Theoretical Physics Dept., Australian Nat. Univ., P.O. Box 4, Canberra, ACT 2600, Australia.
Tomozawa, Y.	Harrison M. Randall Lab., University of Michigan, Ann Arbor, Michigan 98109.
Tyburski, L. J.	Physics Department, University of California, Los Angeles, California 90024.
Tze, H. C.	S.L.A.C., P.O. Box 4349, Stanford, California 94305.
Umezawa, H.	Department of Physics, University of Alberta, Edmonton, Alberta T6G 2J1.
Unger, D. G.	Physics Department, University of Colorado, Boulder, Colorado 80309.

Valanju, P. M.	Ctr. for Particle Theory, University of Texas, Austin, Texas 78712.
Vinet, L.	Centre de Recherches Mathématiques, Université de Montréal, C.P. 6128, Montréal, Quebec H3C 3J7.
Viswanathan, K. S.	Physics Department, Simon Fraser University, Burnaby, B.C. V5A 1S6.
Wadati, M.	Theoretical Physics Inst., The University of Alberta, Edmonton, Alberta T6G 2J1.
Wadia, S. R.	Physics Department, City College of CUNY, New York, N.Y. 10031.
Watson, P. J. S.	Physics Department, Carleton University, Ottawa, Ontario K1S 5B6.
Weidemann, A. W.	Physics Department, University of Maryland, College Park, Maryland 20742.
Weinstein, M.	S.L.A.C., P.O. Box 4349, Stanford, California 94305.
Williams, D. N.	Physics Department, University of Michigan, Ann Arbor, Michigan 48109.
Wilson, W. J.	Physics Department, University of Minnesota, Minneapolis, Minnesota 55455.
Winbow, G. A.	Serlin Physics Lab., Rutgers University, Piscataway, N.J. 08854.
Winn, D. R.	Lab C E310, Fermilab, Box 500, Batavia, Illinois 60510.
Woo, G.	Lyman Laboratory of Phys., Harvard University, Cambridge, Massachusetts 02138.
Yoneya, T.	Physics Department, City University of N.Y., New York, N.Y. 10031.
Young, B.-L.	Physics Department, Iowa State University, Ames, Iowa 50010.
Zachos, C. K.	Caltech, 452-48, Pasadena, California 91125.

SUBJECT INDEX